$60

000688

Global Biopiracy

Law and Society Series
W. Wesley Pue, General Editor

The Law and Society Series explores law as a socially embedded phenomenon. It is premised on the understanding that the conventional division of law from society creates false dichotomies in thinking, scholarship, educational practice, and social life. Books in the series treat law and society as mutually constitutive and seek to bridge scholarship emerging from interdisciplinary engagement of law with disciplines such as politics, social theory, history, political economy, and gender studies.

A list of the books in this series appears at the end of the book.

Ikechi Mgbeoji

Global Biopiracy:
Patents, Plants, and Indigenous
Knowledge

UBCPress · Vancouver · Toronto

15 14 13 12 11 10 09 08 07 06   5 4 3 2 1

Printed in Canada on ancient-forest-free paper (100% post-consumer recycled) that is processed chlorine- and acid-free, with vegetable-based inks.

Library and Archives Canada Cataloguing in Publication

Mgbeoji, Ikechi, 1968-
  Global biopiracy : patents, plants and indigenous knowledge / Ikechi Mgbeoji.

(Law and society)
Includes bibliographical references and index.
ISBN 13: 978-0-7748-1152-1
ISBN 10: 0-7748-1152-8

  1. Patents (International law). 2. Plants, Cultivated – Patents. 3. Traditional ecological knowledge. 4. Plant biotechnology – Patents. 5. Eurocentrism. 6. Biological diversity. I. Title. II. Series: Law and society series (Vancouver, B.C.)

K1519.B54M43 2005   346.04'86   C2005-905369-0

Canadä

UBC Press gratefully acknowledges the financial support for our publishing program of the Government of Canada through the Book Publishing Industry Development Program (BPIDP), and of the Canada Council for the Arts, and the British Columbia Arts Council.

This book has been published with the help of a grant from the Canadian Federation for the Humanities and Social Sciences, through the Aid to Scholarly Publications Programme, using funds provided by the Social Sciences and Humanities Research Council of Canada, and with the help of the K.D. Srivastava Fund.

UBC Press
The University of British Columbia
2029 West Mall
Vancouver, BC V6T 1Z2
604-822-5959 / Fax: 604-822-6083
www.ubcpress.ca

In the course of writing this book, death has dealt me three savage blows, and it is to the collective memory of three people that I dedicate this book:

> *My beloved sister, Eziaha Ocheze Mgbeoji,*
> *who passed away in September 2000*
>
> *My wonderful mother, Victoria Alumugbo Mgbeoji,*
> *who passed away on 6 January 2001*
>
> *My inspirational father, Levi Esiwoko Mgbeoji,*
> *who passed away on 11 July 2004*

Though dead, you all live in my heart and memory forever

> *Fare Well!!!*

Contents

Foreword / ix
Teresa Scassa

Preface / xi

Acknowledgments / xiii

Acronyms / xv

1 Introduction / 1

2 Patents, Indigenous and Traditional Knowledge, and Biopiracy / 9

3 Implications of Biopiracy for Biological and Cultural Diversity / 50

4 The Appropriative Aspects of Biopiracy / 87

5 Patent Regimes and Biopiracy / 119

6 Conclusion / 179

Notes / 201

Selected Bibliography / 280

Index / 305

Foreword

The word "biopiracy" was coined to name a phenomenon that is not new but that has flourished under colonialism, capitalism and, more recently, globalization. By naming it, contemporary scholars have challenged a form of exploitation of peoples and of resources that now poses a threat to both cultural and environmental sustainability. The term "biopiracy" is a direct challenge to the legitimacy of activities that have entered the mainstream of contemporary global capitalism.

This book by the brilliant young scholar Ikechi Mgbeoji is at the forefront of the literature on this important topic. Dr. Mgbeoji provides a trenchant critique of increasingly globalized patent regimes and of how these regimes contributed to the appropriation of plant species and traditional knowledge of the use of plants. He also provides a rich historical, social, and legal context for his analysis and argues that Western ideologies and epistemologies have contributed to a devaluation of indigenous communities and their wealth of traditional knowledge. This devaluation is reflected in those aspects of modern patent law systems that leave traditional knowledge of the use of plants open to exploitation and appropriation. Although he is profoundly critical of the patent regime and of its use by powerful states and multinational corporations to further their own objectives, Dr. Mgbeoji is not without hope for the future. Indeed, his concluding pages contain concrete proposals for change.

The book, which is both rigorously scholarly and profoundly moral, greatly benefits from Dr. Mgbeoji's expertise in intellectual property law and international environmental law. This proficiency is significant because too much of the existing literature lacks a sound grounding in one or both of these fields. In addition to being rich in context, this book is interdoctrinal and interdisciplinary. Dr. Mgbeoji's understanding of the relevant law and institutions, his impressive grasp of the literature across a range of disciplines, and his clarity of vision inform this book and its cogent message.

Dr. Mgbeoji's work is intellectually rigorous and clearly expressed. It is also informed by a passion for justice and equity. I had the privilege of being his supervisor during his doctoral studies. He was a most outstanding graduate student and his brilliance is evident in this book. Dr. Mgbeoji is also a person who inhabits two cultures, whose experience spans both North and South. He is a scholar who can understand our own preconceptions while lifting us outside of them. This book is very much a product of this unique intellect: it is a challenging, engaging, ambitious, and ultimately very important work.

Dr. Teresa Scassa
Dalhousie Law School
Halifax, Nova Scotia
August 2005

Preface

Legal control and ownership of plants and traditional (indigenous) knowledge of the uses of plants (TKUP) is often a vexing issue, particularly at the international level, because of the conflicting interests of states or groups of states. The most widely used form of juridical control of plants and TKUP is the patent system, which originated in Europe. This book rethinks the role of international law and legal concepts, the major patent systems of the world, and international agricultural research institutions as they affect legal ownership and control of plants and TKUP. The analysis is cast in various contexts and examined in multiple levels. The first level of analysis deals with the Eurocentric character of the patent system, international law, and institutions. The second involves the cultural and economic dichotomy between the industrialized Western world, otherwise known as the North, and the "westernizing" world, otherwise known as the South. The North-South divide is untidy, perhaps generalized, but it is used here as a convenient tool of analysis. The third level of analysis considers the phenomenal loss of human cultures and plant diversity.

In examining these issues within the delimited contexts, *Global Biopiracy* argues that the Eurocentric character of the patent system and international law, the cultural and gender biases of Western epistemology, and the commercial orientation of the patent system are implicated in the appropriation and privatization of plants and TKUP. In other words, the phenomenon of appropriation of plants and TKUP, otherwise known as biopiracy, thrives in a cultural milieu in which non-Western forms of knowledge are systemically marginalized and devalued as "folk knowledge" and are characterized as being suitable only as objects of anthropological curiosity.

The implications of appropriation of plants and TKUP in an age of rapid biodiversity loss and homogenization of cultures traverse a gamut of issues, including the sustainability of biological resources, human rights, and distributive justice. Hence, the need to re-examine and redefine the role of patents and international law in the emerging process, bearing in mind the imperatives of a fair and equitable regime on plants and TKUP. Given the interrelatedness of human rights, the commonality of humankind, and the indivisibility of the global environment, these processes are of universal import and ought to be addressed holistically. This book therefore locates the problems of biopiracy as a process of engagement with the question of reconstructing international law, attitudes, and the patent concept. More specifically, the contemporary intellectual property regime, particularly patents, must have a socially mediated core and an ethic of respect, inclusiveness, and diversity of cultures and values.

Acknowledgments

Global Biopiracy owes a lot to the ideas and inspiration of other people. This is particularly so with regard to a subject as complex and controversial as patents on plant life forms and traditional knowledge of the uses of plants. It is therefore proper for me to express my gratitude to those whose ideas I have borrowed or critiqued as well as to express my deep appreciation to the numerous persons who have assisted me in working towards what I have expressed here.

First, I sincerely thank Professor Teresa Scassa at Dalhousie University. From the earliest formulations of the topic, through its gestation and maturation, she provided incredible support, making pertinent suggestions and detailed commentaries. She was splendid, graceful, and resourceful. Her unflagging support also secured me research funding and international exposure, which have contributed immensely in clarifying my views, expanding my horizon, and challenging some of my arguments. I cannot fail to thank Professors David VanderZwaag and Hugh Kindred. David is an encyclopedia of knowledge and his boundless zeal and enthusiasm for international environmental law is better experienced than described. He facilitated my contacts with a lot of useful individuals and institutions, especially the United Nations Environment Programme.

Through his and Teresa's efforts, I was privileged to attend the fifteenth Global Biodiversity Forum in Nairobi, Kenya. From there, I secured the Carl Duisberg Fellowship for research at the Environmental Law Centre of the World Conservation Union, the International Union for the Conservation of Natural Resources (IUCN), Bonn, Germany. In addition, David has been wonderful in shaping my focus on the fundamental problems and theoretical issues of international environmental law. The government of Germany also contributed to the success of this work, especially by awarding me the Carl Duisberg Fellowship. While in Germany I had the privilege of working with several experts at the Environmental Law Centre (ELC) of the World Conservation Union, IUCN: Dr. Francoise Burhenne-Guilmin, Dr. Alexander

Timoshenko, Tomme Young, Dr. Nazrul Islam, Dr. Wang Xi, Isabel Martinez, and Charles Di Leva. The librarians at the ELC, particularly Annie Lukacs, were extraordinarily resourceful. In the course of searching for ideas and materials, I must mention the devoted help of friends such as Dr. Graham Dutfield and Yaw Osafo.

Professor Hugh Kindred and his lovely wife Sheila have been a source of strength, guidance, and comfort. Apart from Hugh's complete mastery of the intricacies of international law, his greatest contribution towards my intellectual development was the human touch that he brought to bear on the principles and ethos of international law. Several other professors and colleagues have been immensely helpful. They include Dick Evans, Aldo Chircop, and Wes Pue. Without the generosity of the Killam Trustees and Richard Owens of the Centre for Innovation Law and Policy, I would never have finished this book in good time.

The librarians at both Dalhousie Law School and Osgoode Hall Law School made an otherwise dreary and frustrating search for documentation an interesting endeavour. I therefore thank Julia Lavigne, Anne-Marie White, Annemarie Hay, Brenda Gillis, Debbie Ritchie, Carla Gobiessi-Lynch, David Michaels, and, of course, Ann Morrison. Sheila Wile has been a friend, confidante, and guide. My thanks to Molly Ross, Sandra Harnum, and Julie Dergal.

Special thanks too must go to the "Nigerian Legion" at Dalhousie University and Osgoode Hall Law School: Chioma Ekpo, Shedrack Agbakwa, Pius Okoronkwo, Annie Brisibe, Bonny Ibhawoh, Ralph Njoku, and Julius Egbeyemi. Special mention must be made of many other friends, especially Professor Obiora Okafor, Ugochukwu Ukpabi, Sonne Udemgba, Chika and Chinwe Onwuekwe, Chidi Oguamanam, Remigius Nwabueze, Chinedu Idike, and a host of other members of the "clan."

Special thanks to my friend Barbara Hinch. I also thank my colleagues in residence during my stay in Halifax for their support and good banter: David Dzidzornu, Gloria Chao, David Parker, Stuart Gilby, and Professor Stuart Kaye. A profound thanks and appreciation to the magnanimity and felicity of Uzoma Nwaekpe, Ogbuagu Odengalasi. As the Igbo say, *Ivu anyi danda, Madu bu uko* (no load defeats the ant, man is strength to his fellow man).

Finally, I must thank my siblings, Ihuoma, Eze, Ebere, and Uchenna. In the course of my intellectual exile, I was devastated by the sudden deaths of my dearly beloved parents and sister. My family and friends have offered solace, inspiration, and encouragement. My brother-in-law, Samuel Ohiara, and my nephews have been wonderful. Of course, I am indebted to the pastor and members of the Seventh-Day Adventist Church in Halifax, whose prayers have been very helpful to me during the most trying moments. My thanks too to my in-laws: Sir and Lady T.T. Onyeaso, Adaeze, Udoka.

To Nkeiruka and our beautiful daughters, Chizaram and Maraelo, I say, I love you all with all my heart.

Acronyms

APEC	Asia-Pacific Economic Cooperation
ARIPO	African Regional Industrial Property Organization
CBD	Convention on Biological Diversity
CCM	common concern of mankind
CGIAR	Consultative Group on International Agricultural Research
CHM	common heritage of mankind
CITES	Convention for the International Trade in Endangered Species
CPC	Community Patent Convention
CSD	Commission on Sustainable Development
DNA	deoxyribonucleic acid
ECOSOC	UN Economic and Social Council
EPC	European Patent Convention
EU	European Union
FAO	Food and Agriculture Organization of the United Nations
FDI	foreign direct investment
GATT	General Agreement on Trade and Tariffs
GDP	gross domestic product
GNP	gross national product
HYVs	high yield varieties
IARCs	International Agricultural Research Centers
IBPGR	International Board for Plant Genetic Resources
ICDPs	Integrated Conservation and Development Projects
ICJ	International Court of Justice
ICJ Reports	International Court of Justice Reports
IFF	Intergovernmental Forum on Forests
ILC	International Law Commission
ILM	International Legal Materials
ILO	International Labour Organization
ILR	International Law Reports
IMF	International Monetary Fund

IPR	intellectual property rights
ISA	International Search Agency
ITTA	International Tropical Timber Agreement
IUCN	World Conservation Union
LMOs	living modified organisms
LNTS	League of Nations Treaty Series
NAFTA	North America Free Trade Agreement
NGO	non-governmental organization
NIEO	New International Economic Order
OAPI	African Intellectual Property Organization
OAU	Organization of African Unity
PBRs	plant breeders rights
PCIJ	Permanent Court of International Justice
PCT	Patent Cooperation Treaty
PFFI	Permanent Forum on Indigenous Issues
PIC	prior informed consent
PSNR	Permanent Sovereignty over Natural Resources
RADIC	African Journal of International and Comparative Law
RAFI	Rural Advancement Foundation International
RNA	ribonucleic acid
TKUP	traditional knowledge of the uses of plants
TRIPs	Agreement on Trade-Related Aspects of Intellectual Property Rights
UNCED	United Nations Conference on Environment and Development
UNCTAD	United Nations Conference on Trade and Development
UNDP	United Nations Development Programme
UNESCO	United Nations Educational, Scientific, and Cultural Organization
UNEP	United Nations Environment Programme
UNFF	United Nations Forum on Forests
UNGA	United Nations General Assembly
UNTS	United Nations Treaty Series
UPOV	Union pour la Protection des Obstentions Vegetales
WHO	World Health Organization of the United Nations
WIPO	World Intellectual Property Organization
WTO	World Trade Organization

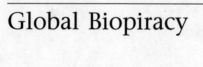

Global Biopiracy

1
Introduction

Various factors have congregated to arouse a huge interest, indeed a controversy, with regard to the legal ownership of plant genetic resources and the knowledge associated with the uses of plants. This controversy has elicited calls for the creation of a regime dealing with access to and equitable sharing of the benefits of plant genetic resources. These issues are often conflated in what has become known as the issue of "indigenous peoples knowledge." The debate has often implicated a variety of issues, such as the imposition of Eurocentric legal concepts (e.g., the imposition of patents on non-European cultures and peoples), the impact of globalization, and emerging norms on legal control of knowledge. In addition, the debate has raised issues pertaining to the prevailing ideology of "civilization" and "development" and its impact on biological and cultural diversity. At the heart of the debate is the political economy and legal control of plant genetic resources and knowledge of their associated uses. Often ignored in the discourse is the fact that the processes by which the dominant cultures and states appropriate the traditional knowledge of Third World peoples are masked in technical, and sometimes diplomatic, understatements. Consequently, it is often difficult to discern the issues at stake or the scale of the disagreements between those involved.

Global Biopiracy explores the contours of that debate and attempts to explain the legal processes and institutional frameworks by which the patent systems of powerful states prey on the genetic resources developed by countries and cultures of the Third World. The main objective is to contribute to a more transparent and open debate, free from the obfuscation and technical shenanigans that have hampered an appreciation of the global forces at play in the appropriation of indigenous peoples knowledge. The debate as presently undertaken marginalizes and underappreciates the role of women and farmers in the development of plant genetic diversity. This is not a coincidence as, since its emergence during the Renaissance, Western science has been masculinist and racist.

Global Biopiracy is divided into six chapters. Chapter 1 is introductory and gives a broad overview of the issues covered. Chapter 2 explains basic concepts in international law, such as indigeneity and biopiracy, and explores the development of national but interlinked system of patent systems. Chapter 2 also provides an anatomy of the structure and process of the relationship between various national patent systems and the patchwork of international instruments seeking to harmonize national patent regimes. It also provides a multidimensional analysis of the evolution and globalization of the patent concept within the context of the North-South divide. Further, it offers a doctrinal analysis of the sources of international law on patents and their status and effect at the domestic level. The underlying theory is that, although the patent system is intrinsically international in an increasingly interactive world, it is, most significantly, aggressively nationalistic. It serves the instrumentalist goals of states, especially powerful multinational corporations capable of influencing and using the machinery of their parent state to influence the domestic patent regimes of other states. Accordingly, powerful states prodded by biotechnology and pharmaceutical firms, tend to favour patent laws that suit and serve their perceived national interests. In this context, the forced globalization of the patent concept owes much more to the influence of industrialized states and the propagation of their national self-interests than it does to altruistic concerns for the welfare of other states.

Chapter 2 is divided into eight sections. The first section traces the origin of the patent system from its roots in the Italian peninsula. The salient point is that the modern patent system is culturally and philosophically Eurocentric. The second section defines the concept of patents, while the third section explores various theories in support of the patent system. The fourth section examines the diffusion of the patent system from Italy to mainland Europe and the eventual imposition of the system of patents on non-Europeans during the age of colonialism and empire. Of course, there a few exceptions, like Japan, which instituted the patent system of its own volition. However, an overwhelming majority of states outside the Eurocentric paradigm and worldview received their patent laws and systems via colonial fiat and imperialist threat.

The fifth section deals with the historical evolution and development of the patent system within the cultural crucible of Europe. The objective is to demonstrate the inextricable link between European normative values and ownership of property vis-à-vis non-European concepts of property ownership. The sixth section briefly explores the influence of industries, especially the life science industries, on the concept and scope of patentable subject matter. Although this issue is dealt with in greater detail in Chapter 5, the objective is to demonstrate that the law on patentability has not stood still; rather, it has moved in directions dictated by industry and the national interests of states. The seventh section takes the argument further by revisiting and interrogating the assumption that patents encourage inventiveness. It is probable that, with or without patent systems, inventions would occur. The eighth section explores the North-South dimensions of the patent system. It shows how the tensions between patent systems and indigenous peoples knowledge have been heightened by the globalization of the concept of patents. It also examines how the domestic applications of international norms on patents create a homogenous regime favourable to the interests of pharmaceutical and agricultural corporations. And it briefly examines recent attempts by the World Intellectual Property Organization (WIPO) to raise global consciousness with regard to the inadequacies of patent systems to protect indigenous peoples knowledge.

Chapter 3 analyzes the evolution, development, and status of the global regime on plants. The objective is to provide a background for subsequent analysis of the methods by which international institutions and the patent system appropriate indigenous peoples knowledge. This chapter has seven sections. The first section examines and evaluates the nature, value, and functions of plant life forms. The second section explores the various religio-philosophical conceptions and how those worldviews have influenced current debates on the patentability of indigenous peoples knowledge. Attention is paid to the philosophical divide between the dominant Judeo-Christian conception of plant life forms and non-Eurocentric philosophies, which often consider certain plants and their associated uses to be sacred.

The third section scrutinizes the influence of a gendered and racist conception of science and "civilization," which forms the subtext to the marginalization and delegitimation of non-European epistemological frameworks. In other words, the process of appropriation of indigenous peoples knowledge is not merely a legal problem; rather, it is a phenomenon that operates within a social structure of inbuilt primordial prejudices and biases against non-Western cultures and non-Western epistemological frameworks. The sum of these processes is the denial of the intellectual contributions of Third World women farmers and healers.

In consequence, domestic standards of patentability, along with the competing priorities, values, and interests of weaker states, have been squelched. Within this milieu of competition and conflicts, the interests of powerful corporations and states, which converge with capitalistic interests, seem to prevail. However, the nature of this conflict affects the domestic status of international instruments and their effect on patents. Weaker states determined to preserve their authority to legislate locally and to protect their threatened national interests may not enthusiastically embrace supranational institutions concerned with the promotion of foreign interests. The principles of international instruments on patents will ultimately pass through the filter of domestic jurisdictions of states and their embedded national priorities, preferences, and values.

The fourth section deals with the phenomenon of plant distribution across the globe and its overall impact and influence on the politics of control of plant life forms. Most of the world's plant genetic diversity occurs in the Third World. This is not solely a function of the whims of geography: it is also a consequence of the enormous efforts of local and traditional farmers and breeders, over the millennia, to conserve and improve plants. However, the diversity of plant life forms is currently under severe threat from the processes of globalization and the consequent homogenization of cultures.

The fifth section deals with the issue of the multiple causes of the modern loss of plant life forms. Certain factors are identified and examined. These include the culture of consumerism, the inequitable global economic regime, overpopulation, agribusiness/bioprospecting and biotechnology, climate change, and the homogenization of cultures. These problems present enormous challenges to the fledgling body of international environmental law. The sixth section explores how international environmental law has responded to the challenges of the erosion and loss of plant life forms. Towards this end, the seventh section examines the provisions of and contributions to the jurisprudence on plants' genetic resources stemming from the Convention on Biological Diversity and the Food and Agricultural Organization of the United Nations' (FAO) Treaty on Plant Genetic Resources.

Chapter 4 deals particularly with the genesis and legal structure of the institutional means for the appropriation and privatization of plants and indigenous peoples knowledge. It is divided into five sections. The first section deals with the concept and mechanism of appropriation and privatization. The major institutionalized process for appropriation of plants involves the International Agricultural Research Centres (IARCs), which function as a conduit for funneling plant germplasm from the South to the North. This mechanism has largely operated under the nebulous and erroneous notion that plants from the South constitute part of the common heritage of mankind.

The second section explores the early beginnings of the appropriation and privatization of plants and TKUP. The third section tackles the question of whether the notion of the common heritage of mankind (CHM) is part of the accepted principles of international law, and, if so, whether it is applicable to plant genetic diversity and TKUP. The conclusion is that CHM does not apply to plants and TKUP. The fourth section examines the appropriative role and functions of the IARCs, while the fifth section explores the role of the FAO in the politics of plant life forms as well as how international law has responded to the question of the legal status of plant germplasm stored in *ex situ* gene banks. As in the preceding chapters, the analysis is conducted within the context of the North-South divide.

Chapter 5 deals with the various methods by which the discordant values and policies of various patent regimes, particularly the United States patent system, have been adjusted and retooled to suit the interests of seed companies, pharmaceutical industries, and biotechnology industries. It is divided into two main parts. Given the differences between the law of patents on plants and TKUP, the first part deals specifically with plants and the development of the legal regime on patents on plants, particularly as it affects the appropriation and privatization of plants through the patent law mechanism. The second part is concerned with the appropriation and privatization of TKUP through the patent process.

The entire chapter is split into eight sections, all of which pay attention to the vagueness of and inconsistencies in the law on the criteria for patentability. All sections examine the various ways in which domestic legislative and judicial activities, especially in the United States and Western Europe, have been designed to facilitate the appropriation of plants and TKUP. Similarly, they examine the loopholes (as well as other juridical curiosities) that exist in the absence of a global standard on novelty and that facilitate the appropriation and privatization of TKUP.

The first section is introductory, while the second examines the changing concept of patentability as it relates to plants. The third section examines the patentability of TKUP. The fourth section examines the nature and quality of international law's response to the challenge of appropriation. Here, the normative thrusts at the soft law level and other emerging hard law commitments are juxtaposed and scrutinized. This section also offers a review and critique of modern ideas on how to stem the tide of appropriation. It gives some examples of how the patent system may be adjusted to accommodate the interests of marginalized cultures and societies that conserve plant diversity and TKUP. It pays particular regard to communal patents and a modified version of Plant Breeders Rights (PBRs).

The questions dealt with in Chapter 6, the concluding chapter, largely relate to the issue of the actual and probable consequences of an appropriative patent regime on plants and TKUP. These questions include, but are not limited to, the law of state responsibility as it affects appropriation and global food security. There are also questions pertaining to biosafety and the influence of patents on modified plants and the general environment.

The core focus of this book is on the sophisticated legal and institutional mechanisms that facilitate and legitimize the body of knowledge possessed by indigenous and traditional peoples concerning the various uses and properties of plants (as well as the derivatives and/or combinations thereof). Generally speaking, until recent normative changes, this body of knowledge "has not been recognized as being either 'scientific' or valuable to the dominant culture and so has been freely appropriated by others."[1] *Biopiracy* argues that, in appropriating plants and TKUP, capital interests across the industrialized world have largely employed two distinct but mutually reinforcing methods; namely, institutional and juridical mechanisms and a gendered/racist construct of non-Western contributions to plant development and use. More important, the legal and policy factors that facilitate the appropriation of indigenous peoples knowledge operate within a cultural context that subtly but persistently denigrates the intellectual worth of traditional and indigenous peoples, especially local women farmers.[2] Cultural biases in the construction of knowledge provide the epistemological framework within which plant genetic resources developed by indigenous peoples are continually construed as "free-for-all" commodities – commodities that are just waiting to be appropriated by those with the cunning and resources to do so. The first mechanism of biopiracy is, arguably, the establishment of the International Agricultural Research Centres (IARCs).

The IARCs have essentially functioned as pipes through which plant germplasm flows from the South to the North. A facilitative instrument of the first mechanism is the patent system. Since the institution of the IARCs and the consequent development of commercial seed businesses in the North, there has been a deliberate relaxation of the traditional conditions for patentability of inventions. This process, which John Frow has described as the "intense process of commercialization and privatization of knowledge,"[3] diminishes the stock of information or resources in the public domain,[4] raising the question of what remains of state sovereignty in the age of globalization. With external forces dictating minimum standards as well as the scope of what may be patented, domestic sovereignty with regard to such phenomena has been usurped by "global" trade institutions whose goals may be incompatible with domestic imperatives and values.

The globalization of patent regimes ignores grassroots concerns about the particularities, perspectives, localisms, and contingencies of societies at various stages of development or with different conceptions of material well-being.[5] Given that patent laws have international implications, there is the corollary issue in international law of state responsibility for domestic juridical and cultural institutions that facilitate the appropriation of indigenous peoples knowledge. As an instrument of Western notions of capital and property, the patent system reveals an intentional socioeconomic and political instrumentality that is alien to many other cultural philosophies.[6] Thus, within the context of the global system, I argue that patent laws and institutions are designed largely to protect and serve the values and interests of propertied states, particularly the so-called North. The pertinent question, however, is whether this regime is compatible with the increasing need for sustainable use of plant life forms, particularly in an age of globalization.[7]

From a strictly philosophical/analytical point of view, especially that proceeding from the Judeo-Christian conception of plants as the property of humankind, ownership of patents on plants, like other forms of property, is about relations between legal persons *inter se* as well as about property relations between plants and human beings.[8] It is doubtful whether patent rights over indigenous peoples knowledge[9] is reconcilable with alternative narratives of the ecosystem and ownership of property.[10]

The unease with conceptions of plants as mere economic units impels the need for a normative, distributive, and efficient assessment of the ramifications of appropriation of indigenous peoples knowledge. In this vein, *Global Biopiracy* serves a prescriptive purpose. Further, I argue that excessive and aggressive patent rights adversely affect the liberty of the public and increase to dangerous levels the power already possessed by those who privatize and appropriate knowledge. Unless this is checked, the overall consequence of this trend is that, with regard to "highly scientific/technological societies guided by cunning, very little is likely to remain free from appropriation."[11] Unless a normative and critical approach is taken to address the issue of appropriation of indigenous peoples knowledge, it is probable that, in addition to plants and knowledge, "all kinds of abstract information in the public domain will fall into private ownership."[12]

Patents, as units of the expression of property rights, must serve some moral value, but they cannot be the basis of moral value itself.[13] It is not enough to say that legal regimes, as legal instruments, exist to serve some ends; it would be better to engage in a vigilant and critical scrutiny of law as a means to an end. As such, international law on patents on indigenous peoples knowledge needs to be reoriented.[14] Legal concepts and institutions that merely constitute an avenue for the maximization of individual wealth at the expense of the overall integrity of the earth[15] and its supporting but marginalized human cultures and societies deserve critical reconsideration[16] and a radical shift in values.[17] As Chistine Frader-Frechette has urged, "we must shift from the individual to the community, from property to common heritage, from uniformity to diversity and poly-culture, from a short-term, quick returns view to a long-term sustainable approach, from exploitation to conservation of nature, from large scale projects to those of human scale."[18]

Global Biopiracy examines the structure and processes by which powerful states, prodded by multinational corporations, have used the norms and processes of the world's dominant patent systems to appropriate and privatize TKUP and various medicinal insights and practices of indigenous and traditional peoples. Beyond the misuse of the patent system,[19] I also look at the causes and effects of the asymmetrical flow of plant germplasm and TKUP from the industrializing states to the industrialized states. I argue that biopiracy is not a matter of aberration or the occasional malfunction of the world's dominant patent systems. To the contrary, the appropriation of TKUP is a predictable and intentional theft of indigenous and traditional knowledge and resources. It is a phenomenon that also implicates plant genetic diversity, global food insecurity, and the various individual and collective human rights of indigenous peoples.[20] Ultimately, at the heart of the debate is what policy direction the patent system should take in order to benefit the global community.[21]

Biopiracy raises serious issues pertaining to the conservation of biological diversity and genetic resources in agriculture, the integrity of plant life forms, a just international economic order, and development. Since the emergence of the biotechnology industry, "biopiracy" has become a lighting rod for activists in the biodiversity and anti-commons debate. Interestingly, a preponderance of commentators on this and ancillary issues are anthropologists, geographers, political scientists, ecologists, agricultural economists, and environmental activists.[22] This trend has greatly enriched the quality and diversity of discourse and has also captured the imagination of many.

However, there is also a tendency for commentators not well versed in the intricacies of international law and the technicalities of domestic patent laws to gloss over or misstate relevant principles of intellectual property law. A prime example of this tendency, as noted above, is the pervading notion that CHM applies to plants within the jurisdiction of states. Similarly, there is a tendency among many commentators to treat the principles or doctrines of patent law as revealed truth – something inflexible and unalterable. Consequently, in this book I attempt to bring a predominantly legal perspective to a complex issue.[23]

Ultimately, *Global Biopiracy* seeks to reconstruct a framework for understanding how the doctrines, principles, and cultural dimensions of patent law facilitate and legitimize the theft and appropriation of indigenous peoples biocultural knowledge.[24] The relationship between patent law and indigenous peoples knowledge is inherently predatory and harmful to the interests, worldviews, and self-determination of the Third World.[25]

2
Patents, Indigenous and Traditional Knowledge, and Biopiracy

The Concept of Biopiracy

Although the importance of plants to human civilization is well recognized, controversy surrounds the legal ownership and control of plant resources.[1] The debate on ownership of plant genetic resources often extends to what may be called "traditional knowledge of the uses of plants," herein referred to as TKUP.[2] The definition of what constitutes TKUP is a little more problematic. TKUP is generally defined in very broad terms. It encompasses a diverse range of tradition-based innovations and creations arising from intellectual activity in the industrial, literary, or artistic fields of indigenous and traditional peoples. Its range includes agricultural products, the medicinal use of plants, and spiritual worldview. TKUP is not a monolithic entity; rather, it is diverse and sophisticated.[3]

Generally speaking, the literature on TKUP largely seems to prefer such phrases as "indigenous peoples knowledge,"[4] "local knowledge," "ethnobotany,"[5] "tribal peoples knowledge,"[6] "folk knowledge,"[7] and so on, some of which are loaded with controversial and sometimes racist insinuations. The term "traditional knowledge[8] of the uses of plants" is preferred for the following reasons. First, although the term "indigenous peoples," with its useful idiom of "indigeneity," constitutes a powerful platform and "moral high ground" in the struggle for human equality and dignity,[9] some definitions of the term in international law may be restrictive. There are several definitions of indigenous peoples in international law. For example, the Martinez Cobo Report commissioned by the United Nations defines indigenous peoples, communities, and nations as "those which, having a historical continuity with pre-invasion and pre-colonial societies that developed on their territories, consider themselves distinct from other sectors of the societies now prevailing in those territories, or parts of them."[10] A treaty definition of indigenous peoples is provided by Convention 169 of the International Labour Organization (ILO): "Those who have descended from populations that inhabited a country at the time of conquest, colonization, or the establishment of present state boundaries, and who irrespective of their legal status, retain some or all of their own social, economic, cultural, and political institutions."[11]

The ILO definition is problematic in that it directly refers "to the histories of the Americas,[12] New Zealand, and Australia, and ignores African and Asian historical realities."[13] At a normative level it constitutes a validation of the discredited "blue water" or "salt water"[14] doctrine in international law on self-determination of colonized peoples.[15] It would therefore be better to adopt a more inclusive concept that recognizes that, within the context of the narrow definition of indigenous peoples offered by some international instruments (e.g., ILO Convention 169), traditional knowledge is not always "indigenous." As the recent World Intellectual Property Organization (WIPO) Report notes, "traditional knowledge is not necessarily indigenous. That is to say, indigenous knowledge is traditional knowledge, but not all traditional knowledge is indigenous."[16] For example, traditional knowledge of African communities regarding the uses of plants endemic to Africa is not indigenous knowledge (within the constraints of ILO Convention 169), it is traditional knowledge.

Second, although the term "tribal knowledge" appears in ILO Convention 169, it is not defined in any international legal instrument; rather, it has pejorative meanings,[17] carrying overtones of primitivism[18] and racial inferiority.[19] The term "traditional knowledge," in spite of its shortcomings, is less offensive than are the aforementioned terms. More important, as the WIPO Report defines it, traditional knowledge embraces a wide range of individuals, communities, and cultures living in a largely non-urbanized setup or, as the Convention on Biological Diversity characterizes it,[20] living "traditional lifestyles." What then are the elements of traditional knowledge? Daniel Gervais has pointed out that traditional knowledge, as represented in knowledge systems, creations, innovations, and cultural expressions, is distinguishable according to how it is generated and transmitted.

A rider must be quickly added; that is, the term "traditional knowledge" should not be understood as denoting or even connoting notions of antiquity, stagnation, and immutability. As the Four Directions Council of the First Nations of Canada has pointed out:

> What is "traditional" about traditional knowledge is not its antiquity but the way it is acquired and used. In other words, the social process of learning and acquiring which is unique to each indigenous group lies at the heart of its "traditionality." Much of this knowledge is actually quite new, but it has a social meaning and legal character, entirely unlike the knowledge indigenous people acquire from settlers and industrialized societies.[21]

Contemporary international law[22] and leading scholars[23] on the subject recognize the dynamism, holism, and diversity of traditional knowledge. To quote from the WIPO Report, "intertwined within practical solutions, traditional knowledge often transmits the history, beliefs, and traditions of a particular people. For example, plants used for medicinal purposes also often have symbolic value for the community. Many sculptures, paintings, and crafts are created according to strict rituals and traditions because of their profound symbolic and/or religious meaning." In light of the above, international law has not yet formulated an adequate definition of traditional knowledge. With regard to this book, however, traditional knowledge pertaining to the uses of plants may be defined as that body of evolving knowledge, including the innovations of individuals and communities, that operates outside the dominant Eurocentric paradigm and that is concerned with the use of plants for social, environmental, medicinal, and therapeutic purposes.

As a body and system for the acquisition, transmission, control, and propagation of knowledge, indigenous and/or traditional knowledge of the uses of plants has moved from the peripheries to the core of modern debates in international intellectual property law. However, the relationship between TKUP and patents remains one of the most widely contested issues in this area.[24] Given that the patent system in and of itself evokes controversy,[25] the addition of indigenous peoples knowledge creates a heated and divisive debate.[26]

This heated debate has been conducted within the context of a growing belief by Third World states, scholars, and their sympathizers in the industrialized world that there is an overwhelming asymmetry in the way major intellectual property systems – especially patents – protect the intellectual property of the industrialized countries while ignoring, and in some cases appropriating, the intellectual creations of Third World peoples and cultures. On the other hand, those who defend the structures, practices, and norms of the dominant models of intellectual property protection in the industrialized world argue that whatever injustices result from the operation of the major intellectual property systems are mere aberrations rather than the systemic predation on Third World biological resources and knowledge. In the words of Naomi Roht-Arriaza, "the appropriation of the scientific and technical knowledge of indigenous and local peoples, of the products of that knowledge ... has become both notorious and contested."[27] The exploitative relationship and alleged theft of biodiversity and associated traditional knowledge has given rise to allegations of "biopiracy" against industrialized states and corporate institutions involved in the business of bioprospecting and the commercialization of indigenous peoples biocultural knowledge.[28] What is biopiracy? Does the term have a relevant and juridical significance beyond its apparent rhetorical[29] and emotive value? With respect to its history, the word "biopiracy" was coined by Canadian activist Pat Mooney. As Graham Dutfield rightly observes the term was devised as

> part of a counter attack strategy on behalf of developing countries that had been accused by developed countries of condoning or supporting "intellectual piracy," but who felt they were hardly as piratical as corporations which acquire resources and traditional knowledge from their countries, use them in their research and development programs, and acquire patents and other intellectual property rights – all without compensating the provider countries and communities.[30]

In order to arrive at a definition of biopiracy one must appreciate the historical context within which the term arose. First, Western intellectual property owners have often accused Third World states and economic actors of "pirating" or unlawfully "appropriating" the intellectual property rights of industrialized entities, especially patents and copyrights. In the wake of biotechnological inventions and the patenting by Western states and entities of indigenous peoples biocultural resources, obtained without their lawful informed consent, Third World States contend that industrialized states, business entities, and research institutions are "pirating" their biological resources. Therefore, the Third World applies the term "biopiracy" to describe what it sees as a misappropriation of indigenous peoples knowledge and biocultural resources, especially through the use of intellectual property mechanisms.[31] If the infringement of patents, copyrights, and trademarks constitutes intellectual piracy, then so does the failure to recognize and compensate indigenous and traditional peoples for the creations arising from their knowledge. Inherent to the biopiracy rhetoric are the notions of unauthorized appropriation/theft of biological diversity and its associated traditional knowledge. The concept of biopiracy concerns law, ethics, morality, and fairness.

Curiously, in defining biopiracy some institutions and scholars posit that the period when the alleged theft was perpetrated is a crucial factor. Thus, the International Chamber of Commerce (ICC) contends that, "in our view, a rational definition of 'biopiracy' would focus on activities relating access or use of genetic resources in contravention to national regimes based on the Convention on Biological Diversity."[32] As Remigius Nwabueze has persuasively argued, there are two serious fallacies embedded in this definition of biopiracy. First, it pretends that national sovereignty over plant genetic resources within the territories of states began with the Convention on Biological Diversity in 1992.[33] Such an argument is untenable. As I have shown elsewhere, states have always had the inherent right to control legal access to resources, including plant resources, within their territorial domains.[34]

Therefore, it is not of any importance whether the theft/appropriation of the biocultural resource in question occurred prior to or after the Convention of Biological Diversity came into effect.

Second, the ICC definition excludes non-members or signatories to the Convention on Biological Diversity. In effect, unless a state is a signatory to the Convention on Biological Resources it can have its biocultural resources stolen by others and not be able to make a case on biopiracy. Without question, this is an absurd argument. A state cannot lose its property simply because it refuses to be a party to a treaty.

Accordingly, biopiracy may be defined as the unauthorized commercial use of biological resources and/or associated traditional knowledge, or the patenting of spurious inventions based on such knowledge, without compensation.[35] Biopiracy also refers to the asymmetrical and unrequited movement of plants and TKUP from the South to the North through the processes of international institutions and the patent system. As Rosemary Coombe has rightly pointed out, this process is characterized by the non-recognition of the intellectual contributions of holders and practitioners of traditional knowledge towards the improvement of the plants or TKUP in question.[36] The question that arises now is, how does the patent system appropriate TKUP?

In deconstructing biopiracy through the patent system, it is imperative to pay attention to both direct and circumstantial evidence as well as to the subtleties of the historical development and internationalization of the patent system, the racialization of knowledge, international economics, and the rise of the biotechnology industry, all of which are implicated in biopiracy. This approach is indispensable because the modern process of appropriation of plants and TKUP is sophisticated and subtle, quite different from the blatant physical bravado of colonial pirates. The appropriation of plants and TKUP through the patent system presents itself as a respectable business fully supported by the paraphernalia of apparent legality. Thus, in order to identify and appreciate the role of the patent system in the appropriation of plants and TKUP, critical attention must be paid to several factors, including the history of the patent system, the original scope of the concept of patentability, the Western biases of the patent concept itself, the circumstances in which the patenting of plants arose and gained global strength, the global imbalance in the distribution of plants, and, of course, the deliberate relaxation of the threshold for patentability of plant inventions and TKUP products.

In addition, it is not enough to analyze what the legal norms of the patent system seek to protect; what they neglect to protect is equally relevant. In short, the patent system must be thoroughly interrogated and its intellectual integrity should not be presumed. In this regard, particular attention ought to be paid to some pertinent US statutes and case law. In addition, the influence of US-based multinational seed and pharmaceutical corporations in shaping patent and plant breeders laws deserves scrutiny. The reasons for this emphasis on the US patent system are as follows: (1) it is common knowledge that Article 27 of the Agreement on Trade-Related Aspects of Intellectual Property Rights (TRIPs), constituting the global minimum threshold for patentability, is an approximation of US jurisprudence and ideology; (2) it is also known that TRIPs is a product of the immense clout of the American pharmaceutical and biotechnology industries; (3) the US patent system accounts for almost half of all patents issued in the world; (4) most of the controversial patents that raise the question of biopiracy were issued by the US Patent Office; (5) the United States has the most appropriative regime on patents; and (6) the pronouncements and decisions of US courts on matters of patent law have immense international influence.

Examples of biopiracy reveal the manifold ways in which legal principles and cultural biases against non-Western forms of epistemology conspire to enable the appropriation of traditional biocultural resources. For example, local farmers in Nigeria developed an insect-resistant cowpea. Needless to say, those local farmers did not "publish" their findings or their results in a "reputable journal" reviewed by their "peers." However, on a trip to West Africa, Angharad Gatehouse, a scientist at the University of Durban, obtained some of these seeds. Using "formal" techniques, he identified in "scientific language" the genetic mechanism that causes the locally developed cowpeas to be insect-resistant. As Buchanan notes, "he [the scientist] promptly left the university and joined Agricultural Genetic Company of Cambridge and they proceeded to apply for a patent on their 'invention.'"[37] The practical result was that local farmers were short-changed by the interplay of patent systems, which erased their hard work and intellect simply because they failed to "publish" their observations in written form.

The double standard inherent in this regime creates a legal order of patents by discovery or importation that is similar to medieval European practices. As William Lesser notes, "those who complain of a double standard regarding IPR [Intellectual Property Rights] protection for genetic materials have a legitimate position."[38] In effect, non-Western contributions to plant development are systematically relegated to an inferior status for appropriation through the patent system.[39] A few concrete examples of this permissive regime may suffice to make my point.

In 1999 two Australian government agricultural agencies attempted to patent chickpeas grown by subsistence farmers in India and Iran. These were unknown to farmers and scientists in Australia until the chickpea germplasm was stored in an international gene bank. What stopped the patent application was a protest by some activists.

In 1997 two professors at Colorado State University were asked to abandon patents on *Quinoa Chenopodium Quinao*, an important food crop among the Andeans. The professors, Duane Johnson and Sarah Ward, applied for and obtained US Patent No. 5,304,718 on a traditional Bolivian variety of quinoa called "Apelawa." This gave them an exclusive monopoly over male sterile plants of this variety and their use in creating other hybrid quinoa varieties. The patent covers both male sterile Apelawa Quinoa and "any" quinoa hybrid that is derived from it. The point here is that male sterility in Andean farmers' varieties of Quinoa had been known for decades among the Andean farmers and peasant farmers in Bolivia, Peru, Ecuador, and Chile.[40]

Furthermore, in 1997 a Texas-based company (RiceTec) acquired US Patent No. 5,663,484 on basmati[41] rice lines and grains. The patent, with twenty claims, covers alleged novel methods of breeding, preparing, and cooking basmati rice. RiceTec's claims are for a specific rice plant (Claims 1-11, 14), for seeds that germinate the patented rice plant (Claim 12), for the grain that is produced by that plant (Claims 13, 15-17), and for the method of selecting plants for breeding and propagating particular grains of rice (Claims 18-20). It should be noted that for centuries basmati rice has been grown and developed in the Greater Punjab region, now split between India and Pakistan. Basmati rice is world-famous for its fragrant aroma, long and slender grain, and distinctive taste. Indeed, the *Oxford Dictionary* defines "basmati" as a "long grained aromatic kind of Indian rice."[42] In 1997 exports of basmati rice constituted 4 percent of India's export earnings. Basmati refers to a particular class or rice, of which there are over 400 varieties in India and Pakistan. Over one million hectares of rice paddy are cultivated in India with basmati rice per annum and 0.75 million hectares are cultivated in Pakistan. In 1998-99 alone India exported US$425 million worth of basmati rice.[43] In this case, the patent on basmati not only appropriated a globally recognized name but also threatened the livelihood of thousands of Punjabi farmers who exported basmati rice.

However, as a result of the re-examination application filed by the Indian government through an NGO named APEDA (Agricultural and Processed Food Products Export Development Authority), RiceTec agreed to withdraw the claims. Further, on 29 January 2002, the United States Patent and Trademark Office issued a re-examination certificate cancelling Claims 1-7, 10, and 14-20 (the broad claims covering the rice plant) and amended Claims 12-13 concerning the definition of chalkiness of the rice grains. However, Claims 8, 9, and 11, covering the specific samples of the plant from the genetically modified seed itself, still stand.

These instances of biopiracy cannot be fully understood outside the context of the historical origins and philosophical foundations of the patent system, the asymmetry in the global distribution of plant genetic resources, and the economic and legal hegemony of the industrialized world over indigenous and traditional peoples. It is therefore necessary to revisit the origins of the patent system and tease out how its philosophical imprints have facilitated the appropriation of indigenous and traditional knowledge.

Origin of the Patent System

Contrary to the popular belief that the patent system originated in industrial England in the post-medieval era,[44] the concept of patent was in fact conceived in Florence, Italy, under circumstances that Owen Lippert has likened to "blackmail."[45] In 1421 the medieval Florentine architect and inventor Filippo Brunelleschi invented an iron-clad sea-craft christened the *Badalone*, which he claimed could transport marble across Lake Arno for the construction of the now famous suspended dome of the cathedral in Florence. Contrary to the existing tradition of disclosure of new scientific discoveries and inventions, Brunelleschi refused to disclose his invention unless the city granted him a limited right to sole commercial exploitation of the sea-craft. Florence yielded to his unprecedented demands and, on 19 June 1421, granted him a public letter to that effect.[46] However, to his and Florence's mutual embarrassment, the *Badalone* sank on its first excursion on Lake Arno. For a long time thereafter, Florence stopped issuing patents. Recovering from this watery debut and failure, the concept of patents drifted to neighbouring Venice, where it became anchored in what is perhaps the first substantive patent statute in the world.[47]

The Venetian patent statute of 19 March 1474 was pioneering in several respects.[48] It offered protection for a period of ten years to all inventions that passed the examinations of the General Welfare Board. In addition, it also provided for punishments for the unauthorized use or infringements of patent grants.[49] Further, in 1474 Venice instituted a registry of patents.[50] As I argue elsewhere, there is little doubt that modern substantive patent law and procedure originated in Venice.[51]

Defining Patents

The term "patent," as an adjective, derives from the Latin verb *patere*, which means "to be open." In relation to a document it means an "open letter addressed to the public."[52] Through such "open letters" European monarchs in the Middle Ages conferred special privileges, status, or titles on individuals or groups of people.[53] In modern times patents have retained their original character as governmental grants or privileges.[54] Machlup has defined a patent title or grant as that "which confers the right to secure the enforcement power of the state in excluding unauthorized persons, for a specified number of years, from making commercial use of a clearly defined invention."[55] What the patentee gets is "the right to exclude other persons for a limited time from making a commercial use of the invention without his/her consent."[56]

Philosophies and Theories of Patents

Since the emergence of patents in Continental Europe, various theories have been advanced to support the existence and imposition of patents on indigenous and traditional peoples. Theories proffered in defence of the patent system clearly show that it is not ideologically or culturally neutral. The philosophy and ideology of the patent concept do not constitute a global value. The patent system is as local, as culture-bound and ethnic, as are comparable legal concepts. The contemporary geographic universality of the patent system should not be mistaken for normative universality. The concept of patents is European in origin as well as in ideology.

Patents constitute and represent a congealed form of a particular cultural ideology, philosophy, and jurisprudence concerning property rights and economics.[57] With respect to so-called intellectual property, early theorists posited that "the labours of the mind and productions of the brain are as justly entitled to the benefit and emoluments that may arise from them, as the labours of the body are."[58] Scholars such as John Locke expounded this theory in terms of possessive individualism, arguing that "every man has a property in his own person. The labour of his body, and the work of his hands, we may say are properly his."[59] Although the Lockean theory has influenced modern patent law, it does not explain the phenomenon of an employer's property being the "inventive genius" of his/her employee.[60]

On the sociocultural plane the metamorphoses and crystallization of intellectual exertions into subjects and objects of ownership[61] covered by patents may be attributed to two socioeconomic factors that have largely defined the Western world – namely, the rise of individualism[62] (and, until recent times, the preeminence of man)[63] and the development of capitalism. Ullman's magisterial work on the former asserts that individualism has some basic elements: the dignity of man, autonomy or self-direction, privacy or private existence within the public world, and self-development.[64] These may be categorized under the concepts of abstract individualism, political individualism, economic individualism, religious individualism, ethical individualism, and epistemological individualism.

The concept of individualism is celebrated and cherished in the Western world. In the words of Oscar Wilde, "for the full development of life to its highest mode of perfection ... what is needed is individualism."[65] In short, man, in the gender-specific meaning of the word, was the supreme being on the face of the earth. It is thus no coincidence that the earliest promoters of the patent system were elitist European men – the gentry and the so-called "gentlemen of science"[66] whose "common interest lay in the promotion and 'professionalization' of scientific activity."[67]

The Calvinist emphasis on the so-called work ethic and the virtue inherent in the accumulation of wealth added to the ideology of individualism.[68] As Jeremy Bentham argued, "[the] individualistic rationalism of capitalism is to be supreme also in the realm of law since egoism is the basic axiom of the legal system even as it is of the economic system."[69] Thus, the primacy of the individual[70] and the dominance of capitalism in Western jurisprudence (as manifested in the patent concept) provide evidence that "intellectual property is a reflection of the Western concern for the *rights of the individual.*"[71]

The notion of a systematized legal entitlement of exclusive and individual rights to the so-called creations of the mind has been "linked to the advent of modern capitalism, individualism, industrial organization, and the technological age."[72] The erroneous yet often unchallenged assumption at work here is that the inventive process is a completely individual enterprise. The reality is that inventions and innovations do not spring *ex nihilo*. Inventors, artists, and other creative people draw from the stock of pre-existing human knowledge and cultures. Isaac Newton spoke well when he noted that, if he had actually "seen further" than the rest of his scientific peers, then it was because he stood on the shoulders of other giants. Put simply, creativity is tied to tradition and the existing stock of knowledge.[73]

In contrast to the idea that the inventive process is an individual effort, most non-Western societies emphasize its "informal," communal nature. And, more important, for the most part they construe information (and, hence, invention) as being a public good.[74] Although Lowie's anthropological study shows that the concept of patents was highly developed among the Andaman Islanders, the Kai, the Koryak, and the Plains Indians,[75] there is a fundamental difference between this and the more generally known European patent system. As Peter Drahos points out, "these societies were, in contrast to Western [societies], more concerned with restricting the transferability of such rights"[76] than with the idea of establishing an exclusive right due to the individual-as-creator. The remarkable thing is that the Western approach, especially after the US entry into the Union for the Protection of Industrial Property towards the end of the eighteenth century, emphasized patents as private property rather than as instruments of public policy.[77] Hence, private "toll-gates," or "fences," were erected around knowledge according to the concept of *ius excluendi* and *ius prohibendi*.[78] *Ex necessitate*, the Western patent system[79] raises cultural and philosophical differences with non-Western societies.[80] However, it has to be recognized that the notion of patents as exclusively private property is contested even in the West. Modern patent laws continue to impose limitations on the exercise of those rights related to ownership of patents. There is, therefore, a combination of both public and private rights over patents.[81]

The patent system was designed to promote the employment of capital and industry in new and profitable directions. Various theories have been posited to rationalize the ideology of the patent system. Like the elusive search for a unifying scientific theory of the universe, the patent system has defied a unified theory.[82] Penrose distinguishes four theories on patents, namely, the natural right theory, the contract/disclosure of secrets theory, the reward theory, and the incentive theory. On the other hand, Oddi has distinguished between the classical and postclassical theories of patents.

Natural Rights Theory of Patents

This theory argues that an inventor has a natural right to his/her invention and that society, represented by the state, has an obligation to recognize, protect, and enforce that right. The natural right theory of patents found expression in the aftermath of the French Revolution,[83] and it has inconsistencies that render it indefensible as a justification for the patent system. First, it requires an acceptance of the notion that ideas can be subjects of exclusive ownership, a position difficult to maintain in ordinary societal relations. Second, it posits that patents are not privileges of governments but, rather, the inherent right of an inventor. Various limitations on patents, such as patentable subjects, duration of patents, and compulsory licensing, render such a proposition unsustainable. Theorists of property have long held that notions of absolute ownership of patents are unrealistic.

According to Edith Penrose international patent law is not based on natural law theory but, rather, is a "social policy" determined by a balance of national costs and benefits.[84] The natural rights theory of patents has also been forcefully rejected by the Secretary-General of the United Nations' Report on the subject.[85] According to this report,

> patent legislation has never been based solely on the concept of the patent as the confirmation of an inherent, rather than the creation of a statutory property right. Such a concept would have left no room for such restraints on the patent grant as its fixed duration, its exclusion for inventions in certain fields ... and the forfeiture or compulsory licensing of patents for failure to work them.[86]

Virtually all patents jurisdictions in the world place limitations on whether or not certain inventions deemed sensitive to national security are patentable.[87] Even the United States, the most vocal advocate of a strong patent system, neither affords nor practices a natural rights theory of patents. For example, inventions in the atomic energy and space/aeronautic fields are not patentable in the United States.[88] Inventions deemed to be detrimental to the national security of the United States are not patentable and may be kept secret or withheld by the government as long as "national security"[89] requires. In sum, the natural rights theory of patents "is virtually dead in learned circles"[90] and neither explains nor justifies the concept of patents.

Reward Theory

This theory argues that inventions are made because the patent system offers a reward to inventors. According to this argument, without this promised reward there would be no inventions.[91] There are some obvious flaws with this theory. First, the emphasis on monetary gain is somewhat generalized: not all inventions are motivated by lucre or expectations of material fortune. Moreover, the commercialization of inventions, on the one hand, and inventiveness per se, on the other, are two distinct phenomena and ought not to be confused.

The better argument for patents would seem to be that they probably serve as a useful incentive for the commercialization and industrialization of inventions. Thus, the crucial role that patents perform is to offer a profit incentive, encouragement, and security to those desirous of commercializing inventions. Inventions would always occur, regardless of the existence or non-existence of a reward mechanism. Another reason the reward theory does not adequately explain the phenomenon of inventiveness is that history is replete with accounts of inventions that were simultaneously created by different inventors in different places. For example, Polzunov in Russia had invented a steam engine before James Watt.[92] Watt got a patent and Polzunov did not. If these inventors had been in the same country, which of them would claim the reward promised by the patent system?

Contract/Disclosure of Secrets Theory

This theory posits that the purpose of the patent concept is to serve as a contract between the inventor and the state. In return for disclosing the secret of his/her invention, the inventor receives from the state a limited monopoly over its use. Critics on several fronts have assailed this theory. First, when secrecy is possible, inventors and industrialists would probably prefer other forms of legal protection (e.g., trade secret laws). Second, even if the inventor kept his/her invention secret, other inventors would eventually hit upon the same idea. This is because, ultimately, inventions are called forth by the needs of society ("necessity is the mother of invention"). Once again, the theory in question confuses commercialization of inventions with the encouragement of inventiveness. Third, there is no way of objectively certifying that the monopoly granted the inventor is equal to the social benefit of the invention. Several inventions that have later proved to be of immense usefulness were "ahead of their time" and, consequently, earned nothing for their creators. A good example is the fax machine, which was invented in 1842 but was not commercialized until the early 1980s.[93]

Encouragement of Invention Theory

There is a general assumption, almost an axiom among patent lawyers, that patents have a causal or organic link with inventiveness and industrialization.[94] An English patent attorney, in attributing the success of the English industrial revolution to the patent concept, declared: "The patent system was our invention, and it gave us the first place among nations in industry for over 200 years."[95] A German patent attorney also vouchsafed that "the countries whose industries take foremost place, also rank highest in patent policy."[96]

Taking these statements *ex facie*, it is tempting to conclude that the patent system is one of the greatest achievements of human ingenuity – something without which humankind would languish in primitive anarchy. Surprisingly, the proof of the causal relationship between patents and inventiveness and economic progress is, at best, very meagre. Upon close analysis, one finds that the hypothesis that patents propel inventiveness is founded upon anecdotes[97] and debatable assumptions.

Thus, in spite of generalized, polemical assertions of a causal link between patent regimes and inventiveness, the "most well-reasoned studies of patent systems"[98] have failed to establish it. Indeed, economists are almost unanimous in their belief that there is no conclusive evidence to show that patent systems have any causal relationship with inventiveness.[99] Surveys of business leaders (with the notable exception of pharmaceutical companies) typically place a low ranking on patents as a stimulant for research and development.[100] The most fundamental difficulty in making any rational claim for or against the alleged relationship between patents and inventiveness is the impossibility of separating out other factors contributing to technological inventiveness, such as "local resource endowment, education of the labour force, availability of capital, and dynamism of the local market."[101]

There are at present no scientific tools with which to compare "what has happened [with] what would have happened"[102] had there not been a patent system. Other motives, apart from the desire for profit, motivate inventiveness. Many scientists and inventors enjoy experimenting with complex things and have a strong interest in gaining academic honours and the respect of their peers. Dr. Salk invented the polio vaccine and gave it to the world freely.[103]

More important, the pervasive phenomenon of serendipity[104] does not fit within the narrow confines of the "patents-propel-technology" hypothesis. As the *Time Magazine* special edition on inventions noted, "in the history of scientific and technological endeavor, there are few if any cases in which the end was exactly what was intended."[105] The list of serendipitous inventions is endless, and in many cases the inventors did not even recognize the full import of their inventions.[106]

Furthermore, as a matter of both historical and modern reality, the expensive nature of the patent process,[107] the complexity of the law,[108] and the radical change in the social structure of inventive activities all cast doubts on the axiom that the patent system is designed for the individual inventor who lacks an arsenal of financial support. In addition, the "historical necessity" argument for patents is erroneous and misconceived. In a classic study of the relationship between patents and the British Industrial Revolution, W.H. Price concluded as follows: "In the mechanical process that took place in the sixteenth and seventeenth centuries, patents were not the leading factors. Some of the most successful mechanical innovations of the period did not enjoy any patent."[109] Similar studies by Ashton,[110] Deane,[111] and Mathias[112] "all hesitate to assert a causal relationship"[113] between patents and the Industrial Revolution.

Indeed, Ashton's studies showed that, on some occasions, the patent system blocked the "way to new contrivance."[114] In his view discovery and innovation might have "developed equally rapidly without the patent system."[115] Schiff's thorough study of inventiveness and patent institutions in eighteenth-century Netherlands and Switzerland led to the conclusion that "inventive activity can be quite vigorous in countries without a patent system."[116] The Netherlands repealed its patent laws in 1869 and the Swiss had no national patent law until 1888, yet during their patentless period both countries witnessed inventiveness and industrialization. Ashton believes that history shows that "patents have not been synonymous with inventions."[117] In modern times brewing, pottery, and the automobile industry have largely operated outside the patent regime yet have made enormous strides in innovativeness and inventions. In sum, it seems that the preponderance of reasoned opinion and empirical research agrees that "the industrialization of a country can proceed smoothly and vigorously without a national patent system."[118]

Interestingly, early European industrial history fails to support the argument for a direct causal link between the patent system and inventiveness. According to Anderfelt, "in a study of the birth and decline of early patent systems in some continental countries ... the conclusion appears to be that the patent institution followed rather than preceded economic and cultural development and later lost its importance when these activities became less intensive."[119] With respect to the Venetian patent system, Anderfelt's study observes that

the existence of the Venetian Patent Act, and of early patent systems in some other continental countries, is interesting from another point of view.

The idea so cherished by the patent advocates, that there exists a more or less organic relationship between the existence of a patent law and the economic and industrial performance of a country, cannot be supported at all by this early experience. In Venice the patent system was introduced when that state was already at the height of its development.[120]

Thus, the argument by patent advocates that the patent system is coterminous with industrialization or that it is a historical necessity is not supported by the history of industrialization and the practice of states.[121] With respect to non-Western societies, it is clear that technological feats have been consistently performed and recorded without the aid of the patent system. As John Needham's epic work on Chinese civilization[122] argues, "Imperial China is an example of a society that achieved spectacular outcomes in science and innovation without the instrumentality of a patent system or a customary equivalent."[123] Similar technological progress was recorded under the Arab and Pharaonic[124] civilizations, respectively, without the aid of a patent system.[125] The import of these irrefutable facts, according to Peter Drahos, is that the "connection to intellectual property, science and economic development is contingent and local rather than necessary and universal."[126] Accordingly, it would be simplistic to attribute modern industrial attainments to the patent system.

A more tenable argument in justification of the patent system has been articulated by Judge Simon Rifkind, co-chairman of the United States President's Commission on the Patent System. In his words,

> the really great, creative geniuses of this world would have contributed their inventions even if there were a jail penalty for doing so. But that in itself would not have been sufficient. *The patent system is more essential to getting together the risk capital which is required to exploit and to develop and to apply the contributions of the genius inventor than to provide a stimulus for the actual mental contribution.* It is to the former that the economic incentive is indispensable. The money will not be risked unless there is some sense of assurance that a benefit will be obtained.[127]

Judge Rifkind's acute observation makes eminent sense. Surprisingly, most advocates of the patent system have consistently confused the commercialization of inventions with inventiveness.

Apart from conflating inventiveness and the commercialization of inventions, the patents-propel-inventiveness hypothesis romanticizes the inventive process and ignores fundamental changes in both the social and legal structure of inventiveness and the ownership of inventions. First, it is necessary to appreciate that the patent system was originally designed for the individual inventor – the proverbial "basement," or "garage," inventor who worked in loneliness and expended his/her scarce resources in the search for socially useful contrivances. However, starting from the era of the British guilds and continuing until the modern age (in which scores of scientists and inventors labour for pay in corporate and privately owned legal entities), the practice of "collective invention" has relegated the legendary lonely inventor to the margins of history.[128] Today, the romantic paradigm of the individual inventor has, as Edith Penrose declared, become more or less an anachronism.

The contemporary reality is about collective inventions and corporate/public ownership of inventions. In arriving at this interesting conclusion, Anderfelt's three-pronged test is pertinent: (1) who is doing research today? (2) who is supplying the money? and (3) who is exploiting the results?[129] With respect to the first question, it is obvious that, for the most part, the inventor remains the individual; however, the context within which she/he now carries out her/his work has changed. Nowadays, inventive activities are carried out by inventors employed by industry or public research institutions.[130] The point is that "these inventors do not live from the particular results of their inventive activities but from a regular income."[131] In effect, the formerly self-employed inventor has to a large extent become an employee of industry.[132]

In a critical sense, the regime of employer ownership of patents marks the ultimate subversion of the romantic raison d'être of the patent system. The dispossesion of the employee from the fruit of his/her own genius through employed remuneration lays bare the capitalist thrust of the patent system. Under this new regime, "few inventions ... can be exclusively attributed to the work of a single individual ... contemporary research is increasingly in the form of teamwork, and often scientists from different fields cooperate on a particular project."[133] The consequence is that now more than 90 percent of all patents are granted to employers rather than to the real inventors. Philosophically, by recognizing an employer's property in the invention of the employee (under some circumstances), the patent system contradicts its original purpose, which was to provide incentives to individual inventors for their "inventive genius." As Soltysinski laments, "the recognition of the employer's right to inventions made by his employee has resulted in depriving the latter of all benefits associated with a patent."[134]

In most cases, apart from the regular salary and the usual raise in remuneration (which other employees with no connection to the inventions also receive as part of their regular entitlements), the individual inventor gets no special personal incentive from the patent system. In the final analysis it is difficult to disagree with Bhupinder Chimni's analogy: "to say that the potential of a patent actually stimulates invention is a lot like saying that you can spur the donkey on by offering a carrot to its rider."[135]

In these circumstances, if individual inventors continue to want their "just reward" for an invention, they may have to look beyond the patent system. To be precise, they would have to depend on their bargaining strength vis-à-vis the corporate and/or public employer. Of course it cannot be denied that providers of infrastructure conducive to inventiveness contribute towards innovation. The fact remains, however, that the patent system was not designed to reward those who provide the infrastructure that enables the emergence of inventions.

Furthermore, it is often argued that states, particularly poor countries that complain of biopiracy but that desire foreign direct investment (FDI), should institute strong patent regimes to attract investment and thus improve their economy and standard of living. Even though this argument seems attractive, it is a simplistic response to a complex problem. Significantly, there is little empirical support for this argument. A study by the United Nations Transnational and Management Division shows that there is little or no empirical evidence to support the view that patents encourage FDI.[136] In fact, other studies have found that:

> the countries with the weakest levels of IPR's protection – the People's Republic of China, Taiwan, Brazil, Argentina, Thailand, etc – over the past decade have routinely been the recipients of the largest net FDI inflows. There has been a significant correlation between the United States Trade Representative's (USTR) list of worst IPRs violators and the highest levels of US foreign direct investment.[137]

Finally, turning to the question of the lack of data to support the patents-propel-technology hypothesis, the near unanimous verdict of intellectual property rights (IPR) economists who have studied the phenomenon is one of doubt, if not polite rejection. According to Frederick Abbott,

> there is a serious gap in economic data and analysis which might demonstrate the positive impact of patent protection as claimed by the patent holders. This gap is well known and accepted among intellectual property rights (IPR) economists and public policy specialists. Since the value of patents to the international economic system has not been empirically demonstrated, there is no concrete basis for analyzing the trade-off in values that patent holders have suggested.[138]

In the considered view of the famous economist Sir Arnold Plant, "the science of economics, as it stands today, furnishes no basis of justification for this enormous experiment in the encouragement of a particular activity by enabling monopolistic price control."[139] Other notable economists and scholars have affirmed this view.[140] Perhaps the last word on this issue should be reserved for economist Fritz Machlup, whose commissioned study on the subject remains a *locus classicus*. In his oft-cited conclusive opinion submitted to the United States Congress, he wrote:

> None of the empirical evidence at our disposal and none of the theoretical arguments presented either confirms or confutes the belief that the patent system has promoted the progress of the technical arts and the productivity of the economy. Scholars must not lack the courage to admit freely that there are many questions to which definite answers are not possible, or not yet possible. They need not be ashamed of coming forth with a frank declaration of ignorance ... No economist, on the basis of present knowledge, could possibly state with certainty that the patent system, as it now operates, confers a net benefit or a net loss upon society.[141]

It seems that, with or without the patent system, inventions would continue to be made. This would seem to be one of the strongest, yet least appreciated, arguments that critics of biopiracy have made against the globalization of the patent system. However, one could argue that the patent system is useful as a guarantor capable of enabling the investment of the risk capital needed to commercialize an invention. Without this form of guarantee, it is reasonable to suppose that investors might be leery about financing inventions. However, this is quite different from arguing that the patent system necessarily encourages inventiveness. More important, even if patents facilitate the commercialization of inventions, does this address the concerns of indigenous and traditional peoples who watch helplessly as their biocultural knowledge is transformed into commercial articles without their consent and without any compensation?

Postclassical Theories of Patents

As earlier indicated, Samuel Oddi's approach distinguishes between the classical and postclassical theories of patents. The theories already discussed may be categorized as classical. On the other hand, the prospect theory,[142] the race-to-invent theory, and the rent-dissipation theory may be categorized as postclassical. Edmund Kitch's well known article drew an analogy between the patent grant and the United States' mineral claims system. This theory has received substantial criticisms, which need not be rehashed here.[143]

Professor Robert Merges and Richard Nelson developed their race-to-invent theory in 1990. According to them, the patent system may be rooted in the belief that society would benefit from granting patents with a relatively narrow scope as this would permit innovations.[144] Samuel Oddi has argued that the most recent theory advanced on patents is the rent-dissipation theory. The essence of this theory is that "patents should minimize rent dissipation at the invention (conception) and innovation (improvement) stages."[145] None of the theories of patents as presently understood offers an unimpeachable justification for patents. Although, with respect to the patent system, scholarship has ranged from faith to agnosticism to absolute dismissal, its instrumentalist character remains the recurrent feature of the practice and institution of domestic patent regimes.[146] It is doubtful whether this trend would yield a more global attitude towards patents – one that seriously considers the values, ideologies, and economics of marginalized states and cultures. Given the absence of a universally accepted justification for patents, how did the concept, from its origins in Venice and Florence, become universal?

The Diffusion and Colonial Migration of the Patent System

The diffusion of the patent concept from the Italian peninsula was not an accident; rather, a combination of certain factors, which led to the decline of Italian pre-eminence in technology, facilitated the diffusion of skilled artisans and the consequent spread of the patent idea from Italy to Central and Western Europe. First, the discovery by medieval seafarers and merchants of a direct sea route to India led to the bypassing of the Italian peninsula and, consequently, to a decline of commerce in the area. Second, the evolution of modern statehood and the rise of Christian doctrines incompatible with Roman Catholicism challenged the temporal authority of the papacy in medieval Italy.

The Church's intolerance of these new Christian interpretations of the Bible led to a persecution of Italian artisans and innovators who were mostly unorthodox in their religious, scholarly, and scientific beliefs. Hence, due to a combination of these factors, Italian artisans started drifting to Western Europe in search of the proverbial greener pasture and for their personal safety. Of course, these Italian migrants took the patent concept with them to Central and Western Europe.

For instance, six of the first nine patents in Brussels were granted to Italians (mainly Venetians). As Maximillian Frumkin noted, "one way or another, Italian influence shows like a thread in all incipient patent systems"[147] of Europe. It is therefore no exaggeration to say that the foundations of most European patent systems were built on the skills and ideas of Italian immigrants. The patent system spread first to Germany and then to other European states (e.g., Russia in 1812, Belgium and the Netherlands in 1817, Spain in 1820, the Vatican in 1833, Sweden in 1834, Portugal in 1837).[148] However, there were some European states, like Switzerland, that detested the patent system. It bears noting that, although the various patent systems in medieval Europe mutually influenced one another, all states designed patent systems and policies in accord with their perceived national interests.[149]

There are three methods by which the patent concept spread from Europe to other parts of the globe. The first method was through the migration of Europeans and their consequent colonization of a host of American, African, Australian, and Asian[150] indigenous peoples. The patent concept was part of the "baggage"[151] that the colonizing Europeans brought with them and imposed on indigenous laws on property and ownership. Needless to say, the concept of ownership and property in European culture and jurisprudence, including distinctions between the corporeal and the incorporeal,[152] are not universal.[153] Indeed, the transplanting of a foreign legal culture has significant juridical, philosophical, and ideological ramifications.

The second method by which the patent concept was spread was through direct, volitional borrowing by independent states. This was particularly the case with Japan: although non-European and never colonized, it independently adopted the patent concept. As Rahn has observed of the Japanese patent law and institution:

> The first legal system introduced from the West was the patent system ... In 1871, only three years after the Restoration [of the Meiji government], it introduced the Summary Rules of Monopoly (*Sembai ryaku kisoku*), a code of 19 articles, which began with the words: he who makes a new invention of anything whatsoever, will from that time on receive permission to sell it exclusively.[154]

The third method by which the patent concept was spread relates to those states that, although politically independent and often disdainful of the concept, were coerced by external political pressures to create and enforce a system of patents within their domestic jurisdictions.[155] Notable in this category are China,[156] Korea,[157] the Netherlands, and Switzerland.[158] Today there is hardly any country that does not have patent law.[159] The patent concept, with its rather dramatic origins in the Italian peninsula, preceded the Industrial Revolution in Europe. Further, in its early manifestation, the patent concept meant different things to different states. More important, the patent concept, whether adopted or rejected, ultimately served an instrumentalist purpose as articulated by the oft-competing interests of states, particularly in Europe. It was also in Europe that the fundamental doctrines of the law of patents evolved. Most of the doctrines and principles of patent law developed in Europe have now been imposed on or inscribed in the laws affecting the knowledge of indigenous and traditional peoples.

Historical Evolution and Development of the Patent System
The contemporary patent concept owes significant debts to the era between medieval times and European industrialization in the late nineteenth century.[160] Until the latter part of the nineteenth century, intellectual property

was a farrago of diverse notions of rights that were presumably derived from intellectual exertions. It was from this idea that the modern patent regime developed. The irony here is that the pioneering clarity and integrity of the Venetian patent system eventually became tarnished with excessive instrumentalism.

Certain substantive principles in the law of patents were already evident by the end of the nineteenth century. First, in shaping the substance of patent law, ideas and discoveries of "natural" phenomena or laws were debarred from patentability. Second, although the terms "invention" and "manufacture" had not yet acquired clear juridical meanings, early patent law and practice was heavily dominated by, if not exclusively restricted to, mechanical inventions and artifices.[161] Third, prior to the evolution of corporate legal personalities and the commercial capitalization of the industrial and inventive process, the patent concept was anchored in the individualism associated with the inventive process.[162] Corporate entities presenting themselves as fictional legal persons could not own patents.

Fourth, given the needs of fledgling industrialization in the Middle Ages, the requirement of industrial repeatability or industrial applicability of patented subjects became a regular feature of patent law and practice.[163] Fifth, the requirement of registration of patents also originated during the period in question, from the provisions of the British designs laws, which regulated the ownership of new textile designs. By this law, it was incumbent on applicants for ownership of new textile designs to specify their designs and to show how they differed from previously registered designs.[164]

Registration thus laid the foundations for its corollary – the requirement for specificity in applications for patents. Thus requirement is judicially associated with the 1778 opinion of Lord Mansfield in *Liardet* v. *Johnson*.[165] The disclosure is a technical outline and description of the invention. The claim delimits the sphere of rights conferred on the inventor by the patented invention. In theory, this specification is a form of consideration for the patent grant. It is in this sense that a patent is said to be as good as its specification. Specifications largely transformed the patent system, at least in theory, from a mechanism for importing existing technologies (in the immediate post-Venetian patent system era) into an institution for rewarding those inventors who created new and useful machines. But, more important, the requirement of writing meant that indigenous peoples knowledge, often recorded or transmitted orally, fell outside the ambit of protected information – a situation that has been instrumental to the appropriation of that knowledge.

However, the political economics of the early patent system obscured the law and procedure of patents. As states deliberately engaged in what J.W. Baxter has aptly termed a regime of "invention by importation, as distinct from inspiration,"[166] the patent system became a pliable tool to advance perceived national interests. For instance, William Cecil (first secretary of state to Queen Elizabeth the First), by granting patent rights to skilled and entrepreneurial Europeans wishing to emigrate in order to establish their businesses in England, applied the patent system as a "strategic international trade policy."[167] This was exemplified in the case of the *Clothesworkers of Ipswich* decided in 1614, where the court held that:

> if a man hath *brought in a new invention and a new trade within the Kingdom* in peril of his life and consumption of his stock, etc., or if a man hath made new discovery of anything, in such cases, the King ... in recompense of his costs and travail may grant by charter unto him that he alone shall use such trade or traffic, not the knowledge of the skill to use it for a certain time because at the first, the people of the kingdom are ignorant.[168]

This practice was not limited to the British Isles. In Continental Europe the early French patent system was a regime of privileges that had no pretensions to originality as a criterion for the granting of patents. For instance, in France article 3 of the original patent law of 1791 provided that "whoever [is] the first to *bring into France* a foreign discovery shall enjoy the same advantages as if he were the inventor."[169] In the same vein, novelty was not a criterion for the grant of patents in early Dutch patent law – one merely needed to be the first importer of the trade or art in question. This is another example of how the early jurisprudence of patent law laid the foundation for the appropriation of indigenous peoples knowledge.

The prevailing motive and underlying policy among the oft-competing states of Europe was the economic welfare of the state and the attainment of a preeminent position in the sciences and technology, especially in textiles, mining, metallurgy, and ordnance. It was the invention rather than the inventor who was the primary focus of the patent system. Ironically, the very states that attained their technological prowess by this national chauvinism (e.g., Britain) were enormously concerned at the "leakage" of their technical prowess to other states (like the United States).

The discretionary nature of patent grants sometimes degenerated into egregious abuses, especially on the part of British monarchs who desperately needed to keep their fawning courtiers or to sustain their grand lifestyle. For example, as Holdsworth records:

> [James] the First was always hard up; and for a consideration he was prepared to grant many privileges both of the governmental and of the industrial varieties. Of the second of these varieties of grants the following are a few examples: grant of an exclusive right to export calfskins; grant of an exclusive right to import cod; grant of an exclusive right to make farthing tokens of copper.[170]

This abuse contributed immensely to the malodorous air of privilege attached to early English patents. In consequence, "Parliament, whose members represented many trades injured by these special privileges,"[171] enacted the famous *Statute of Monopolies* of 1623.[172] For this reason, the *Statute of Monopolies* has been fondly but erroneously referred to as the *Magna Carta* of the Right of Inventors.[173]

In spite of the *Statute of Monopolies* and similar juridical initiatives, by the mid-nineteenth century the patent concept had become a cesspool of bureaucratic bungling, complex and expensive processes of registration, and general juridical confusion.[174] Coupled with a rising public preference for a laissez-faire regime in trade and ideas, there was a widespread call for the abolition of the patent system in England, Continental Europe, and the United States. In 1863 the parliament in Germany condemned the patent system as being "injurious to common welfare" and the Government of Prussia opposed the adoption of the patent system by the North German Federation. Otto von Bismarck, then chancellor of the North German Federation, captured the popular opinion when he noted that it was better to abolish the patent system "than to engage in hopeless attempts to reform [it]."[175] In Switzerland economists of great competence characterized the patent system as "pernicious and indefensible,"[176] and a group of Swiss industrialists argued passionately that, "in the interest of the general prosperity of industry and trade, patent protection, that cup of sorrows"[177] should be avoided. Opinions such as this won the day in the Netherlands and Japan. Hence the Dutch abolished it in 1869, and Japan, which had adopted its patent law in 1872, abolished it in 1873. Across the Atlantic the US House of Representatives followed suit and enacted a bill abolishing the patent system. However, by a handful of votes the bill failed passage at the Senate.

Opposition to the patent system began to collapse in 1873 probably as a result of the great economic depression and the consequent rise of protectionism, economic nationalism, intensive lobbying by manufacturers, and, above all, the willingness of the patent advocates to accept compulsory licensing as part of the "reformed" patent law. Britain relented in 1874, Germany in 1877, and Japan re-enacted a patent law in 1885. Switzerland enacted a patent law in 1887. Since then, the ideology of patents has remained in the ascendant. It is important to remember that doctrines and principles of patent law are never stagnant; they evolve and change in tandem with the demands of industry.

Influence of Life Science Industries on "Patentable Subject Matter"

Patent laws are very technical and sometimes complex. This complexity is further compounded when they are exploited by states as instruments of both domestic and international economic policy. Central to an understanding of how patents enable the theft, or piracy, of indigenous peoples knowledge is an appreciation of the principles pertaining to what is patentable in patent law. Principles of patent laws evolve in directions motivated by the demands of industry, the state, and other powerful stakeholders in the patent system. For example, although patent laws are ostensibly designed to protect and reward invention, key elements of the concept of patent law (such as "invention")[178] have no clear and predetermined meaning.[179] Similarly, other concepts, such as "novelty" and "manufacture," not only lack clear meanings within respective national patent law jurisdictions but also lack a universal standard. Consequently, despite statutory basis, courts have tended to "create" rather than "interpret" patent law provisions. Hence, in jurisdictions such as the United States, pro-business courts have largely expanded the scope of patentable subject matter in ways that appropriate, or facilitate the appropriation of, indigenous peoples knowledge.

The decision whether or not to grant a patent, especially with respect to so-called "natural substances," reveals the vulnerability of patent law to judicial manipulation.[180] The question that arises is whether the courts are designed as institutions for setting public policy or as institutions for interpreting public policy. In the absence of any clear legislative or executive initiative in this regard, it seems that, in most cases, especially in US patent law, the courts have become a policy-making institution in matters related to patentability.[181] What is more interesting is how the deliberate ambiguities and manipulations of the patent system have affected and legitimated the appropriation of plants and TKUP. This observation is most evident in the influence and consequences of US patent law around the globe.

Notwithstanding the attempts at harmonization of patent laws and procedure, there is no international patent system in the strict sense of the word; rather, individual states, while maintaining an essentially domestic patent system of varying degrees of effectiveness, attempt to synchronize their national patent laws and systems with one another. In this dialogue of patent laws, powerful states and important global actors, mainly from the United States, have the capacity to influence the trajectory of patent laws in less powerful states. The overwhelming dominance of powerful states and industry in the formulation of patent law rules has global implications.

Biopiracy and the North-South Implications of Patents
The overwhelming dominance of powerful states and industries in the formulation, enforcement, and globalization of patent law rules has split the world into two antagonistic camps.[182] The contending interests of states,[183] as evidenced by their respective patent laws, reveal an instrumentalist approach towards the protection and sustenance of perceived technological and economic needs. In some cases, patents also function as a juridical anchor for the state's notions of distributive justice and of what constitutes

public morality, or *ordre public*.[184] Given the structure and realities of state-hood, it is in the economic and political interest of every state to maximize the economic returns on its resources, to stay technologically ahead of other states, and to improve the standard of living of its citizenry. Thus, the patent concept is not an instrument of interstate conviviality or camaraderie; rather, it is an instrument for the pursuit of perceived national economic self-interest, even when this leads to the appropriation of indigenous peoples knowledge. As Abbott explains,

> patent holdings worldwide are largely in the hands of enterprises based in a small group of industrialized countries. The claim that patent holders should be entitled to "super-returns" based on the value of invention also is a claim that a higher proportion of global wealth should be allocated to industrialized country enterprises.[185]

In effect, despite all the current emphasis on "international law of patents," national patent systems have always been based on the perceived economic considerations of states.[186] As an instrument in interstate economic and technological competition, the enforcement of patent norms often operates to the economic detriment of weaker states and less privileged cultures. Accordingly, when powerful states of the North[187] use their leverage to coerce and compel states in the South[188] to institute strong patent regimes on the ostensible theory that "patents will promote inventiveness" in the latter, they are engaging in a propagandistic exercise. As Alan Gutowski cogently points out, "for the sake of clarity, if not honesty, developed countries would do well simply to abandon their high moral tone and address issues of IP [intellectual property] as matters of domestic and international economic policy."[189] In economic terms, the South argues that, at best, different levels and subjects of patent protection may be "more appropriate at different stages of economic development"[190] and, consequently, it would prefer a less stringent patent system (if any at all).

Even among the industrialized countries of the North, patents constitute a form of rent paid to the more powerful states. In his research findings O.J. Firestone confirmed that 94 percent of patents in Canada are issued to foreigners – mainly Americans. Thus Canada, which belongs to the category of industrialized states, finds itself in the position of a less-developed economy, tending to import technology and pay rent to American firms, which basically have no interest in working the patents in Canada but are keen on ensuring a continued monopoly in the Canadian market by importing the finished patented products. In regards to patents on indigenous peoples knowledge, the consequence is that poor but gene-rich states are coerced into paying rent on their own biocultural knowledge and products whenever these form the subject of patents. The inescapable conclusion is that, in substantive law and procedure, contemporary attempts at institutionalizing a so-called global patent system usually reflect the congealed national interests of powerful states[191] of the North. The underlying conflict is that the South's level of economic development does not justify a strong patent regime. Of course, the North rejects this.

Ironically, history bears out the position of the South on the patent debate, particularly with regard to indigenous peoples knowledge. As I have already shown, the patent law policies of European countries during their early years of industrialization reveal a clear instrumentalist policy of appropriating the ideas and technologies of other states.[192] Even the United States, which now chastises recalcitrant Southern states unwilling to adopt strong patent laws, was a notorious pirate of foreign technology, flouting patent norms in defiance of then technologically superior European states.[193] As Alford has noted,

> throughout its period as a developing country, and some would argue, even thereafter, the United States was notorious for its singular and, in many regards, cavalier attitude toward the intellectual property of foreigners ... the criticisms that the American government and interest groups level at China today could readily have been applied to this nation a century ago ... The plans for much of the machinery that powered Lowell, birthplace of the American industrial revolution, were shamelessly lifted by none other than Ralph Lowell, himself from Great Britain.[194]

Further, while Britain as a one-time global leader in technology forbade the export of

> engines, parts, and skilled persons, the US imported all three regardless ... *The decision was made in the US that at that stage of economic development, the best policy for the US was lax enforcement of foreign intellectual property.*[195]

As the US Congress itself admitted, "when the United States was still a relatively young and developing country, it refused to respect international intellectual property rights on the grounds that it was freely entitled to foreign works to further its social and economic development."[196] The interesting point here, as Anderfelt has observed, is that,

> when developing countries today are told of the necessity of having an "adequate" patent system ("adequate" meaning accepting at least the minimum requirements of the international patent system) and the "historical examples" of the now industrialized countries are now referred to in sup-

port, the fact that several of the latter at certain periods used their patent systems chiefly as an incentive to "importers" of foreign-made inventions is understandably never mentioned.[197]

Perhaps what makes the contemporary struggle for economic and technological preeminence so intense is the fact that

> the new world economy is increasingly being driven by intellectual capital: knowledge, inventions, expressions of creativity and the accumulated education, training and skills embodied in the scientific, engineering, and professional workforce ... Intellectual capital is surpassing in importance the components of the physical capital-land, natural resources and manual labour – as the main source of wealth creation.[198]

Thus, at the heart of the debate on biopiracy is a struggle over economic profits from patented products, as stringent patent laws ensure the promotion of Northern industries and the creation of an ever-expanding market for patented indigenous peoples knowledge.[199]

As biotechnology becomes a significant component of national wealth, and as increased global economic interdependence expands the frontiers of trade,[200] the economic rent to be collected by a stringent and global application of patents on indigenous biocultural resources becomes a compelling national priority for industrialized states. The North's use of coercion[201] with regard to patent law encroachment on indigenous peoples knowledge shows clearly whose economic interests such a regime serves.[202] As Paul Liu observes, "a strong invisible hand, the influence of industries and industrial organization is evident."[203]

State policies on patents (absent formidable external pressures to the contrary) are dependent on a cost-benefit analysis of the economic ramifications of patents. On the whole, strong patent regimes result in a net economic loss for the South and a net economic gain for the North. For instance, studies show that foreigners hold over 90 to 95 percent of all patents granted in the South, and most of these are not being worked in the countries of the South[204] but, rather, feed monopoly markets for imported goods. The situation becomes more acute in light of current laws, which construe the importation of patented products as the local working of an invention. In the words of J.H. Reichman,

> Evidence shows that foreign patents typically become vehicles for import monopolies in the developing countries, while locally-worked patents seldom produce the desired diffusion of technical knowledge needed to permit future competition by local firms and agencies. Arguably, domestic

recognition of foreign intellectual property can thus inhibit rather than stimulate local innovation, and it can thus enhance the comparative advantages of the industrialized countries in the production of new and old technologies.[205]

In effect, patents, especially on indigenous biocultural knowledge, are deductions from the national wealth of the already beleaguered countries of the South.[206] The inescapable conclusion seems to be that countries of the South "have nothing to gain from granting patents on inventions worked and patented abroad except the avoidance of unpleasant foreign retaliation in other directions."[207] As Amy Carroll has argued, "the United States and other developed countries are winning the North-South patent debate not based on the merits of their argument, but rather because of their economic upper-hand."[208]

Accordingly, the argument that broad patent protection in the South would propel development and technological prowess in that part of the world is of doubtful validity. Frederick Abbott has observed that the "debate stands on firmer ground if premised on the recognition that the industrialized countries are trying to protect an increasingly important component of their national wealth."[209] However, as a special issue of *The Economist* observed, "it is not at all obvious that the developing countries are obliged, either morally or for the sake of sound economics, to meet the rich nations' demands."[210]

Apart from the unpersuasiveness of the argument that technological greatness would follow the adoption of patent regimes in the South, it should be recognized that the argument itself oversimplifies ideas about what might be needed to propel twenty-first-century inventiveness in the South. As Amy Carroll notes, "it would be overly simplistic to conclude that developing countries can be brought in line with developed countries, in terms of economic and social welfare, merely by implementing a stronger legal framework for granting and enforcing IP rights."[211] In the midst of debilitating diseases, illiteracy, political instability, and unspeakable economic exploitation, it would take far more than a compendium of strong patent laws to "induce" inventiveness. Despite weak patent laws, Korea, India, Taiwan, and Singapore have transformed themselves into emerging industrializing giants by making huge investments in education and research and development, thus creating strong public infrastructure and vibrant export-oriented economies.[212]

Indeed, modern scholarship strongly suggests that, until countries of the South reach a certain threshold level of industrial and social "development," patents tend to impede their economic advancement. Hence, Penrose's argument that states that have not attained such minimum level of "development should be exempt from any international patent arrangement."[213] Historically, it was through "piracy,"[214] as the industrial history of the fledgling European states and United States amply demonstrates, that the industrial development threshold was attained.

Given the economic losses that this approach entails for the North, it may seem that Northern countries would utilize persuasion and, if need be, coercion to maintain a regime favorable to them. However, in the words of J.H. Reichman, "imposition of foreign legal standards on unwilling states in the name of 'harmonization' remains today what Ladas deemed it in 1975, namely, a polite form of economic imperialism."[215] Whether economic imperialism is polite or rude, the contemporary pace of globalization has further exacerbated it.[216] The underlying issues, particularly cultural and philosophical differences, must be examined.

Corporate-driven globalization has not yet bridged ideological and cultural differences apparent in the patent laws of states. Forced attempts at the harmonization or unification of disparate national patent laws[217] merely paper over deep cracks, which are part of the notorious North-South[218] schism. Law is a mirror of societal values,[219] and the patent concept is no exception.[220] Interestingly, as Stephen Ladas points out, "discussion on unification or harmonization of law has suffered from an initial false assumption that law is a simple single conception."[221] Under this imperialistic, "one-size-fits-all" assumption, the conception of law that is regarded as material is the Western conception. All other legal cultures, especially those that are opposed to or skeptical about patent systems, are assumed to be retarded, primitive, and backward. Yet all non-Western cultures have laws and law-creating mechanisms independent of the dominant Eurocentric model. It is hardly debatable that non-Western jurisprudence has survived its contact with European jurisprudence.

By ignoring these crucial factors,[222] hegemonic attempts at compulsory unification and harmonization of the international patent system create a phenomenon known as juridical "inter-culture"[223] and the illusion of global legal harmony. Hence, given the inability of the dominant jurisprudential paradigm to explain the effects of sociocultural/legal interfaces, some scholars have argued for "new forms of scholarly representation"[224] to address both the omission and erasure of alternative narratives of law and experience. Although the patent law concept has been displacing indigenous cultures and laws,[225] it must be understood that patent law does not reflect the worldviews of indigenous and traditional peoples.[226]

Left to their own devices, domestic patent laws often reflect the social values and priorities of their respective states, especially where the state in question has the clout to resist foreign coercion. For example, the Japanese patent system, which was volitionally adopted, reflects Japanese sociocultural and philosophical ideas about property and its social role.[227] Although the Japanese patent system is often touted as "strong," a recent critical analysis reveals that it limits the scope of exclusive rights which can be obtained, and thus encourages minimal claiming, licensing, and other "cooperative" behaviors ... it marks a stark contrast with the more individualistic, pioneer-oriented US system. To that extent, the patent system (Japanese) reflects deep-seated cultural differences.[228]

According to Dan Rosen and Chikako Usui, "by all conventional measures, Japan should be a bastion of protection for intellectual property. And yet, compared with many Western countries and particularly the United States, the Japanese version of intellectual property law is porous and the attitude is often ambivalent."[229] All forms of law are value-driven. It would be idle to pretend that the Eurocentric sociocultural values embedded in the patent concept have lost their ethnic flavour simply because that concept has assumed global status. There are inherent conflicts involved in imposing or transplanting foreign legal concepts upon different cultural frameworks.[230] At best, the globalizing but inherently "ethnic" status of a legal concept such as the patent regime may be characterized as "globalized localism."[231] The point is that this globalized localism, as it relates to the patent concept and the latter's expanding role vis-à-vis plants and TKUP, "implicates the discipline of international law and human rights."[232] Regrettably, however, the harmonization and internationalization of patent laws ignores the cultural, environmental, and human rights implications of a forcefully contrived trade-based international framework for patents on plant life forms.

Although patent law pretends that there are no conceptions or regimes of intellectual property protection among non-Western societies, studies and fact-finding missions by the World Intellectual Property Organization (WIPO) and other agencies/scholars consistently show the existence of sophisticated regimes of ownership and control of inventions and innovations among indigenous and traditional cultures.[233] Indigenous intellectual property regimes have been shown to be at odds with Western jurisprudence and worldview on ownership and use of intellectual property.[234] Most non-Western societies construe property interest in the incorporeal in a far more holistic and communal sense than do Western societies, and this usually involves the "right to be recognized as 'owner' *but not necessarily the right to exclude others from use.*"[235] In other words, non-Western societies put great emphasis on usage and management of property, and this serves to militate against the appropriation of such knowledge by persons or entities outside the ambits of the particular locale in question. Ownership of property is thus fundamentally about notions of societal cohesion and distributive justice. These distinctions constitute a formidable gulf between *meum* and *teum*[236] (mine and yours) with regard to property, ownership, and access to the property in question.

Hence, a big problem with the forced imposition of foreign patent standards on all states irrespective of their sociocultural values, priorities, and ideologies of development is that it creates an ideological, philosophical, epistemological, and legal monoculture. As Ruth Gana has argued,

> what the internationalization of intellectual property implies, ultimately, is that there is only one way to participate in the international economy and that is by playing in accordance with prescribed rules, regardless of its impact on a group of peoples. It is a message that is not unfamiliar in the history of world affairs, and yet it is a message, which, so history informs us, has caused devastation of unimagined proportions to human society.[237]

It is a system that also denies legitimacy and efficacy to the creativity of peoples of the South, particularly women involved in the development of plants and the creation of TKUP.[238] Since the patent system cannot or has not yet accommodated the values and worldviews of non-Western societies, there is bound to be tension within the unified internationalizing framework of the contemporary patent regime. The contemporary global divide on the role of the patent system generally and in plants and TKUP specifically involves much more than a struggle over the intellectual "wealth"[239] of states: it also involves the sociocultural and human rights implications of legal transplants and concepts in an age of globalization, where the dominant legal culture preys on other legal cultures.

International Law and the Structural Framework of Patent Systems

Patent protection remains grounded in the national legal systems of each state and, to some extent, in regional arrangements. However, it is also fair to say that, until recent times, the international patent system involved a struggle by states and interested entities to extend and harmonize the reach of the property interest immanent in the patent concept beyond the borders of individual states.[240] During this internationalization process the patent concept has undergone three theoretical and juridical stages of evolution: (1) the primitive era of patents; (2) the development of multilateral treaties on patents; and (3) the linkage of trade and intellectual property rights in the early 1990s as the culmination of the coercive globalization of patent rights over biocultural resources.

Indeed, the internationalization of the patent concept started almost at the same time as the patent concept itself. However, the formal internationalization of patent systems by treaty process probably started over a century ago with the Paris Convention of 1883.[241] The Paris Convention was drafted (after preparatory work in 1873 and 1878) at a diplomatic conference in Paris in 1880 and was ratified by eleven states[242] in 1884. It entered into force one month after the deposit of the instruments of ratification, on 7 July 1884, and it created a union for the protection of industrial property. The convention created four categories of rules. The first category deals with rules of international public law regulating rights and obligations of member states, organs of the organization, and the constitutional character of the organization. The second category deals with those provisions that require or permit member states to legislate within the field of industrial property.[243] This territorial discretion was removed by the norms of the TRIPs agreement.[244] The third category deals with substantive laws in the field of industrial property regarding the rights and obligations of private parties. The fourth and final category contains provisions on rules of substantive law regarding the rights and obligations of parties.

Perhaps the Paris Convention's most important contribution to patent law jurisprudence was the introduction of the principles of "national treatment"[245] and "right of priority,"[246] which created a regime of formal equality between nationals of all member states.[247] However, the Patent Cooperation Treaty (PCT)[248] of 1970 represents the first major step, after the Paris Convention, towards a truly procedural internationalized patent system. It should be noted that neither the PCT nor any other "global" patent treaty or convention creates a so-called global patent. There is no such thing; rather the PCT facilitates the granting of national and regional patents. In other words, the PCT, at least on paper, gives the foreign applicant the much-needed time to protect his/her invention internationally.[249] The essence of the PCT is to streamline the administrative processes involved in filing a patent application in different countries.

The patent concept once again betrays its nationalistic instrumentality, as some countries, like the United States, do not permit their citizens to apply for patents abroad until a national security review of the invention is completed.[250] Notwithstanding the existence of the PCT and similar non-regional international legislative instruments designed to harmonize the procedure for the grant of patents, Europe has obviously remained at the forefront of the internationalization and harmonization[251] of patent laws and procedure. The European patent system is based on two continental agreements, namely, the 1973 Munich Convention on the European Patent or European Patent Convention (EPC);[252] and the 1975 Luxembourg Convention on the Community Patent or Community Patent Convention (CPC),[253] which is an integral part of the agreement relating to community patents signed in 1989.[254] The EPC facilitates patent protection in as many of the signatory states as the applicant wishes.

On the other hand, the community patent introduced by the EPC is intended to bring together the bundle of protection rights resulting from the grant of a European patent and merge them into a single, unitary, and autonomous right of patent protection valid throughout the community of member states of the European Union. This type of patent is governed only ·

by the provisions of the 1989 agreement relating to community patents. This convention has yet to take effect owing to delays in ratification by member states.[255]

The concept of regional frameworks for patent protection is not limited to Europe. In Latin America the Andean Pact[256] has effects and provisions that form comparable multilateral regional arrangements. In addition to the Andean Pact the *Mercado Comun del Sur* (MERCUSOR) is a free trade organization in Latin America[257] that has Brazil, Argentina, Uruguay, and Paraguay as member states. Established in 1991 it seeks to eliminate trade barriers and to create a common market. However, the treaty of Asuncion, which created the MERCUSOR,[258] is unlike its European counterpart in that it does not contain rules relating to industrial property. However, at an Interparliamentary Committee meeting in June 1992 at Asuncion it was agreed that the topic of industrial property should be discussed (as should patent exploitation). Although the MERCUSOR member states have their respective patent laws, there are no uniform criteria for the patentability and scope of inventions.

The North American Free Trade Agreement (NAFTA)[259] is another regional framework on patents, and it also seeks a harmonized patent regime for member states. Another regional actor in the global patent field is the Asia-Pacific Economic Cooperation (APEC), which in 1995 adopted the Osaka Action Agenda on harmonization of patent laws among member states. Other regional actors include the African Regional Industrial Property Organization (ARIPO). This organization was created at a diplomatic conference held at Lusaka, Zambia, in December 1976.[260] At present its membership consists of fourteen countries: Botswana, Gambia, Ghana, Kenya, Lesotho, Malawi, Sierra Leone, Somalia, Sudan, Swaziland, Tanzania, Uganda, Zambia, and Zimbabwe. Membership is open to Ethiopia, Liberia, Mauritius, Nigeria, and Seychelles – all of which participate in its activities.

The objectives of the ARIPO include, *inter alia*, promoting the harmonization and development of a patent regime appropriate to the needs of the members and of the region as a whole and developing a common view on patent matters. It has a patent documentation and information center at Harare. The organization has also developed model laws on patents in order to assist member states.[261] It is significant that, in 1993, the organization's administrative council agreed in principle that PCT applicants may designate for an "ARIPO patent" states that are party to both the PCT and the Harare Protocol of ARIPO. As a result, it has been possible for PCT applicants to designate for an ARIPO patent applications in states such as Ghana, Lesotho, Sierra Leone, Swaziland, Uganda, Zimbabwe, Malawi, and Sudan.[262]

Another African organization on the harmonization of patent laws and procedures is the African Intellectual Property Organization (OAPI). This organization consists of twelve French-speaking African countries. The agreement creating this organization was signed in Libreville in 1962, and subsequent revisions occurred in 1982 in Bangui. The Bangui agreement was also revised on 24 February 1999. The last revision was undertaken so as to bring the OAPI in line with the requirements of the Agreement on Trade-Related Aspects of Intellectual Property Rights.[263] In addition to the multilateral and regional arrangements on patents, there are numerous bilateral and interregional agreements on patents between various states and groups of states.[264]

There are also non-governmental organizations (NGOs) and international institutions whose policies and operations affect the global regime on patents. These include the World Bank, the United Nations Conference on Trade and Development (UNCTAD), the United Nations Environment Programme (UNEP), the United Nations Development Programme (UNDP), the Food and Agricultural Organization (FAO), the International Telecommunications Union (ITU), the World Health Organization (WHO), the World Intellectual Property Organization (WIPO), and the World Trade Organization (WTO).[265] With respect to WIPO,[266] whose headquarters are in Geneva, it is by far the most specialized agency of the United Nations and is responsible for promoting creative intellectual activity and the attendant laws and institutions. Indeed, WIPO has been the most dominant and impressive institution working towards the articulation of issues surrounding the protection of indigenous peoples knowledge. Although it primarily functions as the administrative organ for the Paris Convention and administers a host of other international legislative instruments and agreements that deal with patents,[267] WIPO has been at the forefront of exploring various avenues of redress for indigenous peoples in regard to the appropriation of their biocultural resources. Given the importance of WIPO in the debate on biopiracy, Chapter 5 examines some of its most important contributions to the emerging international law on biopiracy.

In sum, however, there is no coherent international patent system in the strict sense of the term; rather, there is a multiplicity of international, regional, multilateral, and bilateral agreements seeking to harmonize the process of granting patents. These combinations of international patent instruments also seek, to varying degrees, to harmonize various national laws on patents. The desirability of this goal in itself is a matter of debate and legitimate concern. This is particularly important because patents are not founded on human rights, global morality, or obligations *erga omnes*; rather, the patent concept, in the hands of powerful states and entities, is purely a matter of economic self-interests, which may neither mesh with one another nor be respectful of external domestic peculiarities and priorities.

Therefore, the globalization (or forced harmonization) of a minimum standard of patentability through trade-based global institutions, as recent global trends portend, raises fundamental questions concerning the nature of state sovereignty and the domestic implications of foreign usurpation of a state's eminent domain. These concerns are of particular relevance to weak and marginalized states and cultures, and especially to indigenous and traditional peoples. Ironically, these are the same societies and states with a preponderance of the planet's plant and cultural diversity.

The United Nations and the International Networks of Patent Systems
Prior to the emergence of the World Trade Organization (WTO), the politics of the internationalization of the patent concept often played out at international forums for patents, particularly the United Nations and related agencies. These agencies and forums have a preponderance of numerically superior but economically weak states from the South. For example, following the "Brazilian resolution"[268] of 1961, it was clear that a battle line had been drawn between the North and the South on questions pertaining to patents.[269] Subsequently, in the March 1964 final report of the secretary-general, entitled *The Role of Patents in the Transfer of Technology to Developing Countries,*[270] the various differences between the North and the South on the patent question were raised and thoroughly debated.

Despite the inconclusive nature of this report regarding many contested points concerning the nature of international rules on patents, it was generally well received. This may not be unrelated to the ambiguous nature of its conclusions, which left ample room for often contradictory interpretations in line with the North-South divide. Again, the March 1964 report was further dealt with by the Third Committee (on Financing of Trade and Invisibles),[271] which affirmed the findings of the March 1964 report.[272] In subsequent forums at UNCTAD and the the UN Economic and Social Council (ECOSOC), it became obvious that the underlying tensions and conflict between the North and the South concerning the question of patents had not been resolved. Thus, in 1965 Brazil again introduced a draft resolution on the same subject it had addressed in 1961.[273]

A careful observation and analysis of the positions of the various states shows a clear fault line in the approach to the patent question. While the more populous and numerous states of the South wanted a more liberal international patent regime, the North preferred a stronger one. Further, given their numerical superiority over the North, the states of the South were, at the very least, able to prevent further strengthening of patent laws. Indeed, it is fair to assert that the South had a reasonable chance of extracting major concessions from the North concerning the operations of the international patent system.[274]

Following perceptions that the UN forums, including WIPO, were heavily influenced by the preponderant states of the South, there occurred a shift in norm-creating forums for patents.[275] The North relocated intellectual property functions from the UN agencies and forums to the framework of the WTO, where it has effective control of the agenda and norm-making functions.[276] Accounts of the evolution of the Uruguay Round under the GATT[277] framework, which ultimately produced the TRIPs agreement[278] under the jurisdiction of the WTO, have been exhaustively treated elsewhere and need not detain us here.[279]

Article 27 of the TRIPs agreement is probably the most radical and stringent international legal instrument providing minimum standards of patent laws across the globe. It requires the availability of product and process patents for all new and useful products in all fields of technology, without discrimination as to subject matter. Article 27 (1) of the TRIPs agreement provides as follows:

> Subject to the provisions of paragraphs 2 and 3, patents shall be available for any *inventions*, whether *products* or *processes*, in all fields of technology, provided that they are new, involve an inventive step and are capable of industrial application. [emphasis added]

The exceptions stated in paragraphs 2 and 3 are:

> 2. Members may exclude from patentability inventions, the prevention within their territory of the commercial exploitation of which is necessary to protect *ordre public* or morality, including to protect human, animal or plant life or health or to avoid serious prejudice to the environment, provided that such exclusion is not made merely because the exploitation is prohibited by their law.

> 3. Members may also exclude from patentability:
> A. diagnostic, therapeutic and surgical methods for the treatment of humans or animals;
> B. Plants and animals other than micro-organisms, and essentially biological processes for the production of plants or animals other than non-biological and micro-biological processes. However, Members shall provide for the protection of plant varieties either by patents or by an effective *sui generis* system or by a combination of thereof."[280]

Although it allows for the exclusion from patentability of products that protect human, plant, or animal life and health, along with products that are harmful to the environment, opinion is sharply divided as to whether the exceptions are sufficient for the protection of biological and cultural diversity. As Ruth Gana has observed, "it is no secret that the main impetus behind the TRIPs agreement is to secure enforcement of US intellectual property rights abroad."[281] This process is also a reflection of the power and influence of the corporate world in shaping the agenda and law on patents. Further, it evidences the increasing abridgment by powerful multinational corporations of domestic competence in law making.

The implication of Article 27 of the TRIPs agreement on biodiversity, particularly as articulated in the recent Convention on Biological Diversity,[282] is a matter of great importance. Pertinent issues include sustainable use of plants, human rights, global food security, and qualitative environment.[283] This vexed issue is examined in detail in Chapter 3.

Status and Effect of International Patent Instruments at the Domestic Level

Although international law provides the overarching framework for interfacing state and regional regulatory systems, it is at the domestic level of respective states that such norms are played out. International law draws an effective distinction between accession to treaties and conventions and the domestic applicability of such treaties. There are various sources of international law on patents with implications for the effectiveness of international obligations.

Strictly speaking, the sources of international law on patents[284] do not differ from those of general principles of international law.[285] Article 38 of the Statute of the International Court of Justice lists and categorizes primary and secondary sources of international law. The statute provides the following as evidences and sources of general international law:

(a) international conventions, whether general or particular, establishing rules expressly recognized by the contesting states;
(b) international custom, as evidence of a general practice accepted as law;
(c) the general principles of law recognized by civilized nations;
(d) judicial decisions and the teachings of the most highly qualified publicists of the various nations.[286]

Categories (a) to (c) contain the so-called primary sources, while category (d) relates to the subsidiary sources of international law. On the question of customary international law, the most accepted definition is that it consists of widespread state practice[287] and accompanying *opino juris sive necessitates*.[288] Benedetto Conforti argues that evidence of custom may be found in treaties, participation of states in the resolutions of international organizations,[289] diplomatic exchanges, statutes, domestic court decisions, and domestic administrative acts.[290]

On the category of treaties, a treaty by definition requires two or more states bound vis-à-vis each other.[291] A treaty may also encompass an agreement between a state and an international organization.[292] In any event, there is no rule "of international law which might preclude a joint communiqué from constituting an international agreement."[293] The most important principle in the interpretation of treaties is the doctrine of *pacta sunt servanda*. This is codified under Article 26 of the Vienna Convention on the Law of Treaties. It provides that "every treaty in force is binding upon the parties to it and must be performed in good faith."[294] The importance of treaties here is that international law on patents is predominantly grounded on them and, to some extent, on customary international law.[295] Thus, other sources of international law have been subservient to, indeed of marginal utility as sources or evidences of, international law on patents. This preference for *lex scripta* may be due to the specialized nature of patent law or, as Abbott has argued, to neglect by international lawyers.[296]

Domestic judicial precedents, especially from the apex courts of states, may also have extraterritorial persuasive authority. In addition, with the emergence of international arbitrations on patent disputes, it is foreseeable that such decisions will make jurisprudential contributions to the international regime on patents.[297] Furthermore, scholarly works on patents (it would be invidious to name names) and the rules of equity may arguably constitute sources or evidences of international law on patents.[298] The same argument may be made for the category of general principles of law recognized by civilized nations.[299]

In addition, the proliferation of international resolutions and declarations on patents may also constitute sources and evidences of emergent international patent law, particularly on the question of appropriation and privatization of TKUP. Indeed, as Benedetto Conforti has observed,

the proliferation of norms is undoubtedly the most striking development in international relations in recent years. The staggering number of resolutions passed by international organizations ... although non-binding in character, constitute an essential reference point for the assessment of international customary norms and to the interpretation of treaty rules.[300]

Leaving aside the categories of the sources and evidences of international law on patents, the status and effect of such international legal norms, especially treaties at the domestic level, are often a difficult issue to resolve. Thus, in spite of the many centuries of international law, the question of domestic application of international law, especially treaties on trade and patents such as the TRIPs agreement, remains problematic.[301]

The place of a treaty in the internal legal order of a state is determined by its constitutional law.[302] Generally speaking, there are two theories on the application of international treaty law at the domestic level: the monist theory and the dualist theory. The latter conceives of international law as inoperative unless and until the domestic legal system has, by various means and processes, incorporated it as part of state domestic law. This process may also be influenced by state political and economic interests.

On the other hand, the monist theory conceives of international law and domestic law as operating within a unified structure and hierarchy of norms. Of course, on this ladder of norms, international law sits at the apex. Hence international law, at least in theory, has direct effect and applicability in domestic jurisdictions without the need of any transformative legislation. Shorn of theoretical pretensions, the question in monism or dualism concerns the various means whereby international treaty law becomes part of domestic law.[303] Does it do so through a process of transformation[304] or by direct application? In theory, this question should be resolved by reference to the constitutional law of the state concerned. The category of states which require transforming domestic legal instruments for the applicability of international treaty law is comparatively straightforward. What is required in such states is an act or instrument re-enacting the applicable treaty as though it were an act of the domestic legislature.

However, the category of monist states invokes unsettled points of law and, indeed, politics and national self-interest. In monist states like the United States, treaties theoretically have direct applicability once ratified and prevail over previous laws unless subsequently amended by the domestic legislator.[305] Generally speaking, modern constitutions tend to place treaties above the constitutions of states.[306] State practice on monism is not uniform, and academic opinions on the subject lack unanimity. Indeed, contrary to theoretical formulations, the legislatures of some monist states[307] often reserve for themselves the ultimate powers of determining the rank of a treaty vis-à-vis domestic legislation.[308] In making this determination, it is not unusual for considerations of the net gain of the international instrument to be accorded great weight. Thus, some scholars and jurists have argued that there is no hard and fast rule pertaining to distinctions between dualist and monist theories on applicability of treaty law in the domestic legal order.[309]

In addition, some interpretative rules, such as the doctrine of non-self-executing treaties, make hazardous any attempt at definitive theorization on the law pertaining to the purported direct application of agreements such as the TRIPs in the so-called monist states. Theoretically, before an international treaty obligation would have direct effect under the European Community framework, that obligation must be precise and unconditional.[310] This means that the implementation of a trade agreement such as TRIPs must not be subject to any further measures implying "a measure of discretion on the part of the Community organs or the member-states, as the case may be."[311] Thus, apart from some discretionary room in the TRIPs agreement, it would seem that, in theory, if this agreement passes the test outlined above, then it will be directly applicable and binding upon the community as law.[312]

In actual cases, however, at least with regard to the GATT agreement, which is the immediate predecessor of the WTO/TRIPs agreement, one should not rush to any judgment on the presumed direct applicability of the TRIPs agreement in monist jurisdictions.[313] The decision of 22 December 1994 by the Council of the European Community concludes that "by its nature, the Agreement (Uruguay Round) establishing the WTO, including annexes thereto, is not susceptible to being directly invoked in Community or Member State courts."[314] The point is that, under the TRIPs agreement on patent law, patent right holders under the European Economic Community framework may not be entitled to invoke the provisions of the agreement in their private capacity in municipal courts, the purported theoretical postulations of monism notwithstanding.[315]

In effect, the traditional regime of statist sovereignty on patent law sometimes defies or modifies international treaties on patents.[316] Save for the clear rule that states that fail to take necessary measures at the domestic level to allow the application of treaties are in breach of international law, the line between self-executing and non-self-executing treaties on patent law "is a matter of controversy and much confusion ... [and is] vague and volatile."[317] In spite of Article 27 of the TRIPs agreement, it would be premature at this stage of juridical evolution to announce the existence of an international patent law and system.

Hence, state juridical institutions stand at the interface between the domestic legislative framework and international instruments on patents.[318] They (the domestic juridical institutions) may, for a variety of reasons unrelated to the alleged "non-self-executing" nature of treaties, frustrate or modify the intended effect of such treaties on patents at the domestic level. To further complicate theorizing on the matter, the confidential, if not secretive, nature of the negotiation of the TRIPs agreement leaves little or no reliable *travaux preparatoires* to guide domestic courts in precisely determining the intention of the member states regarding the direct applicability (or otherwise) of the agreement.

Thus, although it is good law that treaties should be interpreted in an internationalist sense[319] rather than a parochial sense,[320] the *realpolitik* of the TRIPs agreement compels the prudent view that international law has not yet developed any hard and fast rules on the direct applicability of treaties, especially multilateral trade agreements. In effect, as regards the monist states, no automatic judicial enforcement of the TRIPs agreement is to be presumed. On the other hand, with respect to the dualist states, the TRIPs agreement is subject to the various transformative or adoptive legal regimes of states.

Although there is no international patent system in the strict sense of the term, corporate-driven attempts at standardization of the minimum elements of patent laws have made appreciable progress. Whether this process is desirable and whose interests it serves are matters of legitimate concern and debate. As an economic tool with profound implications, it is within the context of the human environment, particularly with regard to the phenomenon of biopiracy, that the debate on standardization of patent laws finds current urgency. I have traced the gradual development and evolution of the patent system from its Eurocentric basis to its current status as a global instrument for the privatization and commercialization of knowledge. In this quantum leap the most significant fact is that the debate on biopiracy and patents is a redacted version of the larger struggle over control of global biodiversity resource.

In Chapter 3 I examine the status of the plant regime in contemporary international law. As I show, the international law on the conservation and use of plants and indigenous peoples knowledge is no less Eurocentric and gendered than it is with respect to the patent regime.

3
Implications of Biopiracy for Biological and Cultural Diversity

Plant life forms have always been integral to human civilization. In this chapter I examine the nature, function, and value of plant life forms as they relate to the appropriation of indigenous peoples knowledge. I evaluate the various influences that have shaped the nature of legal norms relating to plants and how these have shaped legal regimes that facilitate the appropriation of indigenous peoples knowledge. I examine the present tragedy of plant life forms and the responses of the global community. I explore the texts, contexts, and subtexts of the various legal norms, institutions, and prejudices that underpin the regime on plant life norms with a view towards establishing a deeper appreciation of the nature of the crisis pertaining to the governance of plant life forms. And, of course, I examine how patents are implicated in this phenomenon.

The Nature, Value, and Function of Plants
Plants have always been integral to human civilization. The nature, function, and value of plants are of such importance that the decimation of biodiversity has become significant in recent global discourse. Biodiversity may be defined as the "variability among living organisms from all sources including, amongst others, terrestrial, marine and other aquatic ecosystems and the ecological complexes of which they are part, this includes diversity within species, between species and ecosystems."[1] It encompasses genetic diversity,[2] species diversity,[3] and ecosystem[4] diversity. In simpler terms, biodiversity is the total variety of life on earth. In the midst of the manifold complexity and diversity of life forms, the limitations of the dominant scientific narrative stand in stark contrast.

According to Kalyan, "scientists have only a rudimentary knowledge of biological diversity ... a comprehensive, rigorous, and a general theory of biodiversity is lacking."[5] In addition, there is a fourth layer of complexity to be added to the dimension of biological diversity – human cultural diversity and its implications.[6] For the sake of clarity, I focus on terrestrial plant life forms, which current estimates place at 250,000.[7] However, the arguments made here may be extrapolated to other life forms. The value and utility of plant life forms are varied and immense.[8] Plants have, *inter alia*, utilitarian, aesthetic, and moral functions.[9] In addition, they have an intrinsic value. The utilitarian function of plant resources is invaluable[10] and can hardly be overstated. For example, plants provide food, timber, and medicines; they contribute to balancing the ecosystems, stabilizing soil, and regulating climate.

With regard to health, traditional medicine forms the basis of primary health care for at least 80 percent of the population of the South – a number far in excess of 3 billion.[11] The World Resources Institute estimates that "Indians dwelling in the Amazon Basin make use of some 1,300 medicinal plants, including antibiotics, narcotics, abortifacients, contraceptives, antidiarrhoeal agents, fungicides, anaesthetics, muscle relaxants, and many others."[12] Almost one-quarter of all doctors' prescriptions in the North have their origins in plant species. The dramatic cases of the rosy periwinkle,[13] yew tree,[14] Cameroonian vine,[15] Rauwolfa,[16] and thousands[17] of other miraculous plants[18] merely underscore humanity's absolute reliance on plants for medication since time immemorial. Plants constitute a complex chemical storehouse that contains many actual and undiscovered potential uses in modern medicine.[19] Indeed, of the 250,000 plants known to be in existence, only 10 percent have been tested for medicinal purposes.[20]

On food and global food security, the statistics are awesome. For example, four crops – namely, wheat, rice, maize, and barley – make up 90 percent of the world's annual production of grain. The unpopular or barely used varieties of crops also constitute a formidable insurance for the more commercialized or popular species. For example, seemingly "worthless" species of rice and maize have saved hundreds of thousands of hectares of farms in Indonesia, India, Sri Lanka, Vietnam, the Philippines, and the United States.[21] Further, the 1845-46 blight attack on potatoes, which hit Europe and exiled two million Irish, was only halted when new potato genes were introduced from the South American Andes (the original home of the potato). Genes from cassava in South America have increased yields in Africa and India by up to eighteen times. Southeast Asia's oil palm crop is descended from only four palms taken from West Africa to Java in 1848. Modern yields in palm oil have been improved and increased over threefold by breeding from genetic material from its native home in West Africa. The tomato and sugarcane industries could not have survived without the contributions of their native relatives. The United Nations estimates that only twenty species supply 90 percent of the world's food and just three – wheat, maize, and rice – provide more than half.[22] However, relying on a few species of crops for global food supply is a risky exercise.

The economic significance of plant life forms is staggering. The World Bank estimated in 1990 that agriculture comprised 31 percent of the gross domestic product (GDP)[23] of low-income economies and 12 percent of that of middle-income economies.[24] Similarly, world trade in agricultural products amounted to $3 trillion in 1989.[25] The combined annual global market for plant life forms is conservatively estimated at US$500-800 billion.[26] Yet economic utility alone hardly captures the importance of plants.[27] In addition to their immense economic value, plants play essential roles as regulators of the global climate and environment. The interrelationships between plants and the general environment validate the theory that the entire ecosystem is dynamic.[28] Furthermore, plant resources have immeasurable aesthetic value: many species of plants constitute a source of wonder, inspiration, and joy to human beings as a result of their beauty, intriguing appearance, variety, or fascinating behaviour.[29] It is therefore not surprising that plants have influenced human conceptions of the universe in various ways, particularly in the religious, moral, ethical, and aesthetic dimensions. Consequently, the debate on biopiracy implicates the intense struggle among states for control of earth's plant genetic resources and the ownership of the associated uses of plants. Given the preponderance of plant genetic resources in non-industrialized parts of the world vis-à-vis the genetic poverty of industrialized states (that have an overwhelming edge in technological prowess), it would seem that the commodification of plant life forms threatens global biological diversity and cultural diversity.

Religio-Philosophical Conceptions of Plants

Interestingly, religious views, particularly Judeo-Christian and, of late, Asian, have had and continue to have enormous influence on the development and evolution of international legal norms and institutions pertaining to plants. Often, these varied religious views have affected plants in ways probably unintended but nonetheless significant. Given that these religious-cum-juridical legal norms and institutions are in a constant state of flux, it follows that international law, particularly with regard to the environment, has not remained static; rather, it has moved with the changing interpretations, reinterpretations, and sometimes rejection of hitherto orthodox views. However, the plant regime is not yet a fine synthesis of all diverse religious and philosophical conceptions of plants.

Human conceptions of plants, particularly within the Asian context, may be said to have oscillated between the oft-competitive dynamic between secular politics and religion. This dynamic is evident in the historical competition between religious and non-religious agencies and institutions for temporal power along with the creation of the necessary legal norms to effect dominant ideas.[30] Asia has been celebrated for its holistic conception of all life forms. The ideology of nature as an organic entity and humanity as only a part of an integral whole is a unified theory of life. To quote Indian jurist Krishnan Iyer, it is "a vision of the unity in diversity, and the happy synthesis of materialism and spiritual universality – [that] God inhabits all creations."[31] This theory is the essence of most religious conceptions of nature, particularly within Asian, Native American, and traditional African belief systems. Within this paradigm plants are not capable of being legitimate subjects of private ownership, domination, or control – let alone commoditization. They have a life and juridical existence of their own.

On the other hand, within the Judeo-Christian context plants and other manifestations of "subhuman" life forms are considered as raw materials for the satisfaction, if not indulgence, of human appetite and need. In other words, the raison d'être of plant life forms is to serve the changing needs of humankind. It thus follows that plants can be owned as property and conveyed into the marketplace as objects of trade, that they can function as one of the mechanisms for the acquisition and accumulation of surplus capital. Given the influence of various religious philosophies on the development and evolution of institutions, attitudes, and legal norms that underpin the regime on plant life forms – particularly patents – I now look at some of these belief systems, their conceptions of nature, and their implications for the global regime on access to and control of plants.[32]

First, the ancient Indian Vedic religions conceived of nature as the manifestation of Brahman. In this conception human existence is the highest life form in the long march of the Soul – the *atman* – to the Absolute Infinite, which is posited as the celestial destination of all life forms. Thus, the ancient Indian Upanishads, particularly Hinduism, spoke of the communion of humanity with all creation.[33] In this sense, the ancient Indian *Rig Veda* postulated that God sleeps in the mineral, awakens in the vegetable, walks in the animal, and thinks in the human. In keeping with this theory, a respect for the intrinsic worth and integrity of all life forms was developed.

A caveat is in order here. As Krishna Iyer cautions, "this is not a return to crude nature worship nor [is it] primitive animistic pantheism";[34] rather, it is an appreciation that what is at stake "is consonance with nature and commitment to future generations which are the first charge on our power to use natural resources."[35] It is a conception of the universe as a single entity with only one unitary interest: an abiding and respectful human interrelationship with all manifestations of life irrespective of presumed utility (or lack thereof) to the human life form.[36] Thus, ontologically there is no divide between humanity and nature.

Second, Buddhism,[37] with its ethical imperative of equity and its belief in reincarnation, is equally benign to all life forms. The Buddhist doctrine dwells on the concepts of non-self and impermanence. In its emphasis on the pursuit of enlightenment, it advocates the pursuit of what is "right": right view, right resolve, right speech, right action, right livelihood, right effort, right mindfulness, and right concentration. Thus, Buddhism is clearly non-anthropocentric in that it perceives of humanity as part of a general chain of life. Furthermore, Taoism, like Confucianism and other forms of Asian religious philosophies, abhors materialism and promotes a regime of "social harmony." Similar views and perspectives are echoed in the religions and philosophies of indigenous American nations and peoples.

Third, generally speaking, the indigenous nations of the Americas share a holistic view of all life forms that is strikingly similar to Asiatic views. For example, Andeans do not distinguish "between an individual person, the flora and fauna, or biodiversity in general. All are considered as an indissoluble whole, which within the cosmos achieves a concept of God; what the Incas call 'Pacha Kamaq.'"[38] Anthropological research is replete with the holistic conception of the universe shared by indigenous peoples of the Americas.[39] While the details may differ among the thousands of indigenous American nations, their conceptions and perspectives on the link between humanity and plants share the same broad, holistic pattern. These views also permeate the life philosophy of the Maoris of New Zealand[40] and the Aborigines of Australia.[41] Similar views are manifest in traditional African religions and philosophy.

Fourth, in African indigenous religions it is believed that all life forms are creatures of God.[42] Departed spirits of the dead, the unborn, and even diabolical agents such as witches and wizards are believed to have the capacity to inhabit plants, animals, and sacred groves. A particularly effective method of reverence for, if not conservation of, other organisms, particularly plants and animals, is expressed through the "taboo system"[43] and the notions of "sacred groves" or "evil forests." In fact, in most places in Africa these "sacred groves" "represent the few remaining examples of closed-canopy forests."[44] The taboos are effective because they are associated with religious beliefs. Among the Igbo of southeast Nigeria, plants are inhabited by spiritual entities and are thus imbued with agency.

In Islam it is believed that the almighty Allah created humankind and the entire universe, making the former a steward of the earth. This stewardship has been said to be the "divine rationale for the existence of mankind on the face of this planet."[45] The *Quran* declares that "He [Allah] it is who hath placed you as viceroys of the earth."[46] As noted Islamic scholar Omar Bakhashab has argued, "in the Islamic perspective, people in a community can be compared to passengers on a ship, having a common responsibility. Each passenger has to ensure the ship's safeguard not only for his own safety but that of others as well."[47] With reference to plants, humanity is in a position of divinely mandated stewardship.

The Judeo-Christian conception of nature[48] and the place of humanity in the schemata of life forms may be summarized by the following biblical injunction:

> Then God said, let us make man in our image, after our likeness and let them have dominion over the fish of the sea, and over the birds of the air, and over the cattle, and over all the earth, and over every creeping thing that creeps upon the earth.[49]

It is arguable that the Judeo-Christian philosophy goes beyond merely placing humanity in a stewardship position and asserts a regime of absolute human dominion over plants and other "subhuman" life forms. In effect, "man," in the gender-specific sense of the word, by virtue of a "heavenly and divine" mandate, sits atop a supposed hierarchy of life forms. In this heavenly ordained pyramidal structure, men come first, followed by women, children, animals, birds, fish, plants, and microscopic life forms. On his Olympian pedestal, man's unquestionable authority emanates from the sublime celestial abode of God. In recent times, this flawed theory has been rejected and reinterpreted. Modern interpretation suggests that man is in the role of steward, rather than irresponsible and self-centred lord, of other life forms. The status and function of stewardship carries enormous responsibility, implicating human culture, science, international law, and the environment. Given the profound effects of classical Judeo-Christian philosophy on laws and institutions governing plant life forms, further analysis of the subject is pertinent.

Influence of Judeo-Christian Philosophy on the Laws, Norms, and Institutions Governing Plants and the Environment

Following on the concept of a gendered human preeminence in the scheme of life, and the conception that humanity stands apart from nature, are certain attitudes, institutions, and legal norms on plants and TKUP. First, it is very arguable that this gendered hierarchy contributed to the masculinization of knowledge, particularly the dominant mode of empirical knowledge known as science. Until recent times science was a masculine affair – a discourse among gentlemen.[50] Medieval European women were believed to be incapable of displaying "scientific" abilities, and early scientific forays by women were dismissed as "witchcraft" or "magic."

The vestiges of that era linger on. For example, it is remarkable that, to date, academic degrees in the West are denominated in the masculine gender.[51] A result of this with respect to plants is that scientific contributions by women, especially local farmers and breeders, remain unappreciated, unrecognized, and unrewarded. However, it is becoming increasingly clear that the most important scientific endeavour of all time – the improvement and conservation of plant life forms through farming and selective breeding – was accomplished largely by women who toiled daily in the farm fields, especially in the South.

Further, the Western scientific world, especially in its early days, was an elitist affair with a tight leash on its own orthodoxy and a monopoly on truth. Any deviations from its narrow constructs were construed as quackery, as unscientific, and thus as erroneous. The offending participants stood at grave risk of losing their reputation. This intense tyranny of orthodoxy and peer approval reinforced the unfounded assumption that there was no science among peoples and races of the South. Given this traditional marginalization of both women and non-Western epistemological and narrative frameworks, the intellectual contributions of female traditional farmers in the South were subject to a marginalization that is both acute and multi-layered. Until the emergence of new international norms seeking to validate and legitimize the scientific character of the intellectual contributions of local farmers, breeders, and healers to the improvement and conservation of plant resources, the dominant epistemology denied and erased the input of local women to the construction of plant variety and conservation. This gendered and racist conception of scientific abilities has not been fully vanquished, the redeeming effect of some new treaties notwithstanding.[52]

These cultural factors and biases,[53] particularly the arrogant anthropocentrism in the dominant scientific narrative, contributed immensely to the normative content of international law on biodiversity. For example, Principle 1 of the Rio Declaration on the Environment declares that human beings are at the centre of concerns for sustainable development.[54] The Stockholm Declaration, especially the preamble, is practically an ode to the supposed preeminence of humanity on earth.[55] In this paradigm, other life forms derive their functional utility through how they satisfy human needs and indulge the appetite of humankind. Since the conquest, colonization, and domination of hitherto non-Judeo-Christian cultures, an anthropocentric[56] conception of plants remains dominant in international legal norms and institutions governing plant life forms. Thus, the African Convention on the Conservation of Nature and Natural Resources refers to the manifold diversity of all life forms merely as "assets" to humankind.[57] These unresolved tensions arising from the interface between anthropocentric views and non-anthropocentric conceptions of life forms persist in international law.[58]

However, in recent times international law seems to have witnessed a gradual shift from its anthropocentric conception of life forms to a recognition of the intrinsic worth of other life forms and the role of humankind as trustees or stewards. For example, the World Charter for Nature, although a non-binding instrument, recognizes that "mankind is a part of nature" and that "every form of life is unique, warranting respect regardless of its worth to man."[59] The Convention on Biological Diversity also recognizes and affirms the intrinsic value of plant life forms irrespective of their utility to humankind.[60]

The chauvinism inherent in the postulated divine beginnings of the religions that originated in the Middle East – Christianity and Judaism – influenced early international law. This is true of Christianity in particular; especially after its metamorphosis into the moral fibre of the juridical base of governance in Europe, the emergent states practically adopted the "family of Christianity" as the basis for the fledgling international law.[61] In other words, early international law was both racist and chauvinistic, founded by a religious brotherhood of Europeans professing intellectual and philosophical superiority over all other peoples and cultures of the world.

Notions of racial inferiority thus became the theoretical justification for the acquisition and colonization[62] of the so-called "backward territories."[63] Racist bigotry and economic greed served as "legal" justifications for exterminating the "heathens" who were outside the charmed circle of those redeemed by virtue of their Christian faith and Caucasian heritage.[64] Remarkably, modern scholarship, perhaps due to embarrassment, has barely dealt with the influence of religion, particularly Christian racism, on early international law.[65]

Consequent upon the hierarchical conception of human cultures and races, disdain and contempt for non-Western traditions, cultures, and philosophical conceptions of nature became a feature of early international law. Thus, non-Western conceptions of plant life forms had the status of primitive quackery. The point is that a not-so-subtle racist hierarchy has since become the defining threshold for articulating, evaluating, and rewarding various human scientific cultures and development. Development has thus largely been conceived of and pursued as a lineal process, with the West at the vanguard and the cultural "backward peoples" of the South at the rear.

Epistemologically, Western science denied the plurality of knowledge and relegated most non-Western traditional sciences and narrative frameworks to the "realm of the natural, the mystical and the irrational."[66] Western science was uniquely positioned as empirical, rational, neutral, and universal. In other words, development and civilization, defined as the improved well-being of all, became "equated with the Westernisation of economic categories; Western ideology has been promoted as a universal ideal and attainable by all regardless of differences in culture."[67]

In effect, the definitions of the terms "undeveloped," "underdeveloped," and "developing" reflect a debatable continuum of linear movement by non-Western peoples and their societies towards Western values and worldview. Hence the legal norms and institutions created by European imperialists and colonialists in order to assimilate non-Western traditional societies into the "mainstream" Western culture in the name of "development."[68] The International Labour Organization (ILO) Convention 107 of 1957 is perhaps one of the best known assimilationist international instruments embodying this philosophy of racial, cultural, and epistemological superiority.[69]

However, contemporary scholarship confirms that Western science, like other narratives of science, is in fact "controlled by the social world of scientists and not by the natural world."[70] The implication of this for plant life forms is that the monocultural approach to human dietary preferences, agricultural practices, epistemology, and philosophy that is now in vogue constitutes a formidable threat to the sustenance of the diversity of crops. Similarly, the idea that scientific innovations in plant improvement can only be undertaken within Western empiricism is extremely unfortunate.

In consequence, non-Western scientific contributions to plant improvement have regrettably been perceived as "folk knowledge" unworthy of recognition. Plant life forms, which have witnessed thousands of years of cumulative intellectual interventions and improvements, especially in the hands of women farmers, are denigrated as raw germplasm and wild species unless and until "improved" in Western laboratories by "real scientists."

Recently, however, there has been a movement towards rejecting this policy of assimilation through acculturalization, particularly at the sociocultural level. According to ILO Convention 169 of 1989,

> considering that the developments which have taken place in international law since 1957, as well as developments in the situation of indigenous and tribal peoples in all regions of the world, have made it appropriate to adopt new international standards on the subject with a *view to removing the assimilationist orientation of the earlier standards.*[71]

Yet the underlying values and tensions in this condescending regime continue to surface in modern discourse on patents on plants and plant-related traditional knowledge. In other words, the racist and gendered notion of development has not lost its vigour.

Further, the Judeo-Christian conception of plant life forms as a resource visualizes them as inert, uniform, and mechanistic, separate from and inferior to "man" – that is, it sees them as objects for human domination. In effect, humankind is deemed to stand apart from nature, and the latter, particularly the forest, is considered savage in its primitiveness. And those who live in the forest are savages and brutes. Interestingly, the English word "savage" comes from the Latin word *sylviaticus,* meaning "of the woods."[72] This conception of nature as dormant, unproductive, and wild[73] was not uniformly accepted in early Western thought. Francis Bacon and his reductionist perception of nature and science was critiqued by early Western scientists and thinkers such as Paracelsus, Goethe, and Francis of Assisi. However, their voices were drowned out by the booming majoritarian view of human supremacy and dominion over the earth. As part of the unwritten but dominant philosophical basis of Western science, nature was construed as being made up of complexes of material substances with measurable analytic parameters – ready for human manipulation, analysis, and private ownership.[74] Accordingly, contemporary refrains lauding a better "scientific understanding" of nature may be euphemistic for the ultimate human control and subjugation of nature.[75]

With the attendant Industrial Revolution came a limitless appetite for resource exploitation and accumulation – and profit maximization.[76] While this ideology has brought about an unprecedented level of material progress, a phenomenal consumption of resources, and an enormous accumulation of capital, it has also savaged the environment. Paradoxically, modern concerns about the undesirable environmental consequences of the avaricious exploitation of nature are often construed as mere problems with neoliberal economics[77] or as imperfections of the market.[78]

Another inference from the Judeo-Christian philosophy of human separateness is the juridical demarcation between "nature" and "mankind," as though the two are in a state of gladiatorial conflict. Western-inspired international legal norms draw a line between areas of human habitation and the so-called wilderness. This bifurcation has given rise to the creation of nature parks, often accompanied by forcible ejection of disempowered and marginalized societies from the land where they had thrived and contributed to the diversity of plant life forms. Today, the imposition of Western concepts of wilderness and natural parks is a ubiquitous feature of many continental laws ostensibly designed to protect plants. The African Convention on Conservation of Nature and Natural Resources, for example, defines a strict nature reserve as an area "where it shall be forbidden to reside, enter, traverse or camp, and where it shall be forbidden to fly over at low altitude without a special permit from the competent authority."[79] It also defines a national park as an area

> exclusively set aside for the propagation, protection, conservation and management of vegetation and wild animals as well as for the protection of sites, landscapes or geological formations of particular scientific or aesthetic value, for the benefit and enjoyment of the general public.[80]

In recent times there seems to be a welcome but ill-defined questioning of the assumptions, prejudices, legal norms, and institutional network that sustains the dominant notions of development. Hence, irresponsible anthropocentrism has witnessed severe scholarly attacks, especially by high-profile institutions and individuals, including Prince Charles of the United Kingdom. Human domination of nature unbalanced by a sense of duty, trust, and/or stewardship is increasingly losing favour and legitimacy in juridical and norm-creating circles.[81] The *ancien regime*, as it were, now seems excessively functional, mechanical, and shortsighted. As the philosopher Neil Evernden has argued, "to describe a tree as an oxygen-producing device or a bog as a filtering agent is equally violent, equally debasing to the being itself."[82]

The disenchantment with, and shortcomings of, the anthropocentric conception of plants may also have spawned a new breed of philosophies that has borrowed a considerable body of ideas and inspiration from the South, particularly Asia. For example, "biocentrism," which views humanity as equal to all other forms of life, acknowledges the global morality of its competitive struggle for existence.[83] Similarly, "ecocentrism" focuses on the interaction between living entities and their environment. Even the radical concept of deep ecology developed by Arne Naess was inspired by the Vedantic doctrine of non-duality.[84] Indeed, Japanese Zen Buddhism has strongly influenced Western environmental movements such as Friends of the Earth and Greenpeace.[85]

The danger, however, is that there seems to be an excessive romanticization of the environmental virtues of traditional cultures of both the North and the South – a trend as uncritical as is the glorification of Western empiricism and epistemology. The appropriation and extinction of plant life forms is a hydra-headed problem, and it would be simplistic to assume that a quick embrace of non-anthropocentric views on nature is the only path towards environmental justice and the promotion of plant and cultural diversity. More important, the impact of the recent incorporation of non-Western values into the corpus of laws and philosophies governing plant resources may be exaggerated. As I have argued elsewhere, the process in question merely turns the philosophical bases of present-day international environmental law into a "coat of many colours."[86] Following on this metaphor, the fabric and stitching materials remain Eurocentric. For example, the current loss of plant life forms has generally been perceived as a disaster for humanity: the intrinsic worth of plants is of no consequence. However, the effect of global politics on patents on plant resources, particularly the development of legal norms and global institutions of patents on plants and traditional plant-related knowledge, cannot be understood outside the context of the nature and geography of plant resource distribution across the world.

A Human Geography of Plant Life Forms: The North-South Context
Plant species are not evenly distributed across the face of the earth. Most of

the delicacies today described as "European" or "all-American" or "Kenyan" are not of European or American or Kenyan origin. The potato is indigenous to the Andes; maize is indigenous to Central America; of the twenty major food crops, none originated in North America or Australia and only two – rye and oats – originated in the Euro-Siberian area.[87] Virtually "all of the developed countries' foodstuffs originated in the tropical countries."[88] Corn, rice, potatoes, sugar, citrus fruit, bananas, tomatoes, coconuts, black peppers, nutmeg, pineapples, chocolate, coffee, and vanilla all originated in the tropics. Over two-thirds of existing plant species are located in the South; rice originated in Asia and West Africa; and wheat originated in the Middle East.

In effect, the natural distribution of plant diversity is heavily skewed in favour of the South. More important, however, no state is self-sufficient with regard to plant genetic diversity. The size of states is irrelevant when it comes to measuring species or genetic diversity. For example, Brazil, with only 6.3 percent of the world's surface land area, has 22 percent of the planet's flowering plants. Botanists report that one twenty-acre tract in Malaysia supports 750 tree species. This is more than all the tree species diversity in the United States of America combined.[89] The African state of Madagascar contains about one-quarter of the plant species in Africa. Madagascar is also home to more than fifty species of coffee, "whose caffeine-free beans and disease-resistant qualities could prove valuable to both future consumers and plant breeders."[90] Jack Kloppenburg has captured the stark imbalance in plant distribution:

> Of crops of economic importance, only sunflowers, blueberries, cranberries, pecans, and the Jerusalem artichoke originated in what is now the United States and Canada. An all-American meal would be somewhat limited. Northern Europe's original genetic poverty is only slightly less striking; oats, rye, currants, and raspberries constitute the complement of major crops indigenous to that region. Australia has contributed nothing at all to the global ladder.[91]

According to William Lesser, "the United States, a major food crop producer, is completely dependent on foreign germplasm, including potatoes from Latin America, corn from Central America, soybean and rice from China, and wheat from Syria and environs."[92] In fact, for plant species diversity, North America posts an 85 percent dependency on the South.

Even before the seminal studies of Nicolai Vavilov, this extraordinary phenomenon and its manifold economic and security implications had long occupied the attention of policy makers in the North. Vavilov's travels and studies came to the conclusion that the genetic centres of the world were in the South.[93] Since Vavilov published his findings in 1925, the global politics of plant resources[94] assumed a North-South character. Thus, an array of cultural institutions, legal norms, and mechanisms were designed to relocate the genetic centre of the world and to extract surplus profit by inserting the appropriated plant life forms into the stream of commerce as commodities of trade.[95]

The causes of the profusion of plant diversity in the South are both natural and artificial. As a general rule of geography, moving from the poles to the equator, species richness increases in magnitude. This is largely as a result of the warmer and humid climate near the equator, which is favourable to the multiplication of life forms.[96] Speciation is thus partly induced by geography. In other words, geographical variety and the imperative of evolution facilitate species differences. As a rule of evolution, species mutate and diversify in line with geographical diversity. Hence, part of the explanation for the South's incredible plant species diversity lies in its complex geography.

But nature is not the only agent at work here. Human impact plays a critical role in the multiplication and sustenance of plant species. Plant diversity correlates with cultural diversity. For example, of the nine countries that account for 60 percent of human languages, six also have exceptionally high numbers of unique plant species.[97] Remarkably, of the 6,000 living languages, 1,000 are spoken on a single island – New Guinea. The majority of the world's population speaks only five languages. Ten languages die out every year, and, as cultural globalization sweeps across the earth, it is not a coincidence that the centres of linguistic and biological diversity suffer the highest rate of language extinction. The loss of earth's cultural and linguistic diversity is thus implicated in the debate on biopiracy.

Furthermore, the domestication of plants leads to an increased variety.[98] This is due to the phenomenon of polyploidy; that is, the doubling of chromosome numbers. Since agriculture began 10,000 years ago, selective breeding has resulted in stronger, healthier, and higher-yielding plants, thus increasing the diversity of plant species.[99] The greater the cultural diversity of the local farmers, the more likely they are to breed plants for various cultural purposes (e.g., religious or social festivals), thus multiplying the diversity of plants. Put simply, "agro-biodiversity is not a strictly natural phenomenon but derives from human activities. Indeed, farmers make selections to enrich biodiversity all the time."[100] In this wider context, "cultural diversity and natural diversity are closely linked concepts."[101]

In sum, the immense geographical diversity[102] and cultural complexity of the South practically compel farmers to breed plants suitable for various geographical and cultural imperatives. In dismissing the culturally biased notion that plant diversity in the South is solely a function of geographical whim, Kloppenburg notes that, in fact, the land races of the Third World are most emphatically not simple products of nature. Traditional agriculturists have made very great advances in crop productivity. Domesticated forms of species are frequently very different in form from their wild or weedy relations.[103]

The famous plant breeder Norman Simmonds lent his weight to this considered view when he admitted that, "probably, the total genetic change achieved by farmers over the millennia was far greater than that achieved by the last hundred or two years of more systematic science-based effort."[104] Further, Robert Leffel, the program leader of the United States Department of Agriculture for Oil-Seed Crops, gave credence to this view when he noted that, "in our modest moments, today's soybean breeders must admit that a more ancient society made the big accomplishment in soybean breeding and that we have merely fine-tuned the system to date."[105]

While the link between cultural diversity and plant diversity is irrefutable,[106] it is equally true that modern science and plant-breeding efforts can also contribute towards the improvement of plant species and genetic diversity. However, the threshold that must be established involves determining which human interests and values inspire artificial mediation in plant species and genetic diversity. In this context, it would seem that modern concerns about the industrialization of plants have proceeded on the basis of profit motives alone, thus marginalizing cultural sensibilities and environmental issues and eliminating plants ignorantly considered to be weeds.

In addition, there are concerns that the excessive commercialization of plant-breeding efforts invests the market with awesome powers to determine which plants thrive and which perish. Given the limitations of scientific knowledge and the idiosyncrasies of "market forces," it would seem imprudent to offer additional incentives (like patents) to an already powerful institution such as the market. With respect to the indiscriminate conversion of traditional peoples habitats into parks and reserves, there is a compelling need for care and balance in determining which areas of land are set aside as nature parks. Appearances may be deceptive. As Henry Thoreau advised, "what we call wilderness is a civilization other than our own."[107] According to David Wood:

Virtually all existing environments have in part been shaped by human habitation and use and their continuation requires human involvement ... in fact, forests are cultural artefacts. Present-day biodiversity exists in Central Africa not in spite of human habitation, but because of it. A review of evidence suggests that human influence in tropical vegetation is far greater than previously thought.[108]

Sir Ghilean Prance, the former director of the Royal Botanical Gardens, Kew, observed that

the Amazonian Indians, by an ingenious method of shifting cultivation developed gradually over several millennia, preserved the soils, the wildlife and the ecosystem as a whole ... The indigenous peoples are an integral part of the Amazonian paradise, a fact that has much to teach people from modern industrialized society.[109]

Similar words of caution have been issued with respect to the "impregnable jungles" of the Costa Rican rainforest, hitherto thought to be the heart of pristine jungle.[110] On North America, Jeffrey McNeely, the chief scientist of the World Conservation Union (IUCN) notes that

the wilderness of North America had been thoroughly occupied for thousands of years by a rich diversity of different groups ... the Yellowstone National Park established in 1872 was a territory previously occupied by Shoshone, Crow, and Blackfoot Indians.[111]

William Denevan has also demonstrated that the so-called wilderness, which the early European settlers of North America encountered, was actually the product of the forest management practices of the Native American Indians.[112] The concept of primitive and unspoilt wilderness, which shaped international law on conservation of plants, thus need to be reappraised.[113] It is not every human use of forests that creates disaster.[114] It seems that the better question would concern how human populations use forests[115] and what their ethic of plant conservation might be.[116] Policy makers often lose sight of the blurred interrelationships between human societies and plants.[117]

Early international law instruments[118] on plant conservation pretended that traditional and indigenous peoples were an affliction on the land.[119] While this erroneous attitude may be changing,[120] it formed the basis for the forceful ejection of forest-dwellers from the forests[121] and from their traditional ways of life.[122] However, modern international law now recognizes the benefits of the traditional management style of forest-dwellers.[123] For example, the concept of Integrated Conservation and Development Projects (ICDPs) seems to be in vogue.[124] ICDPs are designed to involve local and indigenous peoples in the management of the forests. Modern international law also seems to have come to terms with the merits of traditional knowledge in its own right and has recognized the inseparability of traditional and indigenous populations from their land.[125] For example, Articles 13, 14, 15, and 16 of ILO Convention 169 of 1989 (although it has not yet come into effect) now forbid forcible removal of traditional and local communities from their ancestral homelands.[126]

Causes of the Contemporary Extinction of Plant Species

The extinction of plant species diversity is often natural.[127] It has been calculated that, since 1600, when the keeping of records on plant diversity started, the rate of extinction of vascular plants may be put at 0.2 percent of the total number of plants estimated to be in existence.[128] Given that only a fraction of the species on earth has been identified, the number of unrecorded extinctions may be higher. Since the age of industrialization and the phenomenal increase of human population, it has been estimated that human-induced plant diversity loss has increased dramatically. Evidence of this is starkly borne out by the loss of forest cover around the globe and the unprecedented conversion of plant habitat for human use. For example, Ethiopia's forest cover and its famous cedar forests have dropped from 30 percent to 1 percent. India's forest cover has shrunk from over 50 percent to 14 percent. Loss of forests is today a matter of global concern.

Given the generally accepted concept of the interdependency of species and ecosystems, the consequences of the extinction of species include the so-called cascade effect.[129] This is the idea that loss of a species may cause a collapse or a malfunctioning of other species within the entire ecosystem.[130] For example, in the tropical moist forests of South America, 900 species of figs provide essential nutrition to spider monkeys, peccaries, and toucans. The existence of the figs depends on their being pollinated by wasps. Without the wasps, it is apparent that spider monkeys and peccaries (and their predators, such as jaguars) would disappear.[131]

Although human influence on plants can be beneficial, it is also true that "market" preferences for various species of plants can yield terrible consequences for plant genetic diversity.[132] John Ryan estimated that, by 2005, three-quarters of Indian rice fields might be sown with only ten varieties of rice as against the 30,000 that have been used in the past fifty years.[133] Similarly, in Indonesia, 1,500 varieties of rice have disappeared within the past fifteen years. Six varieties of corn account for 71 percent of the cornfields and about nine varieties of wheat occupy about 50 percent of the wheat fields in the United States.

The consequences of a narrow plant genetic base can be horrific.[134] The extraordinary rate of loss of biodiversity is a function of a complex web of causes primarily originating from a globalization of the Western concept of development.[135] This manifests itself in consumerism, especially in the North, the high debt burden of the South arising from an unfair global trading system, urbanization,[136] drainage of wetlands,[137] construction of super-highways and dams,[138] biotechnology/bioprospecting and agribusiness techniques,[139] overpopulation,[140] global warming,[141] and global cultural homogeneity.

The prevalent concept and practice of economic development rely heavily on the abundant production and excess consumption of material goods. The attempt to create what Kenneth Galbraith has termed the "affluent society,"[142] with unprecedented amounts of material goods and opportunities for consumption, has profoundly assaulted the environment. Development in this paradigm is largely measured in terms of effective capacity to produce and consume resources.[143] There is a preoccupation with accumulation of wealth as reckoned in gross national product (GNP) and gross domestic product (GDP).[144] Virtually all global financial institutions have propagated this gospel with religious conviction.[145] More often than not, the results have been destructive to the environment.[146] In a sense, this economic and philosophical ideology conflates the exploitation of nature with development.[147] Development so defined and pursued is the rate at which goods and services are commodified and traded. As Shiva contends, the grand "ideology of western development is in large part based on a vision of bringing all natural resources into the market economy for commodity production."[148]

Conversely, other narrative frameworks of living are considered primitive and poor. That is to say, this paradigm perceives precolonial indigenous and local peoples in their self-sustaining environment as poor. Yet most of what is significant in the lives of human beings does not lend itself to quantitative or economic measurement (as in GDPs and GNPs). According to Sara Dillon, "one has a moral responsibility to question the willingness ... to accept GNP and GDP as adequate measures of the material security of most people, despite the fact that these measures tell us little about the way the majority of working people live."[149] As Chakma further notes:

> Sustainable living, allowing a modest consumption of natural resources, is called poverty ... thus the ideology of development views people living within a subsistence economy as poor. They consume organic food grown without the use of chemical fertilizer and high-tech agricultural tools and methods instead of consuming commercially produced and processed food. They are conceived as poor as they live in self-built houses of bamboo and timber.[150]

Apart from the gargantuan resources needed to attain and sustain the standard and style of living promoted through the use of a consumption index as a measure of development,[151] it seems that the crucial question is whether the lineal and consumption-driven model of development is capable of sustaining itself. It is remarkable that, in this context, studies of traditional peoples economic systems conducted by the International Institute for Sustainable Development show that precolonial North American indigenous populations deliberately underproduced and avoided the accumulation of surplus capital so as to increase the resilience of the natural resource base, reduce the risk of resource depletion, and thus ensure the survival of their people.[152] Yet, the lineal notion of development marches on full speed into traditional systems, which sustain plant diversity. As the

Brundtland Report laments, "it is a terrible irony that as formal development reaches more deeply into rainforests, and other isolated environments, it tends to destroy the only cultures that have proved able to thrive in these environments."[153] Of late, there has been a slight rethinking of this consumerist paradigm.[154]

However, international institutions devoted to economic development continue to pursue policies designed to facilitate the production of surplus economic value[155] at the expense of marginalized cultures and peoples.[156] For example, the World Bank, in a July 1991 study by its Office of Environmental Affairs, conceded that its policies and programs have often negatively affected the ecology of native peoples, and it proposed that "the Bank takes a conscious, substantive look at those problems which until recently, development planning has not adequately addressed."[157] In sum, the contemporary ideology of development, which squeezes the last drop of sustenance from the earth,[158] is the root of the loss of plant species.[159] The manifestations and spin-off effects of this regime, which fuel the appropriation of indigenous peoples knowledge and deplete plant genetic diversity, may be broken down as follows.

The Culture of Consumerism

The phenomenon of consumerism is both a catalyst and a product of the prevailing concept of development. Ecological prudence does not characterize the consumption pattern of most of the world, particularly the North.[160] For example, the United States, representing 6 percent of the world's population, consumes 30 percent of the world's mineral production. The consequences of this for plant habitats are devastating.[161] According to the 1991 *World Development Report,* per capita energy consumption in the G-7 nations[162] is at least four times that of the rest of the world. The ruinous consumption pattern in the North and its global implications are probably the greatest threat to the diversity of plants in the world. Using timber as an example, Norman Myers calculated that US consumption of tropical hardwoods increased ninefold between 1950 and 1973. Most of these hardwoods came from forests in Southeast Asia chopped down to produce "wooden floors, snack trays, and salad bowls."[163] The paradox here is that regulation of this unsustainable consumption[164] is practically in the hands of those who actually benefit most from the regime itself. As Morris Cohen warned:

It is only a shallow philosophy which would make human welfare synonymous with the indiscriminate production and consumption of material goods. If there is one iota of wisdom in all the religions or philosophies which have supported the human race in the past it is that man cannot live by economic goods alone but needs vision and wisdom to determine what things are worthwhile and what things it would be better to do without.

This profound human need of controlling and moderating our consumptive demands cannot be left to those whose dominant interest is to stimulate such demands.[165]

Inequitable Global Economic Regime

With respect to the complex problem of the feudalistic global trading system, which results in the high debt burden of many impoverished countries of the South and a ruinous pace of loss of forest cover, the statistics are chilling. In 1990 the cumulative foreign debt of the South was estimated at $1.34 trillion, and it is increasing.[166] According to O'Neill and Sunstein, the ratio of debt to GNP was recently 111 percent for the low-income countries, and interest payments on these debts are crippling.[167] The net effect is that poor countries of the South pay far more than they receive as "foreign aid" in servicing the monumental debts. Much of the debt in question was acquired in the era of illegitimacy of governance and the arms race,[168] which marked the Cold War.

Indeed, some experts believe that, "even after deducting all forms of World Bank and IMF aid and other government and private bank funding, there is a net cash flow of 50 thousand million US dollars each year from the poorest nations of the world to the richest nations."[169] This phenomenon is partly a function of the effects of the colonial and neocolonial structures underpinning contemporary global trade. In this context it should be noted that the colonial powers in Africa, Asia, and Latin America basically designed the colonies as suppliers of plant raw materials such as timber, cocoa, tea, coffee, and other so-called cash crops. The direct effect of this paradigm in the postcolonial era is that the South, which harbours a majority of the earth's plant diversity, was more or less organized to feed the industrial machineries and factories of the North.

Hence, forests are cleared feverishly and mindlessly to plant and harvest cash crops that are sold to the North at low prices (largely fixed by multinational corporations based in the North) in order to purchase finished goods and to service a huge debt portfolio.[170] Interestingly, while the South ravages its forests to expand the scope of monocultural cash crop farming,[171] the industrial entities of the North continue to benefit from the inescapable low price of commodities. In effect, a vicious cycle of environmental massacre continues unabated,[172] with unrelieved poverty and a debt crisis[173] being a couple of its obvious products.[174] It may not be out of place to speak of a new form of feudalism in defining this relationship between the corporate powers of the North and the farming populations of the South. According to Obiora Okafor,

a global neighborhood in which about 20 percent of the population (the North) control and enjoy about 80 percent of its resources, whilst the other

80 percent of the population (the South) control and enjoy less than 20 percent of the said resources ... [is] an international society exhibiting essential forms of medieval feudalism.[175]

Neofeudalism aside, the impact of loss of forest cover on plant diversity is profound. Prior to the era of human pastoral existence, global forest cover ranged from 5 billion hectares to 6.2 billion hectares. Today the original grand figure of 5 to 6 billion hectares has shrunk to 3.4 billion and is still contracting at an annual rate of 0.3 percent. Without a doubt, all pertinent statistics indicate an absolute decline in global forest cover.

The magnitude of the problem has not been reflected in international juridical or normative efforts to address it. In 1983 the International Tropical Timber Agreement (ITTA),[176] a weak "commodity" instrument designed to deal with the conservation and management of forests, was signed. ITTA (revised in 1994), with its bias towards the timber-harvesting industry, has not salvaged global forests since it took provisional effect on 1 April 1985. ITTA established the International Tropical Timber Organization (ITTO), which seems to be polarized along North-South lines with the inescapable inequities. In spite of the weaknesses in and mercantilist bent of ITTA, the 1992 United Nations Conference on Environment and Development (UNCED) "failed to produce a legally binding convention and had to settle with an Authoritative Statement of Forest Principles."[177] The United Nations Commission on Sustainable Development (CSD) and Economic and Social Council (ECOSOC) have, respectively, established the Intergovernmental Panel on Forests (IPF) and the ad hoc open-ended Intergovernmental Forum on Forests (IFF).

These initiatives have culminated in ECOSOC's subsidiary body on forests, the United Nations Forum on Forests (UNFF), established on 18 October 2000.[178] In spite of these international initiatives, it seems that the forest regime, if there is one, lacks policy coherence regarding an enabling international environment. In addition to international juridical paralysis or weakness on the forest front, the influence of ITTO and ITTA is severely limited because extraction of timber does not tell the whole story behind deforestation. Factors responsible for loss of forest cover include ranching, monocultural farming, hydropower, and domestic use of wood for charcoal and firewood.

Overpopulation
Leading population experts Paul and Anne Erlich write that the explosive growth of the human population is "the most significant terrestrial event of the past million millennia ... No geological event in a billion years ... has posed a threat to terrestrial life comparable to that of human overpopulation."[179] The increase in human population is not necessarily evenly spread

across the world. For instance, the average woman in the North bears two children, whereas her counterpart in the South has between six and eight. The result is that 90 percent of the world's population growth is occurring in the developing world.[180]

Population control is so complex politically and economically that mere reduction of numbers per se would not address the issue. For example, serious concerns have recently been expressed in Japan, Russia, Sweden, Canada, and other parts of the North about the declining domestic human population. In some cases, generous economic incentives geared towards reversing that trend have been instituted. Moreover, population increase or decrease does not resolve the question. There are issues of age distribution, skills, land space, and other factors that bear on the desired direction of the growth of various states. More important, there is the question of appetite for consumption of material goods and services, which may impose a greater strain on the environment than mere population numbers. As Henry Steck has argued,

> to focus only on a growing population as the chief villain as many are inclined to do is to ignore the impact of an economic system where both consumer wants and economic incentives for the producer create a combined drive for production at the lowest cost and for higher consumption ... even should a stationary state economy, or zero population growth be achieved, therefore, the question of the distribution of resources would remain on the agenda of uncompleted business between have and have-not nations internationally and have and have-not social strata within nations.[181]

According to Joel Cohen, "the effects of human population on conservation depend strongly on economic and social factors as well as on human numbers, density, and growth rates."[182] The ratio of consumption between the North and the South is 61:1, and studies show that, on the average, a Swiss consumes forty times as much as does his/her Somalian counterpart.[183] Indeed, "15 percent of the world's population in the richest countries enjoy 79 percent of the world's income."[184] Therefore, it is not enough to pursue an aggressive campaign of population control without considering the effects of age distribution and consumption patterns among a given population.

Agribusiness, Bioprospecting, and Biotechnology
The industrialization of agriculture has wrought tremendous havoc on plant species diversity. Industrialized agriculture involves destroying plant habitat to create "a conveyor-belt" mechanism for intensive animal husbandry and agriculture. It also relies heavily on plant monocultures for greater profitability and for the avoidance of "economic inefficiencies" associated

with having a diversity of crop species on a farm. It is estimated that at least 38 percent of the loss of forest cover in the Amazon is due to ranching and agribusiness. The irony here is that, in such circumstances, industrial agriculture *is* in fact an uneconomical method of farming. For example, "raising one million dollars worth of cattle for market requires the destruction of 100 square kilometres of Amazonian forest. In comparison, extracting one million dollars worth of rubber destroys only 6.8 square kilometres of Amazonian forest."[185]

In addition to degrading the land through intensive agricultural farming,[186] this regime encourages the narrowing of the genetic base of agriculture.[187] For example, the excessive commodification of the potato in the United States has reduced the cultivars there to only twelve out of the over 2,000 varieties known to humanity. In fact, 40 percent of the potatoes cultivated in the US are one variety – the Russet Burbank – because industrial food processors such as McDonald's and McCain's need it for its size. Forty percent of all McDonald's "fries" must be two to three inches long, another 40 percent must be over three inches, and the remaining 20 percent can be less than two inches: the Russet Burbank perfectly meets this marketing requirement.[188] This phenomenon drives other species of potatoes out onto the margins, especially when driven by intellectual property rights, such as patents and plant breeders rights (PBRs). A report by the Convention on Biological Diversity (CBD) Secretariat indicates that, since the era of strong intellectual property rights over plant varieties, "vast numbers of traditional crop varieties have disappeared."[189] In effect, intellectual property rights over plants may act as perverse incentives by encouraging the development of only "commercially viable" plant varieties at the expense of plant genetic diversity.[190]

Furthermore, recent studies confirm that the regime of agribusiness, which has a narrow genetic base, is a big threat to plant species diversity. According to *Conservation International*, this factor "is far surpassing logging, mining and slash and burn agriculture"[191] as an agent in destroying plant species and genetic diversity. It seems that the lessons of history have been lost here. It was the rise and pervasiveness of industrialized agriculture that largely drove the plant species diversity of traditional European crops to extinction. For example, 97 percent of the vegetable varieties sold by commercial seed houses in the US at the beginning of the century are now extinct, as are 87 percent of the pear and 86 percent of the apple varieties.[192]

Adding strength to this narrowing of plant species and genetic diversity is the twin process of bioprospecting and biotechnology. Virtually "all the major pharmaceutical firms are already at work screening the genetic resources found in Brazil, Costa Rica, China, Micronesia and other biologically diverse countries."[193] However, it is becoming clear that the screening process used to evaluate the economic and industrial utility of plants exacts a huge toll on plant species and the ecosystem as a whole, sometimes leading to the extinction of rare plant species.[194] This is because most plant-derived drugs are too complex in their chemical structure to be synthesized in laboratories, so the trees that produce them must be "harvested." As a Glaxo spokesperson noted, "it is hard to find a chemist that can compete with nature."[195] In scouring the forests for complex plant-derived chemicals, plant species could be easily decimated. For example, the production of one kilogram of taxol (an anti-cancer drug made from the Pacific Yew tree) requires 20,000 pounds of bark, or 2,500 to 4,000 Pacific Yew trees.[196] Similarly, alkaloids like vincristine and vinblastine derived from the Madagascar rosy periwinkle and used for the treatment of childhood leukemia and Hodgkin's disease must be extracted from the plants that produce them. In this case, fifteen tons of rosy periwinkle leaves yield one ounce of vincristine. This alkaloid sold in 1991 for $100,000 per pound. As a result of the high demand for the plant, the entire native rosy periwinkle habitat in Madagascar has been depleted.[197]

In another dramatic case, a kilogram of twigs, barks, and fruit from a Malaysian gum tree yields a chemical capable of blocking the HIV-1 virus in human immune cells. When this was realized and the scientists went back to Malaysia, the plant species had disappeared.[198] Even the much-publicized Merck-Inbio agreement is a threat to plant species diversity in Costa Rica. As Asebey and Kempenaar have noted,

> Inbio contracted to supply Merck [with] roughly 2,000 natural products extracts for screening, yet Merck's annual through-put[199] is far greater. Merck's screening equipment requires at least 5,000 samples per week to operate efficiently and it is not uncommon for United States pharmaceutical companies such as Merck to have a weekly through-put of 10,000.

Thus, the indiscriminate screening of plant materials for pharmaceutical commercialization, otherwise called the "gene-rush," has depleted plant species and, in some cases, driven them to extinction.[200] With regard to biotechnology,[201] there is mounting evidence that it contributes to the spread of genetic uniformity in agriculture and the concomitant loss of crop genetic diversity.[202]

Climate Change and Plant Diversity Loss

There is mounting evidence that the world's average temperature has been increasing rather rapidly. The atmosphere has been warming up at about 0.3 to 0.6 degrees centigrade on average and 0.8 degrees centigrade on land for the last decade. It is conservatively estimated that the global temperature will increase by 1 degree to 3.5 degrees centigrade over the next 100 years.[203] Although these figures seem fractional, their import may be quite radical. For example, "even a 1 degree centigrade change is a rate unprecedented over the last ten thousand years."[204]

Although the distinction between natural variability of the global climate and human-induced change has not been drawn with mathematical exactitude, there is growing consensus that human activities, indeed, industrial activities and bush burning, have an influence on global climate, particularly in the form of the so-called greenhouse effect caused by some of the gases released into the atmosphere.[205] The consequences of this change on plant diversity may be severe. Apart from the dramatic "acid rain," there is mounting evidence that some parts of Africa and Asia are getting drier while other areas are getting wetter, with the concomitant droughts and floods and eventual devastation of plant habitats. In 1998, for example, *El Niño* spawned forest fires in Indonesia and Brazil, drought in Guyana and Papua New Guinea, and flooding in China, Bangladesh, Ecuador, Peru, and Kenya.[206] Notwithstanding the United Nations Framework Convention on Climate Change in 1992,[207] the impact of global temperature increase on plants is a matter of unresolved complexity.

Cultural Homogenization

As already noted, cultural diversity and plant diversity are linked in numerous ways.[208] Emerging forces of global cultural homogenization are increasingly undermining this linkage. As the dominant concept of development pervades the globe, marginalized cultures and the plants that support them face imminent extinction. Essentially, industrial processes thrive on streamlining methods of production. As Timothy Swanson has argued, industrial human society craves standardization.[209] Hence, the existence of a variety of processes and narrative frameworks is perceived as economically "wasteful."

As market forces determine which plants and human cultures live or thrive, the implicit assumption is that the market is omniscient and therefore knows best. There are serious problems with this logic. First, there is the inbuilt cultural bias of the so-called free market. In other words, when the "free market" is used as the divining rod for plant resources, it is simply a camouflage for a culturally biased judgment about which plants and supporting human cultures will perish. Second, assuming but not conceding that the utilitarian conception of plants is sound, it is a well known fact that modern human knowledge of the utility of plants is at best marginal. A substantial number of plants have not yet been fully appreciated for their utilitarian value. Thus, the "wisdom" of the free market in this case is merely the institutionalization of cultural prejudices and a valorization of ignorance.

Although there is a growing consensus that indigenous and traditional practices are often better attuned to ecological preservation than are Western practices,[210] it is equally true that, in the name of development and civilization, well-meaning but sometimes misguided governmental agencies and missionaries are busy all over the world trying to assimilate indigenous and traditional peoples into a homogenized Western culture. Thus, in addressing the problem of plant diversity erosion and extinction, we need to take a long, hard look at the biases generating cultural uniformity. It is imperative for human systems of knowledge and institutions to become more tolerant of heterogeneity. As Swanson has warned,

> there cannot be plant diversity without cultural and developmental diversity. The West must learn to recognise and accept the fact that there are other viable regimes of development apart from theirs and the South must have the courage to stand up and affirm the merit and credibility of their contributions to human development.[211]

Thus, it is not only plants that are being crushed into extinction by this drive to cultural uniformity; traditional peoples and their time-proven ways of life also face the threat "of being civilized into extinction."[212] The loss of knowledge of the uses of plant resources associated with this homogenization of culture is impossible to evaluate. Local, traditional, and indigenous dietary patterns are rapidly changing, leaving profound negative impacts on the sustainability of various agricultural crops, which, hitherto, supported and nourished traditional and indigenous cultures and civilizations. For example, Africa has more native cereals than any other continent. It has its own species of rice, as well as finger millet, pearl millet, sorghum, tea, guinea-corn, millet, and several dozen wild cereals whose grains have been eaten and cultivated for thousands of years.[213] Over 2,000 native grains, roots, fruits, and other plants of Africa, which have fed people for thousands of years, have been ignored by mainstream agriculture to the detriment of the local species.[214]

International Environmental Law and the Challenge of Plant Species Loss

In response to the rapid diminution of plant life forms, international law has in the past decades attempted to formulate a host of legislative anti-dotes.[215] Critics may label this a postindustrial concern,[216] but the response by international law dates to the 1800s, when the environmental impact of consumerism first engaged the attention of thinkers in the North.[217] In 1929 the health of plants occupied the attention of European states.[218] Apart from the initiatives of 1929 and 1951,[219] which largely focused on phyto-sanitary health, the existence in international law of a general principle of conservation in the field of flora must be inferred from a sequence of international legal instruments.[220]

My hypothesis is anchored on the provisions of such international instruments as the 1940 Convention on Nature and Wildlife Preservation in the Western Hemisphere of 1940, the 1968 African Convention on Nature and Natural Resources, the Convention Relative to the Preservation of Fauna and Flora in their Natural State, and the Convention on International Trade in Endangered Species of Wild Fauna and Flora,[221] all of which provide for and acknowledge the need for the preservation and conservation of plant diversity.

Similar provisions are contained in "soft law" international legal instruments such as the Stockholm Declaration, the Rio Declaration, and the World Charter for Nature of 1982.[222] Therefore, with respect to international law on conservation, there is sufficient evidence of uniformity, consistency, generality, and extensiveness in the acceptance of the legal norms on protection, conservation, and preservation of plant life forms. Furthermore, prior to the CBD era, it is fair to say that the preservation and conservation principle occurred primarily in the preambles and articles of international conventions, thus forming the heart of the given treaties and not an exception thereto.

As Hall has argued, introductory clauses in treaties are sources of international customary law.[223] Article 31 of the Vienna Convention on the law of treaties lends weight to this view as it obliges states to take into account preambular provisions in the context of construing treaties.[224] Moreover, Heijnsbergen has pointed out that "the preservation principle generally concerns treaties of a law-making character rather than contractual treaties, which character makes them more suitable to generate customary law."[225]

R.R. Baxter[226] and Ian Brownlie[227] are also of the opinion that a treaty to which a substantial number of states are parties is powerful evidence of customary law. Constant recitation of principles (such as the preservation principle) in treaties is a material source of international customary law.[228] Accordingly, by a combination of repetitions in preambles and wide state acceptance of the principle of plant conservation, it is correct to say that, prior to the CBD, international law, at least in principle, provided for or recognized the need for conservation of plant life forms.

Although the International Treaty on Plant Genetic Resources for Food and Agriculture, adopted by the Food and Agriculture Organization (FAO) Conference on 3 November 2001, is a powerful complement to existing treaties on the subject, the Convention on Biological Diversity, is without a doubt, the most significant and comprehensive international instrument regulating the use and conservation of plants.

The Convention on Biological Diversity and the Evolving Regime on Plants[229]

The CBD[230] is probably the most detailed and important juridical initiative on the conservation, sustainable use, and equitable sharing of the benefits of plant life forms and plant-based traditional knowledge. Although the true meaning and import of the CBD[231] has not yet congealed into domestic juridical specifics, the general refrain seems to be that this convention marks a historical commitment by all the nations to "conserve biological resources, to use biological resources sustainably and to share equitably the benefits arising from the use of genetic resources."[232] The CBD itself leaves no doubt as to its objectives. Article 1 states:

> The objectives of the Convention, to be pursued in accordance with its relevant provisions, are the conservation of biological diversity, the sustainable use of its components and the fair and equitable sharing of the benefits arising out of the utilization of genetic resources, including by appropriate access to genetic resources and by appropriate transfer of relevant technologies, taking into account all rights over those resources and to technologies, and by appropriate funding.[233]

The character of the CBD, however, leaves much latitude with regard to how states may achieve its goals. As such, the convention imposes few specific obligations on states and few well-defined targets.[234]

The CBD originated from the efforts of the International Union for the Conservation of Natural Resources (IUCN) in the 1980s to articulate a global convention for the protection of biological diversity.[235] In 1987 the United Nations Environment Program (UNEP) Governing Council recognized the need for a convention that would bring together the various treaties on biodiversity protection by offering uniform provisions bearing on biodiversity. This was principally in recognition of the fact that various conventions on the environment and biological diversity were piecemeal, treating only specific aspects of biological diversity.[236] Thus, the CBD is the first global legal instrument governing biological diversity in all its ramifications.[237]

Given the scope, universality, and diversity of the issues raised by plant life forms, a framework convention with room for subsequent specific agreements was preferred. As the Report of the Fourth Global Biodiversity Forum notes, "the Convention on Biological Diversity is a framework for general principles and obligations. There is little of the detailed structure that is necessary to implement its provisions."[238] However, the CBD contains several specific and groundbreaking normative initiatives, and these warrant further examination as they relate to biopiracy.

First, a major contribution of the CBD to the jurisprudence of plant resources conservation and utilization is the recognition of the ecosystem approach to plant conservation.[239] This involves a holistic conception of the human environment as a super-organism in which there are multiple and complex interrelationships between species and habitats.[240] Various resolutions and declarations have, in recent times, incorporated this notion.[241] To a large extent this validates notions of nature as an organic whole rather than as a composition of discrete units. Moreover, this is a quantum leap from the species-based or sectoral approach of pre-CDB international law on plant conservation.

Second, the attempt to reconceptualize biological diversity as a common heritage of mankind (CHM) was rejected at the earliest stages of CBD negotiations. This was perhaps a follow-up to the Stockholm Declaration and other soft law instruments that reaffirmed state sovereignty over plant life forms. For example, Principle 21 of the Stockholm Declaration notes that

states have in accordance with the Charter of the United Nations and the principles of international law, the sovereign right to exploit their own resources pursuant to their own environmental policies, and the responsibility to ensure that activities within their jurisdictions or control do not cause damage to the environment of other states or of areas beyond the limits of national jurisdictions.[242]

Article 3 of the CBD reproduces almost verbatim Principle 21 of the Stockholm Declaration. The Article provides thus:

States have, in accordance with the Charter of the United Nations and the principles of international law, the sovereign right to exploit their own resources pursuant to their own environmental policies, and the responsibility to ensure that activities within their jurisdiction or control do not cause damage to the environment of other States or of areas beyond the limits of national jurisdiction.[243]

This same principle is echoed in Article 15 of the CBD. Given the importance of Article 15 to further discussion on this question of access to plant life forms in relation to biopiracy, the pertinent parts of the article are reproduced *in extenso*:

Recognizing the sovereign rights of States over their natural resources, the authority to determine access to genetic resources rests with the national governments and is subject to national legislation.

2. Each Contracting Party shall endeavour to create conditions to facilitate access to genetic resources for environmentally sound uses by other Contracting Parties and not to impose restrictions that run counter to the objectives of this Convention ...

4. Access, where granted shall be on mutually agreed terms and subject to the provisions of this Article.

5. Access to genetic resources shall be subject to prior informed consent of the Contracting Party providing such resources, unless otherwise determined by that party.

7. Each Contracting party shall take legislative, administrative or policy measures ... with the aim of sharing in a fair and equitable way the results of the research and development and the benefits arising from the commercial and other utilization of genetic resources with the Contracting Party providing such resources. Such sharing shall be upon mutually agreed terms.[244]

The ramifications of the ringing reiteration of Articles 3 and 15 of the CBD – that plants are an inherent part of state sovereignty – has not been lost on scholars and commentators. According to Ranee Panjabi "one of the most important achievements of this Convention lies in its clear endorsement of the fact that biodiversity is a national resource and not part of the common heritage of mankind."[245] Walter Reid adds:

From the standpoint of global biodiversity conservation, the most important thing is that it confirms under international law that biodiversity is a sovereign national resource and that governments have the authority to determine the conditions under which access to that resource is granted. The distinction ... could not be sharper or its implications for conservation more profound.[246]

By emphatically rejecting the categorization of plants as common heritage of mankind (CHM),[247] the CBD's preference for the concept of common concern of mankind (CCM) brings into focus the controversial nature and divisive political economics of plant life forms. While the concept of CHM imposes a global juridical right over plant life forms irrespective of the reach of national boundaries and state sovereignty, the concept of CCM seems to invoke no such legal obligations.[248]

CCM first found expression in international law in the African Convention on the Conservation of Nature and Natural Resources. Chronologically, it seems to have predated CHM. Yet its normative content and juridical weight, if any, remains relatively unexplored and undefined. It seems, however, to contain the seeds for a less nationalistic conception of plant life forms in international law. Although its potentials remain unexplored, it is arguable that, given the global character of plant life forms, the artificiality of national boundaries, and the "global citizenship" of most modern crop plants, insistence on primitive nationalism in relation to plant life forms is no longer either supportable or realistic.

Therefore, the need for a truly global approach to access and use of plant life forms seems compelling. In this regard, the short shrift meted out to CCM as applied to plants may be a huge setback for the CBD's goal to propagate a global approach to resolving questions of access to and use of plant life forms by all humankind. The CBD's preference for a legal regime of near-exclusive sovereign right of states over plant life forms within their respective jurisdictions[249] is understandable, but it does not provide answers to the question of the global significance of plants.

The CBD regime on plant life forms involves creating national strategies and plans, policies, and programs on access to plant life based on terms mutually accepted by "supplier" states and "user" entities. Under modern international law on plant life forms, responsibility for conservation of plants seeks a balance between state sovereignty,[250] international co-existence, and cooperation.[251] To put it rather optimistically, what the CCM principle implies is that the primary state obligation is to ensure that domestic activities related to plants take into consideration the interests, or "concerns," of other states in that resource. Whether this is an extension of the international law principle of *sic uteri tuo ut alienum non laedas*[252] remains an interesting conjecture.[253]

Given that it is a modern international obligation of states to conserve and engage in the sustainable use of plants, it could be argued that a violation of such clearly defined treaty obligations constitutes (at least in theory) an internationally wrongful act.[254] In other words, the obligations created in cases of egregious destruction of plant life forms and habitats are not necessarily founded on the doctrine of *sic uteri tuo ut alienum non laedas*. Indeed, a strong case may be made that wrongful use or wanton destruction of plant life forms, which is contrary to the express obligations created by the CBD, would raise questions of liability under the pertinent rules of state responsibility as articulated by the United Nations International Law Commission (ILC).

The ILC distinguishes internationally wrongful acts from internationally legal acts.[255] While the former give rise to state responsibility, the latter give rise to liability for injurious consequences.[256] In the ILC approach, primary rules constitute the actual obligations. According to Article 2 of the ILC's 2001 Draft on State Responsibility:

> There is an internationally wrongful act of a state when conduct consisting of an action or omission is attributable to the state under international law; and that conduct constitutes a breach of an international obligation of the state.[257]

Given the obligations of states to conserve plants, it would follow that derelictions of such duty would fall under the category of the ILC's primary rules. On the other hand, secondary rules determine the legal consequences of failure to abide by the primary rules.[258] Under international law, three threshold tests are used in determining state responsibility: (1) was there a duty in international law? (2) was the duty breached? (3) can responsibility be attributed to a state for the violation of international law? It should, however, be noted that "acts by non-state entities, such as a citizen or official for whose acts a state is not responsible, do not give rise to state responsibility."[259] There may be difficulties in making attributions of wrongfulness to states where the act or omission in question was not directly performed by organs or officials of that state, particularly in the case of plant conservation.

Further, there is even greater difficulty in asserting or determining liability for infractions of soft law[260] commitments by states. Indeed, this is most pertinent to the CBD given that its obligations are largely couched in generalities, broad objectives, caveats, and other qualifiers having little specificity in terms of state obligation. However, it would seem that the obligation to conserve and use in a sustainable manner plant life forms within the jurisdiction of a state are clear enough to warrant liability for any infraction of that obligation. However, international environmental law has not yet articulated the framework for bringing offending states to justice in cases of wanton and egregious destruction or endangerment of the integrity of plant life forms.

In the absence of a regime of accountability for destruction of plant life forms, Brazil's rebuff of the North's "concern" over its rapid destruction of the Amazon for "developmental" purposes readily comes to mind. The practicalities of implementing a regime of fidelity to international obligations, such as the duty to conserve plant life forms, are complex. Ultimately, it seems that, in spite of international obligations, states have the final say in determining how to balance their national priorities with their obligations to conserve plant life forms under their national jurisdictions.

It is perhaps in recognition of this conundrum that the principles of sustainable development and precaution have been embedded in the CBD and other international environmental law instruments. However, these juridical initiatives are not without their own severe problems, including conceptual incoherence, ideological imperialism, and suspicions among states. The current mantra of sustainable development[261] may be traced to the report of the Brundtland Commission,[262] whose status as part of the emerging body of international environmental law is no longer in doubt.[263] By the same token, academic exertions and international legal literature on the concept are ubiquitous.[264] What is doubtful is the ideological and theoretical coherence of the concept of sustainable development – indeed, its very meaning is doubtful.[265]

First, the notion of sustainable development is vague and, as Sara Dillon argues, "because of its vagueness, sustainable development has fostered diversionary debates over what level of 'development'[266] is in fact 'sustainable.'"[267] Priya Kurian adds that the phrase "'sustainable development' is so ambiguous that it means all things to all people."[268] McCloskey, in despair, concurs that "the idea of sustainable development is a fine phrase without much meaning."[269] Moldan and Billharz[270] have pointed out that sustainability is inherently value-laden and is not applied in a vacuum. Accordingly, in modern times economic development and sustainability have been largely understood within the paradigm of capitalism. This approach has its weaknesses. As Billharz and Moldarn have further argued, "the current concepts of indicators and indicator frameworks [of sustainable development] originate mostly within northern industrial countries and are shaped by their cultural orientation."[271] The point is that, respectively, the concepts of sustainability and development are fundamentally matters of economic ideology and social values, and they are probably in a state of mutual conflict – unless, of course, we provide a fundamental redefinition of the meaning of "economic development."[272]

Given the dominance of the profit motive in the capitalist framework, money plays a major role in the contemporary measurement of sustainability. However, dollars and cents hardly constitute a complete narrative framework for the phenomenon of development. In the words of Billharz and Moldarn, "many natural and social assets and processes are not traded in the market. The use of monetary value as the common denominator for comprehensive approaches thus faces serious methodological difficulties."[273] In this context it is hard to disagree with Sara Dillon, who says that "it would be difficult to imagine a better gift to the multinational corporate order than the concept of 'sustainable development.'"[274] Without yielding to cynicism, it seems fair to say that the concept of sustainable development has largely remained a problematic mantra, indeed, an oxymoron designed to achieve an improbable compromise.[275]

Perhaps in an attempt to remove the sting of some of the above criticisms, the element of equity has become a familiar refrain in the discourse on exploitation of plant life forms. Thus, sustainable use of plants as well as other life forms, as defined by the Global Biodiversity Strategy, means, "husbanding biological resources so that they last indefinitely, making sure that biodiversity is used to improve the human condition and seeing that these resources are shared equitably."[276] However, a survey of the global regime on plant resources shows differences of approach.[277] In addition, the practice of "environmental equity" among and within states[278] shows that the poor bear a disproportionate burden of environmental costs.[279]

It may be an empty ritual to consider the issue of sustainable use of plants in econocentric regimes that concern themselves primarily with issues of monetary savings, investments, excess production of goods and services, and conspicuous consumption without juridical recognition of the legal interests of unborn generations in those same resources. In this regard, the *Children's* case[280] – wherein the Supreme Court of the Philippines granted forty-two children the standing to sue, as representatives of themselves and future generations, to protect their right to a healthy environment – is filled with promise. Speaking for the Supreme Court of the Republic of the Philippines, Justice Hilario Davide held that

> we find no difficulty in ruling that they can for themselves, for others of their generation and for the succeeding generations, file a class suit. Their personality to sue on behalf of the succeeding generations can only be based on the concept of inter-generational responsibility insofar as the right to a balanced and healthful ecology is concerned.[281]

If the concept of sustainability is to rise above being a mere subject of sterile scholarly exertions, it seems inescapable that the domestic laws on standing to sue and access to environmental justice will have to witness a positive sea change. Otherwise, decisions such as the *Children's case* will remain exotic curiosities in the annals of environmental jurisprudence.

With regard to the precautionary principle,[282] as already noted, a state's inability or refusal to adopt this principle in the handling of plant life forms, especially genetically modified plants, may now give rise to certain juridical consequences. However, issues of what actions are to be taken to fulfill the obligation of precaution in international law remain controversial. It may thus be said, in summary, that international liability for obligations created by the CBD is a work in place requiring enormous goodwill from states and other vested interests.[283]

Another significant contribution of the CBD to international jurisprudence on plant conservation is that it makes the domestic order the focal point of action. This is ostensibly a surrender to the demands and sensibilities of the South during the conventions negotiation stages. This posture may be discerned from the rejection of such words as "global" and like terms in the text of the CBD. It is also arguably another manifestation of the South's complete rejection of any connotation of plant resources as a CHM, which the South perceived as a manifestation of neocolonialism.[284]

The benefits of the national focus include the fact that it is at the national and subnational levels that meaningful impact may be made in the conservation of plants. In any case, it is probable that the so-called global values and priorities pertaining to plants may amount to impositions of foreign values. Moreover, of all matters of environmental protection, the conservation of plants is perhaps the least amenable to top-down solutions symptomatic of the so-called global approaches to environmental problems. Of course, there are drawbacks to the instinctive rejection of the denotations of global responses to, or prioritization of values concerning, the conservation of plant resources.

First, the parochial and nationalistic conception of plant conservation is very shortsighted and naive as it ignores the fact that global economic forces are principally to blame for the tragedy of plants at the domestic terrain. Second, the nationalistic approach may create disharmony and inconsistency arising from the uneven abilities of the various states to implement their treaty obligations.[285] The CBD seeks to impose obligations on the governance of plant resources based on the principle that states have uneven capacities and resources with which to prosecute the tall order of conserving plants. Without prejudice to the principle of *pacta sunt servanda,* which underpins treaty obligations,[286] it would seem that, under the principle of common but differentiated responsibilities, states have the latitude to comply with the convention according to their resources and capabilities.[287]

However, state practice amply shows that states may evade their legal obligations, particularly those pertaining to the governance of the environment.[288] Non-compliance may arise from lack of capacity to implement contracted obligations or from a myriad of other reasons. A state is obliged in international law to refrain from acts that would defeat the object and purpose of a treaty to which it has acceded or to which it has expressed the consent to be bound by.[289] Whatever the case may be, a consequence of non-compliance is that international environmental law is weakened and loses legitimacy and efficacy.

The CBD has also contributed to the jurisprudence of plant life forms by creating a regime for fair and equitable sharing of the benefits of plants.[290] One of the three major objectives of the CBD is to institutionalize an equitable access-and-benefit-sharing mechanism as an intrinsic part of the emergent regime on the conservation and use of plants.[291] Without a doubt, the *ancien regime,* as it were, involved a more or less winner-takes-all situation, whereby the North freely appropriated and profited from the South's conservation and improvement of plants. The CBD creates a new regime devoted to sharing the benefits derived from such life forms. As already indicated, this is a fundamental departure from the "former system whereby the South derived no benefit from its genetic resources while Northern companies ... derived ... enormous profits."[292] Naturally, this radical change of the pre-existing regime elicited a considerable degree of opposition.

Perhaps the most innovative contribution of the CBD to the jurisprudence on plants is the unprecedented recognition of the contributions of local communities and indigenous peoples, particularly women, to the conservation and genetic improvement of plant resources.[293] The preamble to the CBD recognizes the

> close and traditional dependence of many indigenous and local communities embodying traditional lifestyles on biological resources ... and the vital role that women play in the conservation and sustainable use of biological diversity.[294]

More important, Article 8 (j) specifically mandates the contracting parties to make laws that will respect, preserve, and maintain such knowledge. The said article states thus:

> Each Contracting Party shall, as far as possible and as appropriate, subject to its national legislation, respect, preserve, and maintain knowledge, innovations and practices of indigenous and local communities embodying traditional lifestyles relevant to the conservation and sustainable use of biological diversity and promote their wider application with the approval and involvement of the holders of such knowledge, innovations and practices arising from the utilization of such knowledge, innovations and practices.[295]

To a large extent, this sea change purges international environmental and patent law of the latent notion that traditional and local communities have made no intellectual contributions of merit to the conservation and improvement of plants. The unfortunate erasure of the scientific and empirical character of traditional peoples knowledge is a wrong that needs to be righted by law and by changes in social attitudes.

Interestingly, vestiges of the premodern view that traditional and local communities were bereft of technological know-how with regard to the conservation and development of plants remain. One commentator has recently postulated that the basis of the recognition of state sovereignty over plant genetic resources rests on the "forbearance" exercised by local communities in preserving plant genetic resources. According to this forbearance hypothesis,

> the philosophy underpinning the convention seeks to reward those who exercised forbearance and thus *preserve* biodiversity ... the sacrifice which developing nations make in forbearing from development arguably equates to the sacrifice of expending labour under the Lockean labour theory from "fruits of one's labour." It should be equally just to recognize a property right resulting from forbearance in this context.[296]

The obvious implication is that, since farming began, traditional farmers, including women involved in improving the genetic base, have done nothing but sit idly by while plants grew or withered away as fate willed. Yet the wheat, tomato, maize, and other numerous plants used as food or medicine are products of millennia of efforts and rational intervention and mediation by farmers and local peoples involved in improving and conserving plants.[297]

Further, the forbearance argument posits that the CBD and local and indigenous peoples are concerned with the "preservation" of plants. Such arguments have no relation to reality. There is a world of difference between conservation and preservation. The dynamic and evolving nature of plant conservation and improvement belies the argument for preservation. The fact of the matter is that there is hardly any unmodified plant genetic resource, especially with regard to food crops. Plants in the farm fields in their *in situ*[298] state are in a state of conservation rather than preservation. It is only with respect to *ex situ*[299] gene banks that the term "preservation" may be used. In any event, the CBD does not deal with plant resources frozen and preserved in gene banks.

Finally, the CBD makes broad inroads in articulating the link between intellectual property rights, particularly patents[300] on plants. Given the controversial nature of this subject, it is not surprising that it was "one of the most divisive issues in the negotiations of the CBD."[301] From both a policy and legalistic point of view, Article 16 (5) of the CBD seems elusive. The article provides thus:

> The Contracting Parties recognizing that patents and other intellectual property rights may have an influence on the implementation of this Convention, shall cooperate in this regard subject to national legislation and international law to ensure that such rights are supportive of and do not run counter to its objectives.[302]

Article 16 (2) is equally pertinent on the question of patents on plant genetic resources. It provides thus:

> In the case of technology, subject to patents and other intellectual property rights, such access and transfer shall be provided on terms which recognize and are consistent with the adequate and effective protection of intellectual property rights.[303]

While the meaning of these provisions remains a matter of academic speculation and conjecture (even within the CBD Secretariat itself),[304] the greater flaw in the CBD seems to be its failure to unequivocally delegitimize the appropriation of "informal plant resources" innovation. This omission drew the ire of some of the countries of the South, particularly Ethiopia. The Government of Ethiopia issued the following statement:

> We express dissatisfaction with the provisions protecting patents and other intellectual property rights without commensurate regard for informal innovations, especially in Article 16, paragraph 2, which opens the way for use by countries with the technological know-how of genetic resources and innovations from countries without the know-how in patents and other intellectual property rights and for taking them out of the reach of even those countries which created the very genetic resources and innovations.[305]

Similarly, the 1997 Declaration of Malta, a non-binding instrument, echoes these views.[306] It calls for

> a re-evaluation of the value of unpaid labour [and notes that] ... current trends to reinforce patent and other intellectual property rights complicate the issue. The concept of intellectual property needs to be re-examined and extended to the traditional knowledge of the poor, local communities and indigenous peoples.[307]

Further, it warns, "the increasing appropriation of knowledge and genetic resources threatens the knowledge systems of the indigenous peoples and local communities, the public right to information, the conservation and sustainable use of biological resources."[308] In the light of dwindling stock of plant genetic and species diversity and the supporting human cultures, it is disturbing that the patent process and some international institutions combine to facilitate the appropriation of plants and TKUP.

In relation to this, the gendered and racist approach of early science to the phenomenon of development has been a factor in the creation of a system that has practically turned itself into an impediment to the sustenance and survival of plant life forms and, by implication, of human life itself. The dominant concept and practice of development is at the root[309] of what Senator George Mitchell has termed, perhaps with some exaggeration, an "ecological holocaust."[310]

In addition, the influence of religion[311] and the concept of development[312] on international law in environmental law remains to be fully explored. Regardless of the outcome of academic and scholarly speculation on these issues, it seems that, in the quest for new solutions to old and new problems,[313] values and priorities will have to be rethought.[314] New ethics may be invoked.[315] These factors wield tremendous influence on the human conception of and relationship with the environment. And, in turn, they affect legal norms in various ways, both intended and unintended.

4
The Appropriative Aspects of Biopiracy

As has been shown in Chapter 3, nature and human mediation have combined to produce the phenomenal preponderance of plant life forms in the South. However, since realizing this fact and appreciating its wider implications there has been a well-designed and executed process on the part of the North to alter this imbalance to its own advantage. A substantial aspect of this process has been the appropriation of the South's plant germplasm and TKUP, with the consequential benefits to Northern agriculture, biotechnology, and pharmaceuticals. The development of industrialized agriculture, biotechnology, and pharmaceuticals in the advanced capitalist economies has been predicated on the systematic and continuous appropriation of plant genetic resources and traditional knowledge from areas that lie principally in the Third World.

Central to the phenomenon of biopiracy are three mutually reinforcing factors. The first is sociocultural and pertains to the cultural and gendered denigration and denial of the intellectual input of traditional farmers and breeders, particularly women, in the improvement of plants and the creation of TKUP. In other words, biopiracy can be understood as part of the cultural war with non-Western peoples, cultures, and epistemological frameworks. It must be recalled that the process of colonization held that traditional and non-European peoples were inferior to Europeans in their intellectual capacity and, thus, their epistemological innovations were regarded as unscientific and inherently inferior to those of Western empiricism. Hence, indigenous peoples knowledge, along with their improvements to plant genetic resources, were deemed by Western scientists to be unworthy of either legal protection or adulation.

To worsen a bad situation, a majority of the innovations and improvements in the farming fields in traditional societies are provided by women, and their enormous contributions are almost invariably ignored within their own cultures. In effect, gender discrimination within indigenous communities adds yet another layer to the external oppression of traditional epistemology. This constitutes the prevalent, albeit unstated and subtle, social context within which appropriation of plants and TKUP is conducted. Consequently, plants and TKUP created or modified within traditional frameworks of knowledge have been systematically relegated to the humiliating statuses of "raw materials," "wild germplasm," "folk knowledge," and "ethnobotany."

The second factor that is central to biopiracy concerns the mechanisms by which powerful and influential states of the North have established international agricultural research centres as research institutions and gene banks for the South's plant genetic resources. This process transfers an enormous quantity, quality, and diversity of plant life forms from the South into gene banks strategically located in several countries of the industrialized North. This ingenious transfer of plant genetic resources is surrounded by a hazy web imbued with the notion of "common heritage of mankind" (CHM), which serves to legitimize what is actually an asymmetrical and illegitimate theft of plant genetic resources. What is striking in this grand international sleight of hand is that, in international law, experts are unanimous in their view that the notion of common heritage of mankind is not applicable to plant germplasm. Accordingly, the phenomenal relocation of the Third World's plant genetic resources without adequate compensation, or even a token recognition of indigenous intellectual contributions towards the improvement of those resources, is probably the most egregious act of biopiracy in international relations.

The third and perhaps most "apparently legitimate" factor central to biopiracy is the patenting of plants and TKUP through the patent system. Interestingly, most of the industrialized world's patent systems have witnessed fundamental changes designed to facilitate and legitimize the appropriation of plants and TKUP. This has been largely made possible by a deliberate lowering of the threshold for patentability and several other forms of judicial and legislative intervention in the patent law system that have resulted in serving the ever-expanding appetite and interests of Western corporate seed merchants and pharmaceutical and biotechnological industries. As Chapter 3 has shown, it would be unhelpful to proceed as though intellectual property rights, particularly patents, have no impact on the plant and TKUP regimes. More often than not this impact has been disastrous for practitioners of TKUP and other indigenous and traditional peoples. As the Bellagio Declaration on the Rights of Indigenous Peoples notes:

> Intellectual property laws have profound effects on issues as disparate as scientific and artistic progress, biodiversity, access to information, and the cultures of indigenous and tribal peoples. Yet all too often those laws are constructed without taking such effects into account, constructed around a

paradigm that is selectively blind to the scientific and artistic contributions of many of the world's cultures and constructed in fora where those who will be most directly affected have no representation.[1]

In spite of increasing global disquiet regarding the appropriation of plants and TKUP, it is in the interests of the North, particularly the United States, to maintain the status quo. Quite simply, it is in the North's economic and industrial interest to ensure free access to the South's plant genetic resources and traditional knowledge.

In this chapter I examine the role of international institutions in facilitating and legitimizing the appropriation of plant life forms and TKUP. My analysis is divided into five sections. The first section deals with the sociocultural background pertaining to the institutional and juridical appropriation of plants and TKUP. I identify such factors as the ethnicization of traditional knowledge and the outright denial of its scientific basis as undermining the scientific legitimacy of traditional knowledge frameworks and thus facilitating the appropriation of its products.

The second section provides a historical survey of the origins of appropriation of plants and TKUP, starting with the so-called Columbian Exchange of the fifteenth century and extending into appropriation through juridical methods, which is a favoured method in modern times. The third section examines the role of the notion of CHM in delegitimizing the scientific basis of traditional knowledge. I argue that there is no generally accepted principle of CHM in international law. Hence, the notion that plants and TKUP constitute a part of CHM has no juridical basis; rather, it is a political and rhetorical tool deployed by both the North and the South as and when it suits their respective interests.

The fourth section presents an analysis of the origins, development, and entrenchment of the international agricultural research centres (IARCs) as an institutionalized mechanism for facilitating the transfer of plant germplasm from the South to the North without monetary compensation and without recognizing the intellectual property interests of traditional farmers, particularly women, who have toiled over the millennia to improve those plants. I argue that, until recent times, IARCs were the largest and most effective institutional mechanism for the appropriation of Southern germplasm.

The fifth section details the role of the Food and Agriculture Organization (FAO) in the global politics on control of plant germplasm and TKUP. I examine the juridical status of the so-called concept of farmers' rights and the role of international law in the vexed question of the legal status of Southern plant germplasm housed and stored in *ex situ* gene banks located in the North. I also examine the creation of a limited common by the 2001 FAO Treaty on Plant Genetic Resources.

A Gendered and Racial Interpretation of Biopiracy

Nature and human mediation have combined to produce the phenomenal preponderance of plant life forms in the South. The advantages that Third World peoples and cultures have in plant genetic resources explain the intense struggle among states to alter the status quo. A substantial instrument in this process has been the North's appropriation of as much of the South's plant germplasm and TKUP as possible, with consequential benefits to Northern agriculture. As Kloppenburg argues, "the development of modern industrialized agriculture in the advanced capitalist economies has been predicated on the systematic and continuous appropriation of plant genetic resources from source areas of genetic diversity that lie principally in the Third World."[2]

Over the centuries, the appropriation of plants and TKUP has been facilitated by institutions and legal concepts specially designed by powerful states of the North. Biopiracy refers to the commercial use of plants and TKUP without compensation and/or without the acknowledgment of the intellectual inputs in the improvement of the plants or in the creation of TKUP, and without gaining the prior informed consent of the owner(s) of the plants or practitioners in question.[3] Biopiracy must thus be seen as part of the legal and institutional processes by which powerful states and corporate interest groups seek to control and dominate Third World plant genetic resources and associated knowledge of the uses of plants.[4] As has already been indicated, the phenomenon of biopiracy[5] is facilitated by three mutually reinforcing factors; namely, the sociocultural, the institutional, and the legal. On the sociocultural front biopiracy derives its vitality from a pervasive devaluation of Third World intellectual contributions. Delegitimizing the intellectual inputs or capacity of indigenous and traditional peoples follows from the racist assumptions of colonialism inherent within the dominant narrative of science. Two methods are clearly discernible in this sociocultural erasure of non-Western epistemologies. First, there is the insidious characterization of non-Western epistemological frameworks as "ethnic" knowledge. Western conceptions of non-Western contributions seek to "other," or "ethnicize," the empirical frameworks of indigenous and traditional peoples across the world. The clear objective is to present non-Western paradigms of knowledge as "culture-specific" in contrast to Western paradigms, which are presented as "decultured" and "universal."[6]

Second, the dominant narrative of Western epistemology continually characterizes non-Western knowledge frameworks as inherently inferior to that of Western empiricism. Although as early as the sixteenth century Francisco de Vittoria argued that the legal principles and concepts of indigenous peoples deserved respect, his views were in the minority.[7] Most of the colonizers of traditional and indigenous non-Caucasian societies and peoples held racist views of the natives, whom they dismissed as "a barbarous race, possessing inferior rational capacities."[8] In such intensely racist conceptions of other peoples and the outright dismissal of indigenous and native knowledge frameworks, it was the opinion of some that "the Indians had an inalienable right to be slaves."[9]

Gradually, however, the doctrine of racial inferiority of colonized peoples was sublimated into benevolent guardianship, which operated on the assumption that the colonized non-Europeans had the mental capacity of children. Given the substitutability of rabid racial discrimination with benevolent, patriarchal protection, it is not surprising that, until recent times, traditional knowledge frameworks languished in the margins, merely serving as objects of curiosity for Western anthropologists keen to explore the supposed "primitiveness" of non-European peoples and societies.

With reference to denial of the internal logic and merit of non-Western empiricism, vestiges of the supposed notion of the "ethnic-ness" of non-Western epistemology persist. Thus, it is virtually a tradition in contemporary discourse on traditional and indigenous peoples knowledge to describe the study and knowledge of plants by Western-trained scientists as "botany," while marginalizing non-Western knowledge and epistemology of plants as "ethnobotany."[10] In the same vein, traditional knowledge is characterized as "folklore,"[11] while Western empiricism is characterized as "scientific." It is immaterial whether users of these discriminatory terms mean any harm or, indeed, whether they understand the racist connotations implicit in such characterizations. Law is about coded and clear rules of social behaviour, and these codes are transported and validated in and by language.

Indeed, it is evident that neither the pervasive and abiding notion of the "ethnic" character of traditional knowledge nor its supposed illogicality and inferiority is supported by reason. It is well known that the products of the so-called ethnic knowledge of traditional peoples and cultures, which ought to have been "culture-specific," command global validity irrespective of where they are used. The examples are legion, but a few will suffice. For instance, the serpent tree, which has been in use in India for thousands of years as a herbal remedy for some types of mental illnesses, is used today to make drugs such as reserpine for epilepsy and high blood pressure. Hallucinogenic and pain-relieving plants such as *papavar somniferum* and *erythroxylon coca*,[12] which were used by Indians for these purposes, are still employed to produce morphine and cocaine.[13] Of course, given their widespread use, especially in Europe and the United States, it would be absurd to suggest that those ethnic plants and their products, such as morphine and cocaine, have efficacy only for people living within ethnic cultural frameworks.

Similarly, quinine, which is used all over the world for the treatment of malaria, is also an ethnic product of the traditional knowledge of the uses of the tree *cinchona officinalis*. It has also been proven that the ethnic plant *withania somnifera* has an anti-tumour property; *gymnema sylvestre* has curative properties in relation to diabetes; *centella asiatica* is anti-leprotic; *embelia ribes* is an antihelminthic; *messua ferrae* is good for respiratory disorders; and *Carum coticum* lowers blood pressure. The list is virtually endless, and all these herbs have pharmacological properties consistent with the insights of the ethnic cultures that developed them.[14]

Over one-quarter of modern drugs prescribed all over the world are directly derived from plant life forms, and most of them are products of the much denigrated "ethnic" and "primitive" traditional knowledge of the uses of plants. Over 80 percent of peoples in the South rely on plants for their medicinal supplies, justifying the efforts of the World Health Organization to involve traditional healers in the dispensation of health care to billions of people all over the world.[15] The absurdity of characterizing TKUP as ethnic knowledge, as a form of knowledge lacking in universal empirical validity, is perhaps best demonstrated in the global use of hitherto ethnic food crops. Virtually every urban neighbourhood in the world has a coffee shop: coffee is an ethnic crop and beverage from Ethiopia and Kenya. Columbus brought maize to Spain in 1496, and today it is a major component of the diets of several countries of the world, especially in East Africa. Similarly, the popular drink Coca-Cola, which is probably the most universally recognizable "face" of global commerce, is partly derived from two ethnic crops: the West African Kola nut and Gum Arabic from the Sudan.

Before Europeans settled in North America, native Shoshone women of Nevada chewed stone seed for birth control purposes.[16] Western scientists initially scoffed at this. Later, modern "science" confirmed that stone seed contains estrogen, which regulates ovulation. Today, stone seed forms the basic ingredient for modern birth control pills. The revolutionary implications for human rights and the empowerment of women wrought by this "culture-specific folklore" are common knowledge. The magnitude of TKUP has not yet been exhausted. As the director-general of WHO, Halfdan Mahler, pointed out in 1977, "let us not be in doubt: modern medicine has a great deal still to learn from the collector of herbs."[17] In sum, the notion that the so-called ethnic knowledge of traditional peoples is culture-specific, while Western science is universally applicable, is inaccurate.

Although modern medicine and science initially scoffed at the practices of traditional medicine, this attitude probably started to change after the end of colonialism, as the ex-colonies began to re-evaluate their own cultures and made some attempts to save them from the patronizing tendencies of the colonizing Europeans. Over the years it has become orthodox wisdom in enlightened quarters in the North to say that non-Western epistemology[18] is no less empirical than is its Western counterpart. It is perhaps in recognition of the narrowness of Western empiricism and its ignorance of the cultural dimension of technological development that

Western science is becoming increasingly multidisciplinary. Of course, most epistemological systems have their limitations and strengths[19] and can complement one another.[20]

Indeed, Vogel and other writers are persuaded that American Indian medicine was at least equal, and probably superior in many respects, to that of sixteenth-century Europe. Professor Brown has observed that

> the impressive knowledge of the Native American peoples about a wide variety of natural phenomena is not however accidental, nor has its acquisition been haphazard. It is based on generations of systematic inquiry. It is the accumulation of and transmittal of repeated observations, experiments and conclusions. Some of the elements of the scientific method were inherent in their processes.[21]

Examples of the empirical basis of TKUP are legion. For instance, Michael Balick notes that, around the Ganges and the foothills of the Himalayas, the natives, by observing how the mongoose ate the *chotachand* shrub before engaging cobras in combat, correctly deduced that the plant offered an antidote to snakebite.[22]

The point here is that non-Western knowledge frameworks are no less empirical than are Western frameworks. The problem is that non-Western epistemological frameworks continue to suffer cultural denigration and humiliation. In effect, as Michael Balick argues, a "deeper reluctance to explore indigenous knowledge systems may be attributed to cultural prejudice dating to the years when the Western powers reigned over colonies."[23] During the period of colonial imperialism, Balick contends that Western medicine

> was taken as a prime exemplar of the constructive and beneficial effects of European rule. This western medicine was, to the imperial mind, one of its most indisputable claims to legitimacy. Since western medicine was regarded as prima facie evidence of the intellectual and cultural superiority of Europeans, the figure of the medicine man or shaman was often viewed as inimical to social and cultural progress. Indeed the pejorative term "witch doctor" has come to stand for savagery, superstition, irrationality and malevolence.[24]

If cultural biases are eschewed, Michael Balick argues, then "Indigenous traditions and science are epistemologically closer to each other than Westerners might assume."[25] Quoting F.S.C. Nortrop, Peter Morley affirms that

> one must seriously ask oneself whether superstition and myth, in the derogatory or non-scientific connotations of these words, are not due to our

judging a given people from our conceptual standpoint, rather than theirs ... when the trouble was taken to find their concepts, then it became evident that everything made sense and that their behaviour and cultural norms followed as naturally and consistently from their particular categories of natural experience as ours do from our own. I believe it is just as much an error to suppose that there were no people anywhere who insisted on empirically, and hence scientifically, verified basic concepts before Galileo. Prevalent as the latter is, it is nonetheless nonsense.[26]

As Morley further argues, "Throughout the vast range of traditional medical systems are many beliefs and practices which contain an element of techno-empirical knowledge."[27] Yet the tendency has been to consider traditional medical practice as "primitive."[28] On the question of the input of the indigenous farmers to cultivating and saving germplasm, William Lesser observes that recent studies clearly show that these farmers and breeders had knowledge similar to the "formalized" findings of Mendel on genetic traits and breeding. In his view, local knowledge is "empirical in a pragmatic sense."[29] In sum, the characterization of traditionally modified plants and TKUP as "raw material"[30] or "ethnic" knowledge, with only parochial validity, is not only objectionable but unfounded.

The Early Beginnings of Institutionalized Biopiracy (1492-1941)

Although the Westphalian paradigm conferred upon states sovereign jurisdiction over their respective municipalities, no state is wholly self-sufficient with respect to its food or medicinal needs. Thus, the irreplaceable and multiple roles, values, and functions of plants have often necessitated a large measure of international interaction and cooperation regarding plants and their derivatives. However, a critical analysis of the direction in which plants flow, and their methods of transfer, from one state or region to the other reveals an asymmetrical and inequitable regime. In other words, the movement and transfer of plant germplasm and products has been largely from the South to the North, with the former having little or nothing to show for the interaction.

Further analysis of the methods of transfer and "exchange" of plant life forms between the North and the South reveals the existence of a potent regime that is a brilliant combination of juridical subterfuge and unequal bargaining power. This situation has been created and sustained by colonialist and neocolonialist structures and the manipulation of international patent law to facilitate the free and unhindered flow of plant germplasm from the South to the North. What is intriguing in this process is the fact that, although North-South relations are notoriously controversial and emotive, especially in light of the colonialist context and neocolonialist tendencies, historians and other analysts have largely tended to focus on the looting, pillaging, and theft of artifacts made of gold, silver, ivory, or wood. Yet land, gold, and silver did not form the only motives for colonialism: exotic plants were also a factor. As Kloppenburg observed, "little note has been taken of the appropriation of plant genetic resources"[31] by the forces of colonialism. Even Karl Marx, known for his revolutionary and profound insight into and analysis of global capital, paid little attention to the role of plants; nor did he explore the asymmetrical movement of plant resources and its radical import in changing the power and political structure across the globe.[32] Other eminent historians, political philosophers, and economists have largely devoted their energies to scrutinizing and explicating the phenomenal haemorrhage of human and mineral resources from colonized territories and peoples.

It may be argued that, unlike gold and silver, plants are renewable. However, it seems probable that the real cause of this understatement of the role of plants is the prosaic nature of plant transfer from one state to the other. Save in exceptional circumstances, the history of plant transfer is no match for the high drama associated with the state-sponsored looting and pillaging of gold and silver by European explorers, pirates, and *conquistadors*. Simply put, plant theft seems benign, if not mundane. It is probable that, if the state-sponsored pirate Sir Francis Drake had presented his Queen with a bowl of cotton seeds, rubber seeds, or peanuts, no knighthood would have gone his way.

Yet plants have always formed the substratum upon which diverse human civilizations have prospered. Since the times of Christopher Columbus

the New World supplied new plants of enormous culinary, medicinal, and industrial significance: cocoa, quinine, tobacco, sisal and rubber. More than this, the Americas also provided a new arena for the production of the Old World's plant commodities (e.g., spices, bananas, tea, coffee, sugar, indigo).[33]

The profound implication here is that the asymmetrical movement of plant life forms from the South to the North largely underpinned and, indeed, redefined the structure and configuration of the global economy, human population distribution, and the cultural, scientific, and international legal order. For example, before the age of iron ships, the United Kingdom had destroyed its own forests to build ships with which it commanded the seas. Had there been no Indian forests to keep Her Majesty's navy and merchant marine afloat, it is doubtful whether British dominance of the seas would have lasted as long as it did.

Similarly, it is remarkable that "a single coffee tree reaching the Amsterdam botanic gardens in 1706 from Ethiopia via Ceylon and Java became the basis for the New World coffee industry."[34] Today, the consumption of coffee is so commonplace in Europe and North America that it is virtually impossible to conceive of a village without a coffee shop. Furthermore, and perhaps more poignantly, the developmental trajectory of the United States, its sustained use of enslaved non-white labour, and the consequential contribution of that disreputable practice to the discourse and law on human rights and dignity would probably not have occurred if there had been no sugar cane or cotton farms. Plants provided the basis for those complex manifestations of human oppression and thus created the institutions and laws that afforded examination of the ramifications and contents of human dignity.

The case of quinine is the perfect example of the process and implications of biopiracy.[35] The grand irony here is that it was quinine that enabled European colonizers to penetrate, survive, and ultimately colonize the malaria-infested parts of Africa, Asia, and Latin America. The point is that plants have always played a critical role in redefining and reconfiguring the global balance of power.

For historical convenience rather than exactitude, the origins of the appropriation of plants may be traced to the "Columbian Exchange" of 1492, when Christopher Columbus's forays into the Americas with some plant germplasm marked the introduction of "exotic" plant species. Since then the face of the earth, in the spread and distribution of human populations as well as in the realignment of geopolitical power, has radically changed. In 1493 Columbus returned to Europe with maize, and in 1494 he returned to the Americas with wheat, olives, chickpeas, onions, radishes, sugar cane, and citrus fruits (for scurvy) in the hope of supporting a European colony.[36]

Subsequent voyages by other European explorers and voyagers added potatoes to the diet of Europe, resulting in a phenomenal increase in European population. Furthermore, the introduction of new plant resources into European diet and agriculture and the settlement of Europeans in the Americans fundamentally reconfigured global economic and political equations.[37] As Kloppenburg observes:

> Maize and potatoes had a profound impact on European diets. These crops produce more calories per unit of land than any other staple but cassava [another New World crop that spread quickly through tropical Africa]. As such, they were accepted, though often reluctantly, by peasantries increasingly pressed by enclosures and landlords, and by a growing urban proletariat.[38]

Since Columbus's sporadic and disorganized transfer of plants from the "New World" to the "Old World," and from parts of the New World to other parts of the same New World, the critical importance of exotic plant germplasm has never been lost on the political leaders of Europe, the Americas, and, subsequently, Australasia. Indeed, in Europe a worldwide network devoted to the collection of germplasm from the South was quickly put in place; hence, the origin of the botanical gardens, particularly in the British Empire.[39] These institutions routinely collected the world's plant resources, of which a decisive majority was tropical or subtropical in origin.

Given that most of these tropical and subtropical territories and peoples were under the European colonial yoke, the asymmetrical transfer of germplasm from the colony to the mother country was perceived as "an internal affair" of the colonial empires. For example, germplasm from British colonies in Asia and Africa was routinely transferred not only to the Royal Botanical Gardens in the United Kingdom but also to other parts of the British Empire, as though the latter were a single juridical entity (if not de jure, at least de facto). Thus the transfer of plant germplasm was not conducted under the notion that plants constituted a free good and a resource for all peoples of the world; rather, the prevailing theory was that a colonial outpost was merely one of several other projections of an imperial state.

Scientists, breeders, and collectors, particularly from the colonizing world, collected and transferred a huge quantity and diversity of economically useful and/or rare plant life forms to botanical gardens, gene banks, research institutions, and breeding programs that were deliberately scattered across the various outposts of the colonial empires. In the absence of the earlier imposition of legal restrictions on this "intrastate" transfer of germplasm, there emerged the unfounded notion, especially among latter-day environmental activists,[40] non-governmental organizations, and some writers,[41] that plants, even in the postcolonial era, come under the rubric of the common heritage of mankind (CHM).[42] This notion is fallacious. A sober analysis of state practice and other evidences of international law in the colonial and postcolonial eras clearly shows that there has never been a CHM regime as applied to plant genetic resources. The false notion that plant genetic resources constitute part of the CHM has significantly facilitated the theft and appropriation of indigenous peoples knowledge. This is a subject that demands close scholarly attention and analysis.

Biopiracy and the CHM Concept in a Postcolonial World

The CHM concept[43] entered the lexicon of international law a few decades ago.[44] Since then, the attempts at defining its scope and meaning have been ambiguous.[45] Notwithstanding the uncertainties surrounding the meaning of its constitutive terms, one major factor remains constant – the narrowness of the scope of the concept of common heritage. The concept of common heritage[46] has attained juridical mention only within the ambit of claims concerning communal rights in areas or resources that lie outside the limits of state jurisdictional authority: a sort of res communis humanitatis.[47] In other words, it is a term applied to the so-called global commons.[48] These include the ocean floor,[49] outer space,[50] the moon,[51] and Antarctica.[52]

It is thus apparent that the notion of common heritage is the opposite of principles of international law governing access to or control over assets or properties, particularly natural resources, that fall within the jurisdiction of a recognized state. In effect, sovereignty and jurisdiction over a territory is an indefeasible aspect and character of statehood, and whatever falls within the boundaries of a state is subject to the amplitude and magnitude of state jurisdiction.[53] This is a well known principle of international law, and it need not detain us here.[54]

Ideologically, the notion of common heritage is a political and rhetorical tool of convenience used by both the North and the South whenever it suits their respective interests. Assertions of the applicability or lack thereof of the principle of common heritage to any resource, by either the North or the South, should be critically examined before being accepted as a correct expression of the law. For example, the concept of common heritage was a counterpart of the doomed attempt by the South to establish a New International Economic Order (NIEO). In the movement to establish a new international economic order, the common heritage concept was primarily designed to deny a technologically advanced group of states (from the North) the legal right to exploit and lay claims of rights of ownership over the last frontiers of the world, such as the international seabed and Antarctica.[55]

Conversely, the industrialized states that have largely rejected the notion of common heritage as a general principle of international law, particularly with respect to the South's claim for a new international economic order, have been quite enthusiastic in proclaiming that this concept applies to plant genetic resources, which are found mainly in the South. Needless to add, the North's argument is the product of self-interest. As Kloppenburg further explains:

> Common heritage and the norm of free exchange of plant germ plasm have greatly benefitted the advanced capitalist nations, which not only have the greatest need for and capacity to collect exotic plant materials but also have a superior scientific capacity to use them.[56]

Yet again, when the industrializing states believed that their agricultural outputs could be dramatically improved by adopting the intensive method of farming developed by the industrialized states and by appropriating the so-called high yield varieties (HYVs) created by the latter from germplasm originally collected from the South, the industrializing states enthusiastically declared all plant life forms (including the HYVs) to be a common heritage.

Naturally, the industrialized states rejected this characterization of genetically modified HYVs as common heritage. Henry Vogel has analyzed North-South polemics and posturing vis-à-vis the applicability of common heritage to the conflict between privatization of the benefits of plant re-sources and the socialization of the cost of access to those resources:

> Genetic resources are a prime example of privatization having more to do with power relationships in the contemporary world than with neo-classical economic science. Until quite recently, Northern industry has been able to privatize the benefits of bio-technologies that derive from genetic resources while at the same time socializing the cost of access to those genetic re-sources. Genetic resources were free under the doctrine known as the "com-mon heritage of mankind." Being on the opposite side of the trade, Southern countries have long wanted to privatize genetic resources but socialize ac-cess to biotechnologies. Rather than arguing for a symmetrical reform and the privatization of profits and costs, both the North and the South would like asymmetrical reform: the privatization of just their profits and the so-cialization of just their costs. For the North this would mean that the South gives up its genetic resources but recognizes its intellectual property rights (IPRs); for the South this would mean that the North gives up its IPRs but recognizes a Southern claim on the use of its genetic resources. In the struggle for inefficiency and inequity, the North is winning.[57]

Accordingly, the common heritage notion, as espoused by both sides of the global economic and industrial divide, has been a barely disguised ideologi-cal tool in the struggle for control of plant genetic resources across the globe. Leaving ideology aside, the question remains whether, in international law, there is a settled principle of CHM, and, if so, whether such a principle governs the regime on plant resources where they are located within the boundaries of sovereign states. In answering the second part of this ques-tion, one must make reference to the pertinent sources of international law, particularly the primary sources – namely, treaties and custom.[58] In answer-ing the first part, it seems that, notwithstanding the substantial confusion that has afflicted the concept of CHM, five major characteristics may be said to delimit it under contemporary international law.[59]

First, the area to which the concept of common heritage may apply must be free from appropriation of any kind and, hypothetically, must be man-aged by all states.[60] Second, under the proposed common heritage regime, it follows that all peoples would be expected to co-manage the common space in their capacity as representatives of "mankind." In other words, there can be no supervening national interests wherever the concept of common heri-tage is deemed to be applicable. Third, whatever economic benefits accrue from this global management of a common space would be vested in the global community. These are the necessary inferences from the quality of the term "common" as used in the notion of common heritage. Fourth, the area of common global ownership must entail a completely demilitarized zone where only peaceful activities are conducted.[61] Fifth, scientific research must be freely and openly permissible, and the physical environment and ecology of the area in question must not be impaired. Even a cursory examination of these elements, and an extrapolation to the principles of state sovereignty, clearly shows the inapplicability of the notion of common heritage to plant life forms within the boundaries of states.

Although the concept of common heritage has enjoyed recognition in some treaties, especially treaties dealing with the deep seabed,[62] the moon,[63] outer space and celestial bodies,[64] and the continent of Antarctica,[65] some scholars doubt whether it is now a generally accepted principle of international law. In other words, there seems to be a debate, perhaps semantic, concerning whether recognition of the concept of common heritage in treaty law is synonymous with the status of a "generally accepted principle of international law."[66] "Strict constructionists," or purists, will readily argue that common heritage is not yet a generally accepted principle of international law. On the face of it, there are some arguments that may be made for this doctrinaire, perhaps sterile, point of view.

Strictly speaking, for a concept to be considered as a generally accepted principle and a part of international law, it must be distinct.[67] Given the problematic meanings of the constitutive words "common," "heritage," and "mankind," it may be doubted whether any coherent clarification of the concept exists in international law. The word "common," for example, refers to something that belongs to all. Explicitly and implicitly, management of such entities or resources requires the consent and representative mandate of all who have property in the thing held in common. The term "heritage" refers to property that has been inherited. It is impossible to conceive of the relevance of this term to plants, which may well be unknown to humanity, never mind being capable of being passed on as a heritage. As already indicated, the term "mankind" has not yet acquired any juridical meaning in international law. Accordingly, scholars such as Wolfrun, Gorove, and Joyner are on solid ground when they argue that the concept of common heritage is afflicted with internal inconsistency. However, these arguments are not airtight.

First, it is a matter of common knowledge among international lawyers that there are principles of international law, which, although not known for their clarity, are nonetheless generally accepted. In other words, conceptual clarity is not a condition precedent to the emergence of legal concepts as generally accepted in international law. Ready examples of legal concepts that are not the epitomes of conceptual clarity but that have recently assumed the character of generally accepted principles of international law include "sustainable development" and "precaution." Thus, although conceptual clarity is a virtue and a desirable value in the evolution of legal norms, its absence is not necessarily fatal to the status of a concept in international law. After all, in the development of international law, vague terms and phrases often ripen into coherent principles of law.

Second, it would seem that the crucial factor in determining general acceptance of principles of international law is that the state practice resulting from compliance with that concept (regardless of its clarity), in this case, common heritage, must be demonstrably evident and accompanied with the requisite *opinio juris*. A corollary to this requirement is that the custom of acceptance of that concept must be widespread. Here, the common heritage concept stands on shaky ground. It is remarkable that, apart from the UN Convention on the Law of the Sea (UNCLOS), treaties that recognize common heritage have witnessed the lowest numbers of ratifications. This phenomenon is particularly significant in the context of the global implications of the common heritage concept. For example, the moon treaty has only the barest number of ratifications for its becoming effective – five. Apart from this miserably poor number, none of the five parties to the moon treaty – Austria, Chile, the Netherlands, the Philippines, and Uruguay – is a space-faring state. When this fact is juxtaposed with the universal significance of the moon and the importance of space in daily life (satellite television, telephony, weather forecasting, and so on) the low number of ratifications of those treaties that promote the common heritage concept leaves it in a weak position in its claim to be considered as a generally accepted principle of international law. The inescapable conclusion is that, while the concept of common heritage may be a principle of international law, whether it is "a generally accepted principle" of international law is open to debate.

Assuming, especially in relation to UNCLOS, that the common heritage concept is a generally accepted principle of international law, the second part of the question – whether the "principle" of CHM is applicable to plant resources – deserves further examination. We must look at the sources and evidences of international law as well as the principles of state sovereignty and how they apply to plant life forms. Article 38 of the Statute of the International Court of Justice details the general sources and evidences of international law.[68]

First, even if the concept of common heritage has been considered to be a generally accepted principle of international law prior to the Convention on Biological Diversity (CBD), no international treaty or convention characterized or designated plants as part of CHM. Indeed, it is striking that, unlike with other "emergent" concepts and principles in international law, particularly regarding the environment, there is not a single declaration or resolution by the UN General Assembly that refers to plants located within a state's jurisdiction as constituting part of CHM.[69] Even the FAO International Treaty on Plant Genetic Resources for Food and Agriculture reiterates that plants are part of the national sovereignty of states. Article 10 thereof provides:

In their relationships with other States, the Contracting Parties recognize the sovereign rights of States over their own plant genetic resources for food and agriculture, including that the authority to determine access to those resources rests with national governments and is subject to national legislation.

More significantly, all references to common heritage in treaties and declarations of the organs of the United Nations have consistently been in the context of the remaining frontiers on earth, along with celestial space and objects. None of the treaties that mention common heritage pertains to spaces traditionally under state sovereignty and jurisdiction, such as plants and plant habitats.

The absence of a treaty law basis for the purported applicability of CHM to plants is neither a remarkable omission nor a coincidence. Since the emergence of the Westphalian paradigm of international law, the latter has been state-centric, founded on the control of each state over its own territories, subject to principles of international law.[70] Article 8 of the Montevideo Convention is clear on the point as it provides that no state has the right to intervene in the internal or external affairs of another. Indeed, all aspects of international law, especially treaty law and state practice, on state sovereignty implicitly and expressly recognize the undoubted powers of states to regulate access to plant life forms within their respective jurisdictions.[71] Contemporary international law instruments are similarly unequivocal in their assertion and reiteration of domestic state sovereignty and the inadmissibility of external interference in internal state affairs. Clearly, prior to the CBD or FAO Plant for Food Treaty, there has been no treaty law support for the notion that plant germplasm from the Third World (or elsewhere, for that matter) is part of the global commons.

In the absence of treaty law support for common heritage pertaining to plants in the pre-1992 era, attention may be shifted to customary international law during that same era – 1492 to 1992. Here again the notion of a common heritage regime on plants finds no support. First, the concept of common heritage entered global discourse in the late 1960s, while the Columbian age of colonial transfer of plant genetic resources across the globe and the transfer or appropriation of plant life forms through colonial instruments, including "botanic gardens" and "research" institutions, began in 1492 and lasted until the late 1960s. Thus, assuming (but without conceding) the common heritage concept to be a principle of customary international law, such a recent principle could not have governed activities that took place hundreds of decades before it came into existence. Customary international law does not operate retroactively.

Furthermore, an examination of relevant state practice during the period between 1492 and the 1960s shows that, indeed, the concept of common heritage has no roots in customary international law. That is to say, state practice or custom[72] accompanied by evidence of *opinio juris* clearly shows that the notion of a common heritage pertaining to plant life forms was not part of customary international law during the period in question. A careful analysis of state practice shows that states have always sought to protect and sustain their monopoly of and hegemony over economically useful plants. Thus, even though states, particularly the gene-rich states of the South, were made to yield their plant life forms to their colonial masters as contributions to international agricultural "research," such practices lacked the requisite character of customary international law. This is because the element of *opinio juris sive necessitates*, which Anthony D'Amato, in his classical disquisition, has beautifully reformulated as the articulation of reciprocal international behavior with legal consequences,[73] is conspicuously missing.

Here, the psychological element is the articulated expectation in state relations that a particular act or omission will have legal implications. The *opinio juris* is the voluntary choice to be bound in law by a free act. In the absence of the psychological element of articulated expectation of reciprocal juridically significant behaviour, otherwise known as *opinio juris sive necessitatis*, the transfer of plant germplasm from the South to the North through the instrumentalism of international research centres or colonialist institutions lacks the legal sense of obligatoriness that is the essential characteristic of customary international law. As such, those transfers of plant life do not amount to an expression of customary international law of common heritage on plant life forms.

In international law repetition of the colonial practice, no matter how frequent, would not yield a legally binding obligation unless it could be shown that the practice was articulated and carried out in the belief that there was a legally binding obligation to do so.[74] This distinction is crucial as it constitutes the divide between a mere social usage and a legally binding principle of customary international law.[75] Indeed, a close study and analysis of relevant state practice shows clearly that states have always sought to keep economically useful plants out of the reach of other states. This practice is particularly evident in the case of states with the requisite enforcement mechanisms for such an exclusionary policy or with the ability to police their territorial borders and to control the inflow and outflow of plant resources. Of course, the reproducibility of plants makes it difficult for such policies to achieve absolute success.

Notwithstanding the challenges associated with controlling the outflow and inflow of plant life forms across international borders, states largely conducted themselves in a manner clearly supportive of their sovereignty over plants within their jurisdiction and were under no legal obligation to grant free access to such resources. For example, in the colonial era economic and military powers like France, the Netherlands, and the United Kingdom adopted elaborate and often stringent measures to "keep useful [plant] materials out of competitors' hands."[76] Further examples of state control over plants are legion. The French were so determined to sustain their monopoly on the indigo dye trade that the export of indigo seeds from French Antigua (a French colony) was made a capital offence.[77]

Similarly, prior to the CBD era, the Government of Ethiopia (one of the rare African states to escape the full rigours of colonialism) embargoed the transfer of coffee germplasm from Ethiopia. In the colonial era it was very difficult to obtain black peppercorns from India. Ecuador does not freely supply cocoa germplasm to other cocoa-producing states. Peru and Bolivia once made trade in quinine (an extract from the bark of the Cinchona tree native to those countries) a government monopoly. Of course, no state could have seriously argued that such actions violated international law.

Desperate and draconian measures were sometimes taken by states to maintain control over plants in their territories. For example, the Dutch, in order to maintain their global monopoly on the supply of nutmeg, destroyed all nutmeg and clove trees in the Moluccas except those on the three islands where they located their plantations. It was this tight control of the transfer of plant germplasm in the period before the CBD that compelled states to engage in or support audacious attempts to break the monopoly of other states on some key, economically significant plants. In some cases the audacious activities were acts of theft or brigandage, such as the smuggling of plants and black market trade in rare or important plants. For example, in order to break the Brazilian monopoly on the global supply of rubber (Brazil controlled 95 percent of global trade), in 1876 British authorities encouraged and aided Henry Wickham when he successfully smuggled out 70,000 rubber seeds in a boat, which eventually reached British colonies in Asia. Wickham's theft of the rubber seeds and eventual escape literally sowed the seeds of the collapse of the Brazilian monopoly on rubber production.[78]

It is evident from the above instances that the undoubted powers of states to regulate access to and the use of plant life forms within their domains has always remained an inherent aspect of statehood. In addition, domestic legislation restraining or controlling access to forests, wildlife, parks, and trade in certain plants has always been part of the exercise of state sovereignty over its plant resources.[79] Most countries of the world, if not all, had national quarantine laws regulating the importation of diseased or potentially diseased plants. Yet it has not been suggested that such domestic legislative powers pertaining to plants are dependent upon external permission or the pleasure of an external transnational authority or even the so-called concept of common heritage on plants. Such authority predated the 1992 CBD and 2001 FAO treaty. For example, pursuant to the power to regulate access to plant resources, the International Covenant for the Protection of Plants, signed in Rome on 16 April 1929, mandated the contracting parties to "establish relevant machineries for the regulation of the import and export of plants."[80]

States have always had as an intrinsic part of their status as sovereign entities the legal authority to regulate the inflow and outflow of plant life forms within their own domestic jurisdiction. [81] In effect, the concept of state sovereignty over plants falls within the more extensive concept and principle of Permanent Sovereignty over Natural Resources (PSNR).[82] In the governance of the same subject matter, the concept of PSNR[83] and CHM are thus mutually exclusive.[84]

The Amazon issue best dramatizes the conflict between CHM and PSNR. The industrialized states have long argued that Brazil *should not* (this is clearly distinct from "does not") have absolute sovereignty over the Amazon region.[85] Brazil controls approximately three-fifths of the Amazon. The Amazon itself comprises some 42 percent of Brazilian territory. It produces 50 percent of the world's oxygen[86] and a substantial part of the world's fresh water and biodiversity.[87] Given its universal importance, it is very tempting to misconstrue the Amazon as a CHM.

However, the concern expressed by the industrialized states over Brazilian (mis)use of the Amazon has never really impressed the Brazilians, who insist that "we are masters of our destiny and will not permit any interference in our territory."[88] Thus, concerns over Brazilian (mis)management[89] of the Amazon would not entitle any state or group of states to assert a right of individual or collective extraterritorial jurisdiction over Brazilian Amazonia.[90] At best, other states may express the requisite amount of "concern" over the use or misuse of such resources occurring within the boundaries of sovereign states. By the same token, arguments by Third World countries that technological products protected by national intellectual property regimes constitute a part of CHM and, thus, should be freely available to all humanity has been resisted by robust rebuttals from industrialized states.

The extent to which states may allow such concerns to influence their style of management or control of such plant resources depends on myriad factors, such as the need for international cooperation, but no legal right may be exercised or presumed by other states in respect of plant resources within the sovereign jurisdiction of other states. However, no state is self-sufficient in matters of biological resources and, more important, issues of conservation, use, and commercialization of plant resources are often intrinsically international in character. Hence, international cooperation is virtually unavoidable with respect to plant germplasm.

In sum, states have always had the right to determine, regulate, and control access to plant life forms within their respective jurisdictions. If there were any juridical doubts on this issue, it is fair to say that the CBD has laid them to rest. Accordingly, it is an overstatement, indeed an error, to argue, as some commentators[91] have done, that the CBD created a "new" regime of state sovereignty over plant life forms. The CBD merely reaffirmed an inherent, pre-existing right of state jurisdiction over plant life forms. What has changed in recent times is the stridency with which states, particularly from the gene-rich South, are reasserting their perceived right of sovereignty over plants within their respective jurisdictions.

Biopiracy and the Role of International Agricultural Research Centres
In addition to the misconception surrounding the CHM concept in relation to plant germplasm, another factor that has facilitated the appropriation of Third World germplasm is the role of international agricultural research centres (IARCs) in the massive transfer of plant germplasm from the Third World to the industrialized world. Indeed, any analysis of the phenomenon of biopiracy that fails to shed some light on the historic transfer of Third World germplasm by the IARCs misunderstands the multiple factors that fuel biopiracy.

Following a meeting in 1941 between Vice President Wallace[92] of the United States and Raymond Fosdick, the president of the Rockefeller Foundation, "it was thought that a program of agricultural development aimed at Latin America in general and Mexico in particular would have both political and economic benefits"[93] for the United States. By 1943 the Rockefeller Foundation had started its Mexican Agricultural Program, with the ostensible primary focus being on the improvement of wheat and corn.[94]

However, the idea was not simply to "improve" wheat and corn; rather, "from the very first, the collection of indigenous germ plasm was an important component of the Rockefeller Foundation's Mexican Agricultural Program and of the other Latin American Initiatives."[95] What is equally significant here is the perspicacity in the timing of this paradigmatic shift regarding the appropriation of plants. That is to say, the Rockefeller-Ford initiative displayed profound business acumen in that it realized the imminence of the demise of colonialism and the consequential loss of the colonial apparatus for funnelling plant germplasm from the South to the North. Hence, a new mechanism for appropriating plant life forms was imperative if the crucially important transfer of plant germplasm was to continue unhindered. As William Lesser has argued, with the "decline of the empire system ... governments lacked the military presence and legal authority to compel sovereign nations to yield valuable germ plasm."[96]

Thus, the Rockefeller-Ford initiative probably marked the beginnings of a postcolonial institutionalization of the global network for the appropriation of the South's germplasm on the ostensible grounds of fostering research on agricultural crops. The Rockefeller Foundation, in conjunction with the US government, spawned IARCs in the South. The location of the IARCs in the various Southern countries was not a product of fortuitous occurrences or a series of coincidences; rather, they were established in each region that was known for its phenomenal stock of indigenous germplasm (i.e., the so-called Vavilov centres). For example, the International Rice Research Institute is located in the Philippines,[97] the International Center for Agricultural Research in Dry Areas is located in Syria,[98] the West African Rice Development Association is located in Liberia,[99] the International Potato Center is located in Peru,[100] and so on.[101] In justifying the choice of Guatemala, the Rockefeller Foundation argued that "the Tropical Research Center has been located [there] to search for genes or characters that will improve our corns."[102]

Through this process the United States amassed a large collection of corn and other plant germplasm subsequently released into the farm fields of the United States and Europe without any economic benefits for the South or any recognition of the intellectual property rights of the local farmers, again particularly women, who had spent millennia improving the plant germplasm in question. Thus, with regard to each crop of global significance, such as corn, wheat, rice, and other cereals, an IARC was established at its centre of origin and funnelled the plant germplasm to the North. In little time an enormous quantity, quality, and diversity of plant germplasm had been collected, and it is no coincidence that it was also during this period that the idea of expanding and reconfiguring the patent system to allow for patents on plants started putting down roots.

The US then turned to the next phase, how to store the collected plant germ. By 1956, this problem had largely been solved by the construction of the National Seed Storage Laboratory at Fort Collins, Colorado, United States.[103] It should be noted that the plant germplasm could easily have been stored in the countries of origin but this did not happen. Instead, while over 80 percent of all economically useful plant germplasm and varieties indigenous to the South was collected by the IARCs, most of the actual storage or gene banks of the collections (called accessions) are located in the North.[104]

The FAO has identified over 1,200 plant genetic resource collections worldwide, held in more than 160 countries and territories.[105] Overall, governments hold 83 percent of the accessions, IARCs hold 11 percent, and the private sector holds 1.27 percent. The IARC collections contain about 35 percent of the unique samples, making them "the world's most significant collection"[106] of Southern plant germplasm. As Naomi Roht-Arriaza notes, "most genetic materials collected in Southern countries – 68 percent of all crop seed, 85 percent of all livestock, and 86 percent of microbial culture collections – are held at the IARCs or in Northern countries."[107] In effect, the advanced capitalist nations, though poor in naturally occurring plant genetic diversity, are as rich in "banked" germ plasm as the developing nations of the Third World. Indeed in a number of crops (wheat, barley, food legumes, potato) the advanced capitalist nations possess more stored germ plasm accessions than do those nations that are the regions of natural diversity for the crop.[108]

Thus, removed from its areas of origin, the collected plant germplasm was released to the public in the North without either economic payment to the traditional farmers who developed it or recognition of the intellectual contributions of local farmers and breeders of the South. Without question, the activities of the IARCs constitute the most far-reaching institutional appropriation of traditional and indigenous-owned plant germplasm in recent times.

Until recently the IARCs were coordinated by the Consultative Group on International Agricultural Research (CGIAR), with the mandate to "coordinate the disparate and haphazard network of germplasm collection around the World."[109] The CGIAR was established in 1971. According to Kloppenburg, it "spearheaded and sustained what has come to be known as the Green Revolution."[110] But that is hardly the end of the story. Again, according to Kloppenburg:

> The IARCs perform a dual role in the processing of plant germ plasm. They necessarily collect and evaluate indigenous land races and primitive cultivars that are the raw materials from which HYVs (high-yield varieties) are bred. And because their "imported" agricultures are based on the very species that the IARCs are mandated to improve (i.e., corn, wheat, potato), such collection and evaluation are of direct value to the developed nations. *The IARCs are ... vehicles for the efficient extraction of plant genetic resources from the Third World and their transfer to the gene banks of Europe, North America, and Japan ... The CGIAR system is, in one sense, the modern successor to the eighteenth- and nineteenth-century botanical gardens that served as conduits for the transmission of plant genetic information from the colonies to the imperial powers.*[111]

In effect, the CGIAR functioned as an institutional channel for funnelling plant germplasm from the South to the North free of charge and without acknowledging the intellectual contributions of traditional and local farmers of the Third World:

> Seed banks and gene banks collect Southern germ plasm and distribute it to gene-poor Northern countries; thus a large proportion of commercially used genetic material moves to the Northern countries via the IARCs. Studies

estimate, for example, that 21 percent of the US wheat crop was derived from material stored at the International Maize and Wheat Improvement Center, the IARC for wheat.[112]

It is equally noteworthy that storage of the germplasm *ex situ* in the North often renders it inaccessible to the original providers from the South.

However, it was the success of the CGIAR in providing the germplasm for the development of the so-called HYVs that ultimately brought into focus the dangerous consequences of excessive and mercantilist reliance on these varieties.[113] In other words, driven by the forces of profit, most crops in the agribusiness world, particularly in the North, are based on very narrow genetic diversity. This raises the real potential for disastrous vulnerability to pests and diseases. The Green Revolution and the HYVs, enthusiastically promoted by the corporate and political interests of the North across the globe, showed the limitations of using the shallow and whimsical criteria of "market forces" to determine which plant species should perish and which should thrive.[114] The problem of genetic uniformity in plants thus brought to the fore, especially in the 1960s and 1970s, the economic importance of Southern germplasm as "insurance" against the ravages of any pest or disease attack on industrial crops or farming as an industry. In turn, this realization fuelled yet another round of aggressive collection of plant germplasm by the Northern-controlled IARCs. The FAO organized two conferences at which a decision was reached to the effect that "a coordinated global program of collection and conservation was necessary to ensure that the essential raw materials of plant improvement were not lost."[115]

Similarly, in 1972 the UN Conference on the Human Environment in Stockholm issued a resolution calling for an international program to preserve the germplasm of tropical crops. In consequence, in 1972 the Beltsville Conference in Maryland recommended the establishment of the International Board for Plant Genetic Resources (IBPGR). The significant fact here is that, instead of locating this international program under the FAO, which is a UN agency, the Northern-controlled CGIAR argued that it should be designated the research arm of world agricultural development.

The anomalous compromise in this situation was CGIAR's creation in 1974 of the IBPGR.[116] Although it was physically placed in the FAO offices in Rome,[117] to all intents and purposes it was constituted as a CGIAR institution and thus operated under the political and financial control of the industrialized states of the North.[118] For example, the IBPGR's budget is not provided for by the FAO but, rather, by twenty of the most industrialized states who have little or negligible indigenous plant germplasm of commercial significance but who are the greatest commercial users of plant germplasm. In fact, 69 percent of the IBPGR's 1984 budget was underwritten by just six of these donors; namely, Canada, Japan, the Netherlands, the United

Kingdom, the World Bank, and the United States Agency for International Development.[119]

Further, the policies of the IBPGR are not set by debate among the global community constituted in the wider FAO but, rather, through decision-making processes internal to the CGIAR. In Kloppenburg's scathing and irrefutable words, "the IBPGR may cloak itself in the 'internationalist' legitimacy provided by its association with the FAO, but the board is not subject to the control of the United Nations. The financial heart and soul of the IBPGR lies elsewhere."[120] The loyalty of the IARCs to the North is also reflected in their ideological and political stance on the non-release of plant germplasm to states perceived to be ideological enemies, especially during the Cold War era. In addition, until recent times powerful Northern states made legal claims to ownership of plant accessions taken from the South but stored in the North's gene banks.

Interestingly, in some cases, those gene banks, for political and/or economic reasons, have been known to refuse to release or grant free access to the stored germplasm to those states who were the original donors of the plant accession in question. For example, the gene banks located in the United States have been known several times to refuse free access to requests from such original donors as Afghanistan, Albania, Cuba, Libya, Nicaragua, and the defunct Soviet Union. According to the administrator of one of the gene banks,

> We are willing to accept selected collections for long-term maintenance at Fort-Collins. *They would become the property of the US Government*, would be incorporated with our regular collections, and made available upon request on the same basis as the rest of the collection ... As you know, it has been our policy for many years to freely exchange germ plasm with most countries of the world; however, *political considerations have at times dictated exclusion of a few countries.*[121]

Naturally, as patents and plant breeders rights exacerbated the overt politicization and appropriation of plant germplasm taken from the South and "improved" in the North for resale to the South, the latter began to dissent.[122] The North dismissed the arguments of the South and observed that there were no legal barriers stopping the South from "improving" the plant germplasm itself and reselling it in the global market.[123] The discontent of the South was soon to find expression in the framework of the FAO and ultimately led to an aggressive reassertion of national and state sovereignty over plants. Ultimately, the juridical solution was the emergence of an international legal order that recognizes the peculiar character of plant genetic resources as both a national and global asset. This is apparent from the texts of both the CBD and the newly adopted FAO International Treaty on Plants. The importance of the FAO treaty warrants further analysis and inquiry.

The FAO and the Politics of Plant Genetic Resources

At the FAO's twenty-first biennial conference in 1981, a resolution was passed instructing the director-general to prepare a draft of an international agreement that would provide a legal framework for controlling the flow of plant genetic resources across the globe.[124] The 1983 draft agreement was designed to be a legally binding convention, but, in a classic display of diplomatic manoeuvring by both sides, the document was reduced to a hortatory and ineffectual treatise,[125] a non-binding instrument full of platitudes and ill-defined, turgid exhortations on CHM and plant germplasm. In hindsight, it seems that the South and the North overreached each other in the struggle for juridical and political control over plant life forms.

In sum, the 1983 undertaking seems to be an attempt by the South to redress perceived inequities in the prevailing regime. The South, probably influenced by the controversial book by Canadian activist Pat Mooney,[126] wanted a regime of common heritage on plants, including the so-called elite varieties produced in the laboratories of the North from germplasm originally collected from the South through the IARCs. The shortsightedness here involves the erroneous notion that these "elite" versions of plant germplasm are desirable when, in fact, they require enormous agricultural inputs and are antithetical to sustainable agriculture as known to hundreds of millions of local farmers in the South.

This misperception of the utility of elite plants led the South to attempt to establish a regime of global free access to those varieties modified and marketed by the North under patent regimes. Thus Article 2 of the 1983 undertaking defines "plant genetic resources" to include the reproductive or vegetative propagating materials of (1) cultivated varieties (cultivars) in current use and newly developed varieties; (2) obsolete cultivars; (3) primitive cultivars (land races); (4) wild and weed species, near relatives of cultivated varieties; and (5) special genetic stocks (including elite and current breeders' lines and mutants).[127] As Tilford has observed, "by the terms of the 1983 Undertaking, the common heritage blanket spreads over not only the Vavilov centres and Third World farmer's fields, but over Northern agricultural laboratories as well."[128] Needless to add, the North considered this challenge to the status quo to be unacceptable.

The response by the North, particularly the US seed industry, was immediate and intense. The American Seed Trade Association fulminated that the 1983 international undertaking struck "at the heart of free enterprise and intellectual property rights."[129] The US government refused to sign the undertaking or to join the Commission on Plant Genetic Resources, and it promised no financial support. Other countries of the North followed suit.

What is quite remarkable, indeed ironic, is that, while the North enthusias-tically supported the erroneous characterization of plants from the South as common heritage, it steadfastly maintained that inclusion of elite or com-mercial varieties under the CHM umbrella was absolutely unacceptable. In the North's view, it could not freely give away the right of commercial plant breeders.[130] Yet it expected the rights of local and traditional farmers and breeders to be sacrificed at the altar of an unfounded notion of CHM.

Although the movement and transfer of plant germplasm from their *in situ* sites in the South to *ex situ* facilities in the North (such as botanic gar-dens, arboreta, seed banks, tissue and cell culture collections, etc.) was pro-moted as an aid to scientific research for the benefit of humanity, later developments, particularly relaxed patent regimes for plants, clearly revealed the appropriative functions of those institutions. It was therefore inevitable that Third World states would eventually question some of the acts of ap-propriation carried out under the ostensible cover of "scientific research." As already evident, the Third World seemed to have operated under the naive assumption that the germplasm under the control of the IARCs could not enter the global market as a patented commodity of trade. This assump-tion was dispelled when biotechnological corporations and other research institutions in the industrialized countries began to obtain patents on im-proved plant genetic resources originally stored in the IARC gene banks. The South began to complain loudly that its plant germplasm, which had been transferred to the North by the IARCs for research purposes, more often than not was modified, patented, and resold in the global market for enormous profits and with little regard to the countries and local farmers and breeders of the South.

These concerns agitated the South, and one of the questions to be ad-dressed involved the legal status of the gene banks and control of the IARCs. Given that these gene banks hold at least one-third of the unduplicated samples of the world's plant germplasm, it is not surprising that an acrimo-nious legal debate ensued between the North and the South as to the exact legal status of the IARCs and the plant accessions held by them. However, not all the IARCs were mired in confusion. For example, the CIMMYT-International Maize and Wheat Improvement Centre, and the IRRI-International Rice Research Institute seemed to have had clear policies on ownership of stored germplasm. Both institutions clearly recognized that genetic materials in their custody deposited before the CBD were "held in trust for the world community" and that plant genetic resources obtained by them after the CBD were governed by the terms of deposit and CBD guidelines.[131]

With respect to the other IARCs the position was far from clear. A 1986 FAO legal opinion, nebulous in conception and filled with equivocations, concluded that the CGIAR gene banks existed in a unique world between national and international law. They are not created by a formal treaty concluded among States or other international legal persons, and their activities are not directed by States or such other international legal persons. The gene banks maintained by the IARCS are neither under the control of any given State or national authority, nor in the private sector. Their status is, in fact, *sui generis*.[132]

Another FAO legal opinion, this one prounounced in 1987, hardly added any clarity to the issue:

Ownership of genetic material held in government gene banks or those of public institutions was in most cases, for practical purposes considered to be vested in the State in which these gene banks are located. However, for material held in the International Agricultural Research Centres, the legal position was unclear.[133]

This obfuscation was not to stand for long. Following agitation by the South and threats of restricted access to plant germplasm, a clearer legal answer to the question emerged in 1995. Thus, by virtue of an agreement between the CGIAR gene banks and the FAO, control of the former was vested in the latter.[134] The new FAO treaty on plant genetic resources provides a legal framework for access to the IARC germplasm collection but hardly stems the flow of appropriation of plant germplasm.

However, a regime of compensation seemed futuristic, and in the cases where the origin of the IARC accession is unknown, the FAO suggested that "compensation might be provided to developing countries collectively."[135] The FAO/IPGR agreement thus brought the global network of the Rockefeller-founded IARCs within the framework of the FAO, which, in turn, became incorporated into the framework of the newly adopted FAO treaty on plant genetic resources. In a bid to exorcize the ghost of CHM from plant life forms, the FAO treaty on plant genetic resources recognizes the "sovereign rights of states" over those plant resources within their territorial jurisdiction.[136]

It seems that the definition of the legal status of the IARCs has not resolved the North-South differences on access to plants. Indeed, following the inability of the South and the North to arrive at a compromise on the question of farmers rights, the international undertaking came under review at the Leipzig conference in Germany, where a global plan of action was adopted.[137] This Global Plan of Action (although non-binding) recognizes the intellectual input of local farmers in the improvement and conservation of plant germplasm.[138]

However, its platitudinous and patronizing praises offered little comfort to local farmers whose contributions to plant development continue to be appropriated through the juridical framework of the patent system. Suffice it to note that, in an ostensible attempt to stem the tide of appropriation of plant germplasm, the FAO concluded a draft international code of conduct for the collection and transfer of plant germplasm.[139] In addition, in 1993 the FAO passed Resolution 7/93 requesting the Commission on Genetic Resources to provide a forum for negotiations among governments to harmonize the 1983 undertaking with the relevant provisions of the CBD. According to the Commission on Genetic Resources, the revised undertaking was designed to become a legally binding instrument when finalized. As earlier noted, these FAO initiatives culminated in the adoption, on 3 November 2001, of the FAO treaty on plant genetic resources and the draft article on traditional knowledge and equitable sharing of the benefits derived from it.[140]

In retrospect, if the 1983 undertaking contains any redeeming element, it is its Article 7. The provisions of Article 7 practically sought to rewrite the well-orchestrated Northern institutionalization of the appropriation of the South's plant diversity. The said article mandates the development of

an internationally coordinated network of national, regional, and international centres, including an international network of base collections in gene banks, *under the auspices or jurisdiction of the FAO*, that have assumed the responsibility to hold, for the benefit of the international community, and on the principle of unrestricted exchange, base or active collections of the plant genetic resources of particular species.[141]

Article 7 thus mandated a transfer of the IARCs to the FAO. As already pointed out, the IARCs functioned as the pipeline for funnelling the South's plant germplasm to the North, without economic rewards or recognition of local farmers and breeders intellectual contributions towards improving the plants in question. The North initially opposed the incorporation of the IBPGR into the FAO structure. Following severe North-South disagreements on the letter and spirit of the 1983 undertaking, subsequent attempts were made to "interpret" the undertaking in a manner suitable to all sides. Such interpretations include Resolution 4/89, which recognized symmetry between plant breeders rights (PBRs), as provided for by the International Union for the Protection of New Varieties of Plants (UPOV) and the undertaking. Similarly, Resolution 5/89 defined the concept of farmers rights, and the sovereign rights of nations over their genetic resources were reaffirmed in Resolution 3/91. Finally, in 1993 the FAO Conference adopted Resolution 7/93 for the revision of the international undertaking and its harmonization with the CBD.[142]

The harmonization, which occurred in 1994, led each CGIAR centre to sign an undertaking with the FAO putting its plant genetic resources in trust under the latter's auspices. The agreements are largely the same and describe the roles and responsibilities of the centres as trustees. These stipulate that:

1 the CGIAR hold the designated germplasm in trust for the benefit of the international community, in particular, the Third World;
2 neither the CGIAR, or any recipient, will seek intellectual property rights protection over the designated germplasm or related information;
3 the CGIAR will undertake to make samples of the designated germplasm and related information available directly to the users for the purpose of scientific research, plant breeding, or genetic resource conservation without restriction.

Arguably, the term "related information" in relation to germplasm as adopted in the agreements between the centre and FAO refers to indigenous or traditional knowledge of the uses of the plants. However, this phrase is far from clear. In addition, there is a legal grey area pertaining to when a variety is sufficiently different from the original in-trust germplasm to give rise to a legitimate claim for intellectual property rights protection. The effects of the agreements on the jurisprudence on biopiracy remain unclear.

The emergence in 2001 of the FAO Treaty on Plant Genetic Resources has not silenced debates on the aforementioned issues. The centerpiece of the treaty is its multilateral system for access and benefit sharing. It is generally acknowledged that access to the Third World's wide genetic base of plants will allow the further development of improved varieties. The sticking point, however, has remained the lack of compensation to the Third World and recognition of the intellectual contributions of its peoples and cultures towards the improvement and conservation of plant varieties. The FAO treaty is in many respects a comprehensive instrument, and thus its Article 9 would seem to remedy the extant shortfall in international law pertaining to the denial of indigenous contributions to plant development. Article 9 states that parties are to

> recognize the enormous contribution that the local and indigenous communities and farmers of all regions of the world, particularly those in the centres of origin and crop diversity, have made and will continue to make for the conservation and development of plant genetic resources which constitute the basis of food and agricultural production throughout the world.

In the context of the treaty, recognition of indigenous intellectual contribution to plant conservation and development improves on "farmers rights."

It should be noted that farmers rights was a concept originally introduced into international discourse in 1988, when the Colorado-based Keystone Center mediated the dispute between several states and itself under the aegis of the Keystone International Dialogue on Plant Genetic Resources.[143] The Keystone mediation accepted the concept of farmers rights proposed by Mexico in exchange for a regime of free access to Third World germplasm under the 1989 FAO Undertaking.[144] As articulated in the FAO Undertaking of 1989 as per Resolution 5/89, farmers rights are defined as

> rights arising from the *past, present and future contributions of farmers in conserving, improving and making available plant genetic resources, particularly those in centres of origin/diversity. These rights are vested in the international community as trustee* for present and future generations of farmers for the purpose of ensuring full benefits to farmers and supporting the continuation of their contributions, as well as of the overall purposes of the International Undertaking.[145]

It is instructive that, under the concept of farmers rights as articulated under the 1989 undertaking, "future" intellectual exertions and input of farmers from the South in improving plant germplasm are vested in the "international community." In sharp contrast, intellectual exertions by "scientists" in the laboratories of the multinational seed companies of the North are construed as private property and are secured with patents and PBRs. While these international instruments are non-binding agreements and are now dated and obsolete, they provide a nuanced understanding of the politics and economics of the struggle for control of plants through the mechanism of patents, legal concepts, and international institutions.

The FAO treaty makes three substantive contributions to the concept of farmers rights. These are (1) the protection of relevant traditional knowledge (echoing Article 8 [j] of the CBD); (2) recognition of the right of farmers to participate equitably in sharing benefits arising from the utilization of plant genetic resources for food and agriculture; and (3) recognition of the right of farmers to participate in making decisions at national levels. These are significant improvements on and contributions to the jurisprudence pertaining to protection of indigenous peoples knowledge. In fact, the FAO treaty, in this respect, goes further than Article 8 (j) of the CBD. However, Article 9 (3) of the treaty specifies that

> nothing in this Article shall be interpreted to limit any rights that farmers have to save, use, exchange and sell farm-saved seed/propagating material, subject to national law and as appropriate.

While this provision is seemingly neutral, given the limitation of responsibility for implementation to national governments, it is arguable that this is a loss for farmers, especially, indigenous and traditional farmers.

Another important aspect of the FAO Treaty on Plants is the creation of a limited commons in certain food-related plant genetic resources. Thus, despite its multilateral system, the scope of the treaty is limited to a list of crops. The reason for this is that certain countries generally rich in biodiversity – even if they are not equal in plant genetic resources for food and agriculture – wanted to limit the application of the multilateral system, thereby leaving some room for future bilateral arrangements. Similarly, it was agreed by virtue of Article 12 (3) (a) that material made available through the FAO multilateral system should be "provided solely for the purpose of utilization and conservation for research, breeding, and training related to food and agriculture." In effect, chemical, pharmaceutical, and/or other industrial uses beyond food and animal feed are excluded.

There is debate as to whether the thirty-five crops listed as belonging in the global commons are sufficient or appropriate. These include most major food crops, such as rice, wheat, maize, sorghum, millet, beans, peas, lentils, chickpeas, cowpeas, potato, cassava, and yam. Article 11 of the treaty provides that the criteria for drawing up the list are food security and interdependence. Critics of the list have argued that there is no completely objective means of constructing a list according to these criteria.

It can, however, be argued that there are reliable indicators of which food plants are most often used by the global community. The absence of such food crops as soybean, groundnuts, sugar cane, and the many unknown relatives of the genus *manihot* and tomato from the list suggests a lack of objectivity in drawing up the list of food plants included in the global commons. On the other hand, it may reflect a desire by the Third World states to keep as much of their food plants as possible away from the global commons. Interestingly, the list does not include some of the major staple food plants of the Third World, such as taro, cocoyam, and grass-pea. It also omits African forage grasses and Latin American forage legumes. It is arguable that the absence of industrial crops like tea, coffee, rubber, and oil palm is a result of prevalent bilateral and/or commercial arrangements for these commodities, notwithstanding their significant roles in food security and interdependence. Be that as it may, the list is fairly representative of globally and regionally important crops. It would, however, be simplistic to assume that food security can be guaranteed by a reliance on cultivated crops: harvested landraces (close relatives of cultivated species) and wild species often offer greater protection from diseases and environmental problems.

With respect to the controversial issue of biopiracy, Article 12 (3) (d) of the treaty constitutes the most contentious point. It has been surmised that the refusal of Japan and the United States to vote for the adoption of the treaty stemmed from their opposition to this provision.[146] During the negotiations all parties agreed that intellectual property rights such as patents and plant breeders rights would not be applied to all plant genetic resources in the form in which they are actually received from the multilateral system. A similar agreement was reached with respect to plant genetic resources held by the IARCs under the CGIAR.

It must be noted that neither the texts of the treaty forbidding intellectual property rights over plant genetic resources held in trust by the CGIAR nor the negotiating documents of the treaty affords reasonable guarantees against appropriation and biopiracy of plant genetic resources held in trust by the FAO or its agencies. For example, as I argue in greater detail in the next chapter, it is possible to patent DNA sequences that have been isolated from plant material without any structural modification. In effect, Article 12 (3) (d) does not in any serious manner protect Third World plant genetic resources held in trust by the FAO from appropriation through lax patent regimes. A patent holder could easily restrict use of such protected material by others, including the donor Third World country, simply by enforcing the patent on the isolated DNA sequence. This would clearly constitute a misappropriation of plant genetic resources. In respect to this provision, the Conference of the Parties to the CBD has noted that "intellectual property rights might, under certain circumstances, constrain access to and use of genetic resources and scientific research."[147] The most worrisome aspect of this provision is that it does not adequately appreciate that appropriation of plant genetic resources is a contemporary phenomenon carried out through the mechanism of DNA patents on sequences that have not in any way been "modified" by the patent holders.

5
Patent Regimes and Biopiracy

As I have argued in the preceding chapters, the process of biopiracy of plants and TKUP is multiple, involving racialization of knowledge, international institutions, and juridical mechanisms, especially patents and PBRs. In this chapter I examine the appropriative function of the interplay between patent systems of industrialized states and, in a limited capacity, PBRs. The interplay of various national patent laws, particularly of the US patent system in fostering and legitimizing the unauthorized use and commercialization of indigenous knowledge about the uses of plants, has inflamed the passions of many activists, scholars, and indigenous peoples.

In addition, I analyze not only what the legal norms of the patent system seek to protect but also what they omit to protect. In short, the patent system must be thoroughly interrogated and its intellectual integrity should not be presumed. In this endeavour I pay particular attention to some pertinent US statutes and case law. In addition, I scrutinize the influence of US-based multinational seed and pharmaceutical corporations in shaping patent and plant breeders laws.

I divide this chapter into eight sections. The first section deals with the concept of patentability and how it has evolved over the years. I critically examine the juridical regime created by the CBD, TRIPs, and the case law on plants and TKUP. Given that, for all practical purposes, Article 27 of the TRIPs agreement elevates US patent law into a set of global legal norms, I focus close attention on US statutes and case law. I conclude that the changes in the concept of patentability owe much more to judicial assertiveness and the rise of corporate control of the seed industry than they do to legislative boldness.

In the second section I examine the expansion and relaxation of the various criteria for patentability as they relate to plants. I divide this section into four parts. The first part deals with the criterion of novelty and how it has been changed to suit the peculiar needs of inventions relating to plants. The second part examines the criterion of specification and the various methods by which some patent law jurisdictions have sought to overcome the problem of the inherent indescribability of certain qualities of plants. The third part deals with the criteria of inventive step and utility of invention, and the fourth part examines the criterion of uniform reproducibility.

In the third section I focus on TKUP and how it relates to the concept of patentability. I contend that the fundamental elements of the patent system have been rewritten and watered down in order to accommodate the special interests of the seed and pharmaceutical industries. The result is a set of double standards and a permissive patent system – one that facilitates, condones, and legitimizes biopiracy. I divide this section into three parts: the first deals with the debate on "products-of-nature" exceptions to patentability; the second examines the relaxation of the criterion of novelty as applied to TKUP; and the third deals with the issue of utility of inventions and the determination of what constitutes inventive step in relation to TKUP.

In the fourth section I examine the various ways in which international law has attempted to meet the challenges of appropriation of plants and TKUP. Regrettably, international law on state responsibility for the appropriation of alien property is preoccupied with real property and, thus, leaves traditionally generated intellectual property rights unprotected. In light of this lacuna, it seems that states from the South, NGOs, the World Intellectual Property Organization (WIPO), and academics have paid enormous attention to soft law instruments and other alternative options designed to counter the rising wave of biopiracy. I examine the adequacy of such measures and argue that, in addition, we need to rethink domestic laws on patents. Furthermore, regional juridical responses to the challenge of biopiracy should be adopted. Thus I explore the options of registration of TKUP and modification of the domestic laws and institutions on patents. I divide the section into three parts: the first critiques the concept of state responsibility as it relates to appropriation of plants and TKUP; the second examines the law on access to plants and TKUP; and the third part exposes the weaknesses of soft law approaches to the problems of appropriation.

In the fifth section I revisit the option of registration of traditional knowledge as a means of avoiding the appropriation of plants and TKUP. In the sixth section I make a case for modifying domestic patent regimes in order to deal with the problem of biopiracy. Here I evaluate some of the objections scholars make concerning the potential of positive changes for patent systems and also outline the features of a communal patent concept. Finally, in the seventh section I look at the issue of PBRs and conclude that this form of private property rights over plants is as problematic as are patent grants.

Appropriative Functions of Patents

Although the process of appropriation of plants and TKUP is a multiple one involving international institutions and juridical mechanisms, the most effective and controversial process of biopiracy involves the interplay between national patent systems,[1] particularly the US patent system, and other patent systems.[2] The consensus of most scholars, activists, and indigenous peoples is that the patent system has not been sensitive to the dignity, rights, and worldviews of indigenous and traditional peoples. Most critics argue that the patent system is incompatible with the values and cultures of traditional and indigenous peoples. The root cause of this problem is the epistemological and cultural prejudice against non-Western frames of knowledge. Biopiracy is thus a cultural, institutional, and juridical phenomenon.[3] It follows that biopiracy cannot be fully or properly understood without reference to the inherent biases of patent law, Western cultural supremacy, and the inadequacy of international institutions governing plant germplasm and use of indigenous peoples knowledge.[4]

As an epistemological phenomenon, biopiracy delegitimizes the profound intellectual input of local farmers into the improvement of plants. Within the paradigm in question, centuries-old efforts of indigenous and traditional farmers are diminished by being referred to as "informal" or "unorthodox" or as lacking a scientific basis. The consequence is that seeds and plant germplasm improved over the centuries by local and traditional farmers are construed as free goods and appropriated as part of the global CHM, whereas their counterparts from the North are seen as "improved varieties" deserving patent protection and, to some extent, PBRs.

As fallout from the insensitive and predatory patent law regime, biopiracy cannot be fully understood unless one pays significant attention to the subtleties of patent law, its history, cultural biases, and mercantilist instincts. This is indispensable because the modern process of biopiracy is sophisticated and subtle, differing completely from the bravado of sea pirates in the age of empire. Biopiracy through patents is more or less an art form, fully supported (or at least facilitated) by the paraphernalia of apparent legality. As one recent commentator observed, with an admixture of historical and contemporary imagery:

> Today's pirates don't come with eye patches and daggers clenched in their teeth, but with sharp suits and claiming intellectual property rights. So those rich countries which take seeds away from their poorer neighbors and then try to patent them are guilty of theft – plain and simple: biopirates by another name.[5]

In addition, it is not enough to analyze what the legal norms of the patent system seek to protect; it is also necessary to analyze what they omit to protect. The patent system must be thoroughly interrogated and its intellectual integrity should not be presumed. In this endeavour particular attention must be paid to some pertinent US statutes and case law. The reasons for this relative emphasis on the US patent system are not far to seek.

First, it is common knowledge that the TRIPs agreements, especially Article 27, which sets the global minimum threshold for patentability, is an approximation of US jurisprudence on patentability. Second, the TRIPs agreement is a product of the immense clout of the American pharmaceutical and biotechnology industry. Third, the US patent system accounts for almost half of all patents issued in the world. Fourth, the American Patent Office issued most of the controversial patents that raise the question of biopiracy. Fifth, the United States has the most robust regime on patents. Sixth, the pronouncements and decisions of US courts on matters of patent law have immense international influence, often affecting the jurisprudence of other courts in several countries.

Biopiracy and the Concept of Patentability in Patent Law

In analyzing how patent law facilitates the appropriation of indigenous peoples knowledge, the first legal concept that should be scrutinized is the criterion for patentability. The definition of patentable inventions was originally intended, by the first patent laws in Continental Europe, to be limited to machines or processes involving new "art," "manufacture," "compositions of matter," and "designs." The general interpretation by the courts and patent offices was that only inventions pertaining to mechanical[6] products, artifices, and processes were patentable subject matter. However, as Graham Dutfield has pointed out, with the rise of the chemical and pharmaceutical industries, the scope of patentability gradually encompassed the products and processes of those industries.[7] Life forms, including plants, were excluded from patentability. Today, however, the scope of patentable subjects includes artificially modified life forms and DNA sequences. Some critics of this template shift in the patent system argue that life forms are ill fitted for the patent system. There is enormous force in this criticism, but what is important is that the concept of patentability has expanded rapidly and now encompasses a variety of subjects that hitherto it did not.

A close study of the process by which the patent system was reinvented to accommodate the interests of the seed merchants and commercial plant breeders reveals that the appropriation and privatization of plants through the patent law system followed a careful, gradual, and methodical expansion of the concept of patentability. This process has been pursued at three distinct levels: the state, the regional, and (lately) the global. At all three levels the influence of the global seed and pharmaceutical industries is unmistakable.

It is significant that, at the state level, soon after Vavilov's groundbreaking studies, the American seed industry made efforts to extend property rights on plants through patents and, to a lesser extent, by granting PBRs on plants. Although the initial plan was to have patents on seeds and plants of agricultural importance, the industry sensed that such a bold proposition would be drowned in public outrage and indignation. Therefore, it embarked upon a gradual process that began with establishing the principle that plants were capable of being privately owned. In order to outflank public outrage, the first legislative proposal for patents on plants was made by a seemingly benign US group: rose breeders and horticulturists.

Since a significant number of important agricultural crops are sexually reproducing, the threshold on patents on plants was first established through asexually reproducing plants. At the initial stage, sexually reproducing crops, which constituted staple diets for millions of people (e.g., wheat, rice, barley, oats, maize, etc.) and formed the basis of several sensitive industries (e.g., brewing), were deliberately excluded from the initial grasp of patentability. Commenting on this shrewd manoeuvre, Kloppenburg notes that

> in 1930, Paul Stark advised the American Seed Trade Association's Plant Patent Committee to drop their efforts to have sexually reproducing species included in the proposed Plant Patent Act. He suggested that it was best to let the establishment of patent rights to asexually reproducing species *set a principle that new plant forms could be considered patentable*.[8]

It must, however, be noted that gaining this initial threshold was no mean feat and, ultimately, required a fundamental reconfiguration and rewriting of both substantive and procedural patent law.[9] Given that the United States and other champions of patents on plants are dependent on the South for a variety of plant germplasm, the political and economic motives behind this template shift, and the implications of this must not be underestimated.[10] Ordinarily, one would expect that states rich in plant germplasm and diversity would be at the forefront of the campaign for privatization and commercialization of plant genetic diversity.

Further, in the United States the *Plant Patent Act*,[11] *Plant Variety Protection Act*,[12] and the courts have often weighed in to expand the boundaries of the patent system in favour of the industries that rely mainly on plants and TKUP. According to John Golden, "whether as a result of a pro-patent judiciary or as a consequence of the natural extension of prior legal doctrine, by the early 1990s patent law had resolved many fundamental issues in favour of biotechnology's patentability." [13] The practice of extending patents to plants and TKUP originally faced strong opposition, especially in Europe. For example, in Germany patents on plants started only in 1934. Prior to that, the German patent office held that plants could not be patented as they were not inventions in the strict sense of the word. It would seem that, in tearing down the walls of opposition to plant patentability, the courts of powerful states with huge investments in the seed industry, particularly in the North, have often seized the initiative from the legislature.

Perhaps the most celebrated judicial pronouncement in the United States on the rationale for including life forms as patentable subjects is the majority decision of the US Supreme Court in *Diamond* v. *Chakrabarty*.[14] Here, the court held that, since the new bacterium in question was an artificially created "composition of matter" hitherto unknown to humanity, it was patentable. In addition to the pre-existing concept of PBRs created by the US *Plant Variety Protection Act* (*PVPA*), the notion of granting intellectual property rights such as patents on plants was further amplified in *Ex Parte Hibbard*,[15] where it was held that plants, seeds, and tissue cultures can be covered by patents.[16] The initiatives in the United States also found support in similar laws in many states of the North, where legislative responses designed to cater to the interests of commercial seed breeders have been made.[17]

Given the importance of judicial influence in expanding the frontiers of patentability,[18] it is interesting to look at the Canadian case of *Pioneer Hi-Bred Ltd.* v. *Commissioner of Patents*.[19] In this instance a patent application was filed by Clark Jennings in respect of a new soybean variety known as "Soybean Variety 0877." The claims related to the soybean plant, pod, and seed. According to the specification filed alongside the patent application for this new soybean variety, its novelty rested on the fact that it had a high oil content, matured early, provided stable high yields, and had a resistance to seed shattering and certain diseases. Moreover, as the claimant argued, these characteristics could not be achieved by natural breeding. Only "artificial intervention" would make it possible to hybridize two different lines of soybeans to produce "soybean variety 0877." In determining whether this new variety of soybean was patentable, the Canadian Federal Court of Appeal considered the problematic issue of patentability of plants.

In the opinion of the Appeal Court plant "inventions" do not qualify as "manufacture" within the purview of the *Patent Act of Canada*.[20] As Marceau J. noted, the legislature did not contemplate plants as patentable subject matter. If they did, such words as "strain," "variety," or "hybrid" would have appeared[21] in the legislation. In a concurring opinion, Pratte J. further argued that, since a complete and accurate disclosure of the soybean variety was not possible, plants were not capable of being patented. On further appeal to the Supreme Court of Canada, the Court had to deal with the question of whether a new variety of soybean resulting from artificial cross-breeding represented an invention within the meaning of the Canadian *Patent Act*. In grappling with this question, the Court opined that the level of "human intervention" required in the invention of plants for patent purposes must be such that it alters or defies the natural laws of reproduction.

In other words, artificial changes in plants that only reflect natural laws of heredity or Mendelian principles do not rise to the level of patentabililty.

Accordingly, while plants were not necessarily debarred from patentability, the standard of patentability had to ensure that plant inventions crossed or defied the boundaries of natural laws. On the question of specification of plant inventions, neither Canadian nor European courts were persuaded that plant inventions could be sufficiently disclosed, even by a deposit of the invention in question. However, the laws of various American states do not display the same rigorous scrutiny of purported plant inventions and, consequently, permit their patenting.

Domestic legislative initiatives, particularly in the United States, laid the groundwork for international conventions for patent-like rights over plants, such as the UPOV[22] Convention,[23] which, interestingly, started off like the US *Patent Act* by initially limiting its scope to asexually reproduced plants. In order to be eligible for protection under the UPOV, new plant varieties must be (1) distinct from existing commonly known varieties, (2) sufficiently uniform, (3) stable, and (4) new (in the sense that they must not have been commercialized prior to certain dates). In creating this right, the breeder's authorization must be obtained with respect to the use of the propagating material of his or her protected variety for any of the following acts: (1) production or reproduction (multiplication), (2) conditioning for the purpose of propagation, (3) offering for sale, (4) selling or other marketing, (5) exporting, (6) importing, and (7) stocking for any of the purposes mentioned from 1 through 6.

At the regional and continental level, treaties such as the *European Patent Convention of 1973*[24] were concluded and put into effect. For example, Article 53 (b) of the EPC states, "European patents shall not be granted in respect of plants or animal varieties or essentially biological processes for the production of plants and animals."[25] This treaty provision formed the basis for member states of the European Union to modify their laws accordingly.

In comparison with member states of the European Union, the United States has pursued an aggressive expansion of property rights over plants. For example, although the UPOV Convention recognizes an optional protection of plants by PBRs *or* patents, most European Union states opted for PBRs *instead* of patents. Second, until the European Patent Office made its famous *volte face* in policy in the *Plant Genetic Systems* decisions, it was believed that Article 53 (b) limited the scope of patent protection for plants to plants *simpliciter* instead of plant varieties.[26] However, it now seems that the EU law on the matter has reverted to the pre-plant-genetic-systems regime.[27]

Article 27 of the TRIPs agreement marks the culmination of the unrelenting campaign by powerful Northern seed traders to extend the frontiers of the patent system by expanding the concept of patentability at a global level.[28] In the creation and enforcement of a global regime of patents on life forms, the ubiquitous influence of the seed merchants and pharmaceutical giants is undeniable.[29] As Valentina Tejera has critically observed:

Seeking the enactment of the TRIPs agreement,[30] executives from large US pharmaceutical companies exerted strong lobbying to shape United States policy affecting the Uruguay Round GATT negotiations. In fact, many of the transnational companies served as advisors to the GATT Agreement because of their interest in having an "even playing field" in the international market ... For the establishment of this uniform playing field the Uruguay Round Agreement compels Third World signatories to adopt the patent policies and laws of the United States.[31]

While business executives from powerful Northern states, particularly the United States, shaped the content and structure of legal norms on the patenting of plants and TKUP, it is interesting that most of those contributing to the improvement and sustenance of plant diversity were left out of the law-making process. With particular reference to the African delegation, the OAU Draft Declaration on Community Rights and Access to Biological Resources laments, "a smaller part of humanity, represented by 40 States concluded the negotiations for the creation of the World Trade Organization (WTO) in 1994. African countries had negligible or no inputs into the negotiations."[32] As the expansion of the concept of patentability rolls ahead with barely explored consequences,[33] one of the things that stands out in the globalization of the industrial model of development is how plants are being incorporated into this gigantic process for purposes of private profit.[34] As Mark Ritchie argues:

The TRIPs Agreement embraces an industrial model whereby the products of scientific research become the private property of its corporate sponsors. The new rules developed during the Uruguay Round are in conflict with many existing national laws and the traditions of many agricultural and indigenous communities where knowledge of the nutritional and medicinal uses of plants and the results of plant breeding are shared as a community resource.[35]

As the moral, ethical, and legal debates on the patentability of plants and TKUP rage on, legislative and judicial efforts designed to expand the concept of patentability deserve to be scrutinized. It is arguable that the shift from mechanical inventions and compositions of inanimate matter to life forms has opened the way for the industrialization and privatization of life forms, especially plants.[36] To achieve this end the elements of patentability have had to be readjusted and lowered to fit the demands and requirements of the seed-breeding and pharmaceutical industries of the North.[37] Such juridical changes have performed and continue to perform appropriative functions.[38]

Strictly speaking, prior to the WTO era, there was no global minimum standard for patentability of inventions. Article 27 of TRIPs[39] may thus be said to constitute an approximation of what may be crudely regarded as a global standard on patentability. Article 27 (1) of the TRIPs agreement offers the modern minimum yardstick for patentability of all inventions. The said provision provides as follows:

Subject to the provisions of paragraphs 2 and 3, patents shall be available for any *inventions*, whether *products* or *processes*,[40] in all fields of technology, provided that they are new, involve an inventive step and are capable of industrial application.[41]

The exceptions stated in paragraphs 2 and 3 stipulate that

2. Members may exclude from patentability inventions, the prevention within their territory of the commercial exploitation of which is necessary to protect *ordre public* or morality, including to protect human, animal or plant life or health or to avoid serious prejudice to the environment, provided that such exclusion is not made merely because the exploitation is prohibited by their law.

3. Members may also exclude from patentability: diagnostic, therapeutic and surgical methods for the treatment of humans or animals; plants and animals other than micro-organisms, and essentially biological processes for the production of plants or animals other than non-biological and micro-biological processes. However, Members shall provide for the protection of plant varieties either by patents or by an effective *sui generis* system or by a combination of thereof.[42]

The regime established by these provisions stipulates that member states must establish patent systems or an effective *sui generis* system or a combination thereof for plants. The conditions for patentability of plants are fourfold; namely, the article must be an invention, the invention must be new, the novelty must involve an inventive step, and, finally, the novel invention must be capable of industrial application and must also be useful.

These criteria deserve closer treatment and analysis, and it must be pointed out that, in the process of biopiracy, appropriation of plants ought to be treated separately from appropriation of TKUP. Another point worthy of note is that, under the TRIPs agreement, states must have a patent regime for plants, or a *sui generis* system, or a combined regime of patents and *sui generis* system. Most scholars have pointed to the PBRs of the UPOV laws as the obvious *sui generis* option. Given that the latest revision of UPOV grants rights similar to patent grants, there seems to be an illusion of choice for states ill-prepared or unwilling to grant patents on plants.

Two questions have often been posed with respect to the impact of TRIPs on indigenous peoples knowledge. The first question concerns whether there is any conflict between the CBD and TRIPs; the second question concerns whether TRIPs has any implication for indigenous peoples knowledge. With regard to the former, there is little doubt that the TRIPs agreement marks a radical transition from the national character of patents to its linkage to global trade, with an ideological and global superstructure (such as the WTO) to ensure its implementation. On the other hand, the conception of the imperatives of the CBD, especially Article 8 (j) thereof, is to set up national initiatives pertaining to protection of indigenous peoples knowledge and plant biodiversity, even when it is obvious that such national initiatives would have global ramifications.[43] As Chidi Oguamanam has presciently noted, the CBD paradigm is "expected to set the stage for a global framework for the protection of indigenous knowledge on the basis of indigenous conceptions, and as a global plan of action how to conserve biological diversity."[44] Although TRIPs is a global instrument, intellectual property rights are ultimately national grants, and thus, to the extent that a state's laws on intellectual property rights meets the minimum standards set by TRIPs, it is arguable that there is no inherent conflict between the CBD and the latter. However, there would be a conflict between Article 27 of TRIPs and the CBD if, in pursuit of the latter, states enact laws that detract from the omnibus scope of TRIPs. What happens when there is a conflict between two pertinent treaties addressing the same subject?

The Vienna Convention on the Law of Treaties, which codifies customary international law on treaty law, may not be very helpful on the issue.[45] The closest aid from the Vienna Convention is Article 30, which provides as follows:

(a) If a treaty says that it is subject to, or is not to be considered as incompatible with, another treaty, that other treaty will prevail.
(b) As between parties to a treaty who become parties to a later, inconsistent treaty, the earlier treaty will apply only where its provisions are not incompatible with the later treaty.
(c) As between a party to both treaties and a party to only one of them, the treaty to which both are parties will govern the mutual rights and obligations of the States concerned.[46]

A careful scrutiny of the Vienna Convention yields the impression that the above-stated rules are essentially residual; that is to say, they are meant to apply in the absence of express treaty provisions regulating priority. A scrutiny of both the TRIPs and CBD agreements shows that neither of them makes a clear provision on the issue of priority. The closest to a rule on priority in this instance are the provisions of Article 1 of the TRIPs agreement. The said article provides that

> members shall give effect to the provisions of this Agreement. Members may, but shall not be obliged to, implement in their law more extensive protection than is required by this Agreement, provided that such protection does not contravene the provisions of this Agreement. Members shall determine the appropriate method of implementing the provisions of this Agreement within their own legal system and practice.[47]

Given that the TRIPs agreement is specifically focused on intellectual property, it seems that its provisions would override the CBD should there be any conflict on such issues as patentability. On issues that pertain to the protection of indigenous peoples knowledge through intellectual property regimes, it seems that Article 8 (j) of the CBD and Article 27 (3) of TRIPs are reconcilable. This issue raises the question of whether there are implications for traditional knowledge arising from the operation of the TRIPs agreement. Prominent skeptics such as Pat Mooney of the Canadian activist group Rural Advancement Foundation International (RAFI) and the Indian eco-feminist Vandana Shiva, activist scholars, indigenous groups, and many others, have argued that TRIPs poses significant threats to the economic and social well-being of indigenous peoples. RAFI and many others have been quite effective in drumming up global support at such diverse forums as ethnobiological conferences, events organized by indigenous groups, meetings of the Conference of the Parties (COP) of the CBD, and intergovernmental meetings of WIPO.

RAFI and other activist groups believe that TRIPs facilitates the piracy of indigenous knowledge. The problem is that TRIPs represents an alien jurisprudence. Indigenous peoples have their own normative structures governing the acquisition, transfer, and use of knowledge. To the extent that TRIPs does not recognize indigenous regimes, there is little question that it violates the equality and humanity of indigenous peoples across the world. The point is that states that have acceded to the TRIPs agreement rarely pay any heed to indigenous intellectual property regimes. It is hardly debatable that, while patent law, especially the concept of patentability, has been modified in ways that are highly supportive of pharmaceutical, biotechnology, and myriad other industries in the industrialized world, powerful states are impervious to the concerns of indigenous peoples. It is therefore arguable that the TRIPs Agreement is a form of European and American "neo-colonialism."[48] Beyond the changing concept of patentability and the negative impacts of TRIPs on indigenous peoples knowledge, biopiracy is also facilitated by the way in which Article 27 of the TRIPs agreement defines or fails to define the concept of novelty in patent law.

Plant Biopiracy and the Criterion of Novelty in Patent Law

Article 27 of the TRIPs agreement does not offer any definition for inventions. Remarkably, since the early beginnings of the patent system, the definition of what constitutes an invention has been subjected to endless scholarly and judicial analysis. In effect, there is no unanimity as to what an invention in patent law means, and domestic exertions on this issue, while seemingly eclectic, are indeed oriented towards appropriation of plants, especially by the powerful states of the North. As the United Nations recently noted:

> The TRIPs agreement contains no definition of invention and therefore leaves member countries relatively free to draw the line between patentable "discoveries" and actual inventions in the biological field ... The lack of consensus concerning biological patents thus allows countries considerable leeway in fashioning their policy options.[49]

Early patent law jurists proposed that an invention must be a tangible thing, a manifest result of an intellectual exertion. This threshold theoretically debarred scientific discoveries from the patent regime. Second, until the rise of the biotechnology and pharmaceutical industries, patent lawyers were of the view that patent law disallowed patent protection for the so-called products of nature. These distinctions held sway for some time, but, with the emergence of a strong pharmaceutical lobby, legislative changes, and court decisions, the distinctions between products of nature and refined chemical compounds no longer hold. Arguments that inventions may be distinguished from the so-called discoveries and products of nature are no longer either clear-cut or supportable by case law and state policy (particularly arguments that come from the North).

Realistically, scientific discoveries contribute immensely to technological development, and most inventions are based upon theoretical principles established by scientists and philosophers. However, the general perception is that the patent system has long established the legal doctrine that only tangible inventions are protectable.[50] In reality however, the law on patentability of inventions is not well settled and, indeed, some scholars have argued that it is the "arch-problem of patent law."[51] Patents are today granted to "inventions" that are, strictly speaking, discoveries of the laws of nature or isolates of natural chemical compounds.

Apart from the issue of absence of global standards on what constitutes inventions, there is the issue of the blurred distinction between invention of plants, as it were, and discovery of new plants. On this point, certain interesting issues arise, particularly with respect to US law, which is the most extreme on this subject. For example, the definition of patentable subject matter under the US *Plant Patent Act* of 1930 is quite curious. Section 161 thereof provides that

> whoever invents or *discovers* and asexually reproduces any distinct and new variety of plant, including cultivated sports, mutants, hybrids, and newly found seedlings, other than a propagated plant or a *plant found in an uncultivated state*, may obtain a patent therefore, subject to the conditions and requirements of this title.[52]

The practical significance of this legislation, particularly the emphasized portions, is that "the PPA ... extends patent protection not only to inventors but also to 'discoverers' of eligible subject matter."[53] The combined consequence of these provisions is to permit the appropriation of foreign plants through permissive and geographically biased legislation. Indeed, when the provisions cited above are read in conjunction with Section 102 of the US *Patent Act,* a far more disturbing regime of biopiracy becomes clearer. Section 102 provides that:

> A person shall be entitled to a patent unless:
> (a) the invention was known or used by others in this country, or patented or described in a printed publication in this or a foreign country, before the invention thereof by the applicant for a patent, or
> (b) the invention was patented or described in a *printed publication* in this or a foreign country or in public use or sale in this country, more than one year to the date of the application for patent in the United States, or
>   ...
> (g) Before the applicant's invention thereof the invention was made in this country by another who had not abandoned, suppressed, or concealed it.[54]

The US Supreme Court *dictum* in *Gayler* v. *Wilder*[55] brings out the oddity of this law. According to the Court:

> If the foreign invention had been printed or patented, it was already given to the world and open to the people of this country, as well as of others, upon reasonable inquiry ... *but if the foreign discovery is not patented, nor described in any printed publication, it might be known and used in remote places for ages, and the people of this country be unable to profit by it.* The means of obtaining knowledge would not be within their reach; and as far as their interests is concerned, it would be the same as if the improvement had

never been discovered. *It is the inventor here that brings it to them, and places it in their possession. And as he does this by the effort of his own genius, the law regards him as the first and original inventor, and protects his patent, although the improvement had in fact been invented before, and used by others.*[56]

In effect, under section 102 of the US *Patent Act*, prior knowledge, use, or invention in the United States can be used as evidence to invalidate a US patent for lack of novelty;[57] however, as Shayana Kadidal points out, "almost all similar foreign activity cannot be used against a US patent."[58] In effect, what may be construed and upheld as a new plant invention in the United States may in some circumstances refer to plants used by and well known to peoples and societies outside of the United States.

This is not to say that all Northern states have parochial and state-centric notions of novelty of inventions. Most states, including Canada, apply a global standard of novelty.[59] Indeed, of the major patent-granting countries, only the United States and Japan operate regimes of limited, state-centric, and geographically specific notions of printed publications in the determination of prior art.[60] The radical difference and import is that both states issue more than half of all patents operative in the world. The obvious consequence is that, for states and peoples who do not have strong formal structures for patenting plants and/or publishing their knowledge in journals,[61] their domestic plant resources could be "discovered" and taken to a country like the United States or Japan for the purposes of patent protection. Considering the huge information gap between the North and the South, the enormous global ignorance of plants located in many Southern hinterlands, and the limited knowledge of the diversity of plant life forms available to and in common use among local and traditional peoples in the South, it hardly takes more than a visit to a Southern village for someone from the North to become a "discoverer" of a "new" plant.

Further, the notion of printed publication has not been given any clear meaning, particularly in the United States and Japan. The courts in the United States have, for instance, tended to consider the words "printed publication" as halves of a two-tiered standard.[62] In other words, the prior art must not only be printed but also published. Interestingly, the meaning of the word "printed" could sometimes yield bizarre judicial interpretations. For example, some Northern states (like the United States), which have deemed it convenient to jettison the central requirement of written descriptions in plant patents, have often turned around and insisted that typescripts of prior art are not "printed" matter when they come from Southern countries.[63] This emphasis on printed publication places "oral and other evanescent sources clearly outside prior art for the purposes of determining inventions."[64]

Given that most of the peoples and cultures in the South, where plant diversity thrives, tend to rely on the oral transmission of knowledge, the cultural and economic damage that this regime wreaks on them is enormous. Biopiracy cannot be eradicated unless this dichotomy between oral knowledge and published knowledge is removed.

Often, biopiracy thrives on the exploitation of the hospitality and naiveté of local and traditional peoples. For example, in May 1986 a chief from the Secoya community of Ecuador exchanged some specimens of the rare and useful plant, *Banisteriopsis caapi* (otherwise known as "yage" in the local language), for two packs of Marlboro cigarettes. This exchange, as it were, occurred between the chief and a person whom he would later simply describe as a "gringo." The gringo was Loren Miller of the International Plant Medicine Corporation. Miller had heard of the psychoactive properties of yage as a hallucinogenic (the variety of Banisteriopsis he took had been domesticated by the Indians for hundreds of years). Shortly after the exchange, Miller returned to the United States with the "discovery," applied indigenous methods of breeding, and applied for and obtained plant patent no. 5,751 from the US Patent Office on the "new" breed.[65] In another case, Larry Proctor of POD-NERS in the United States patented (US Patent No. 5,894,079) yellow beans he collected in Mexico. Mexican farmers have known this species of beans for centuries.[66] These examples, which have raised considerable controversy, hardly exhaust the long list of instances of biopiracy.

The appropriative function of section 102 of the US *Patent Act* appears to have been mitigated by the US Uruguay Round Agreements Act[67], which purports to limit prior art on inventive activities to member states of NAFTA and the WTO. For the purposes of clarity, I reproduce the said amendment *in extenso*:

S.1042. Invention made abroad.
(a) In proceedings in the Patent and Trademark Office, in the courts, and before any other competent authority, an applicant for a patent or a patentee, may not establish a date of invention by reference to knowledge or use thereof, or any other activity with respect thereto, *in a foreign other than a NAFTA or a WTO member country*, except as provided in sections 119 and 365 of this title.[68]

It is curious why the United States and the member states of the WTO should not have a simple global standard of absolute novelty rather than relying on the questionable and exploitative national and/or regional limitations regarding what constitutes novelty and prior art.

Plant Biopiracy and the Requirement of Specification in Patent Law
The second aspect of the intentional weakening of patent laws and standards in order to facilitate biopiracy involves the lowering of the threshold

on specification. Under general patent law an applicant is required to submit a detailed description of her/his invention. This criterion epitomizes the supposed social raison d'être of the patent system – that is, the enrichment of the stock of knowledge in the public domain in exchange for a limited monopoly. Indeed, one of the major theories on patents is that the state grants the privilege of patents in exchange for the inventor's full disclosure in a manner that would enable those as skilled in the art as the inventor to replicate the invention freely after the term of the patent. It is for this reason that patent law experts agree that specification lies at the heart of the whole patent system.[69]

Thus, absent strict enforcement of the criterion of full and frank disclosure, the entire edifice of the patent system collapses into a disreputable regime of privileges, like the unsavory medieval royal monopolies in the United Kingdom. In relation to plants, it is obvious that a faithful adherence to this rule of complete description is impossible because morphological characteristics and features such as "the taste of a fruit, the smell of a flower, the baking power of a cereal or the brewing power of barley"[70] cannot be reduced to documentary specification capable of enabling a skilled person in the art to replicate the plant. In other words, differentiating one plant from another by way of depicting their literal characteristics is virtually impossible. It was for this reason that classical patent theorists opined that, in the absence of radical legislative changes, the patent system could not be applied to plants.

Indeed, attempts to cure this juridical black hole by requiring the deposit of the "new" plant hardly ameliorate the radical defect in specifying the purported new plant. The major advantage in written specification is that it enables the public, with minimum hassle, to have access to the information contained in the disclosure. This need is hardly resolved by depositing the new plant with the patent office. It is difficult to conceive of how depositing a sample of the new plant could be of any scientific value to interested members of the public or how it would enable thousands of other persons skilled in the art in question to have easy access to what is meant to be a novel addition to knowledge in the public domain. In the *Pioneer Hi-Bred* case the Supreme Court of Canada considered the issue and reasoned that, since a specification

> lies at the heart of the whole patent system[71] the test to be applied in determining whether disclosure is complete is that the applicant must disclose everything that is essential for the invention to function properly. To be complete, it must meet two conditions: it must describe the invention and define the way it is produced or built. The description must be such as to enable a person skilled in the art or the field of the invention to produce it using only the instructions contained in the disclosure.[72]

In considering the case at hand, the Court found that, in the absence of any special legislation on deposit of samples, it could not unilaterally expand the traditional tests for patentability. Accordingly, since "deposit of the seed by itself [did] not comply with the applicable law"[73] the patent application on the new variety of soybeans was refused. At the moment, it is not yet clear whether the courts would accept the deposit of biological material other than microorganisms as sufficient disclosure and specification. In the *Tetraploide Kamille* case involving chamomile plants, the German Federal Supreme Court, like its Canadian counterpart, explicitly left the question unanswered.

The requirement for special legislation to bring plants within the scope of patentability has been met in some jurisdictions, such as the United States, through the enactment of the *Plant Patent Act*. This type of law attempts to lower the traditional requirement of complete and accurate specification through written description. Even with such special laws, which lower traditional standards of specification and/or elevate the deposit of samples to the status of actual and complete specification, there are still conceptual and practical problems. For example, it is still difficult to comprehend how certain indescribable qualities of plants would be reduced to drawings and writings, as is the case with non-life-form inventions. As some of the earliest critics of patents on plants have asked, "how will a plant breeder describe his new product? It is almost impossible to describe in words what a violet smells like, or a Jonathan apple tastes like."[74] As Robert Allyn asks, "pray tell me, what does an onion taste like?"[75] Plants are inherently incapable of falling within the scope of what may be specified under traditional patent laws.[76]

However, this obstacle did not prevent countries from forcing plants into the patent regime. After the publication of Vavilov's findings on the nature of the global distribution of plant varieties, the United States took the lead in expanding the patent system to accommodate plant and seed merchants by deliberately "relaxing the written description requirement in favor of a description as complete as is reasonably possible."[77] In 1930 identical bills were introduced in both Houses seeking to remove the written description requirement, which had for centuries constituted a pillar of the patent system.

For example, section 161 of the US *Plant Patent Act* provides that "no plant patent shall be declared invalid for non-compliance with section 112 of this title if the description is as complete as is reasonably possible."[78] The American initiative was followed in the Netherlands in 1942 and Germany in 1953 and then by other European countries.[79] In justifying this rewriting of a hitherto fundamental pillar of patent law, the US Congress admitted that the template shift was designed to "afford agriculture, so far as practicable, the same opportunity to participate in the benefits of the patent system as has been given industry."[80] By way of contrast, no mechanical invention would be patented if it were not accompanied, *inter alia*, with a full and frank specification. The inescapable inference is that "the requirements for obtaining a plant patent are substantially more liberal than those mandated for a standard utility patent,"[81] and this has had an appropriative impact on plants from the South.

Plant Biopiracy and the Criteria of "Inventive Step" and "Utility of Invention"

Two concepts close to invention are novelty and utility of invention. However, as is already evident, the concept of novelty is relative and sometimes arbitrarily determined. The determination of whether a particular inventive step is worthy of patent protection is subjective. Furthermore, as Reichman has noted, "there is no international agreement or uniform set of guidelines for implementing the now universal eligibility criterion of 'non-obviousness.'"[82]

Cultural bias and the subtext of gender discrimination, especially for plants that are largely improved by Southern women farmers, cannot be easily discountenanced given that the latters' farming has been erroneously regarded as mere drudgery. The ambiguities and biases surrounding these concepts facilitate a Northern free ride on the South's plant germplasm, which the North generalizes as raw materials for its pharmaceutical and biotechnology industries. In order to appreciate how the obfuscation of the meaning of novelty and inventive step facilitates the appropriation of plant life forms, it is pertinent to briefly examine judicial and scholarly debates on the issue.

Unlike the copyright regime, which protects both "the most impassioned poetry and the sheerest doggerel"[83] alike, patent protection is available only for inventions that are deemed to be prior art.[84] The problem lies in determining what quantum of innovation amounts to an invention.[85] There is thus a theoretical difference between mere improvements, which could be made by the hypothetical unimaginative person "skilled in the art," and changes that rise to the level of invention. Needless to say, there is a measure of discretion between these two poles, and the fundamental problem of the patent law system lies in this grey area between inventions and improvements.[86] As Tomlin J. noted in *Samuel Parkes and Co. v. Cocker Bros., Ltd.*,

nobody, however, has told me, what is the precise characteristic or quality the presence of which distinguishes invention from a workshop improvement. Day is day and night is night, but who shall tell where day ends or night begins ... it is, I think, practically impossible to say there is not that scintilla of invention necessary to support the talent.[87]

It is not only judges who have grappled with this slippery aspect of patent law; scholars have also found the concept of inventive genius elusive. In his analysis Richard Gardiner despaired that,

> in the light of uncertainty as to what it is that is protected by patent law (both in cases of what required element of inventiveness is central to patentability and the extent of what the patent actually protects) readers of the Reports of Patent Cases might well reach the conclusion that the state of the law in this field depends on how key concepts strike the judge hearing a cause or fit the line of reasoning ... Invention ... idea ... ingenuity and discovery are used by the courts in conjunction with novelty and the notion of what is inventive or not obvious in unpredictable ways.[88]

A careful perusal of both case law and scholarly writings on the question of what constitutes the appropriate quantum of novelty to justify a patent grant shows that there are no *a priori* thresholds; rather, patent examiners and judges have, in spite of protestations to the contrary, essentially relied on their values, critical faculties, and hunches.

In effect, although all legal authorities are agreed in theory that there is no invention when the alleged improvement is obvious to the hypothetical person skilled in the art, the fact is that the determination of that "requisite level of inventiveness" is a subjective process that evaluates the gradation of innovations and affixes value on the patent application in question. This is what the German jurists who have examined this phenomenon have termed a *Werturteil*, a judgment of value.[89]

In the exercise of this judgment of value, a close scrutiny of judicial decisions and a quick comparison with judicial attitudes to plant patents vis-à-vis mechanical inventions clearly show that the juridical system of patents leans towards mechanical inventions rather than plant inventions or biotechnological products. In this process of double standards and a lenient approach to purported plant inventions, the level of "inventiveness" required is significantly low.

For example, although the US *Plant Patent Act* (which creates special provisions for plant inventions) requires that, for a plant to be patentable, it must be "be distinct and new,"[90] the novelty required is not necessarily of the standard provided for under section 102 of the US *Patent Act* (which provides the rules for general patents excluding plants). The courts have interpreted this to mean that the "new" plant did not exist previously in a capacity in which it could reproduce itself.[91] In addition to this lowered threshold, the term "distinct," which was ostensibly intended to substantially distinguish one patented plant from the other, has not received such a purposed intent from the courts; rather, subtle and frivolous biochemical distinctions (instead of the usual morphological and utilitarian distinctions) have dominated the jurisprudence on patents on plants.[92]

It is within this context that the requirement of utility of inventions has been abused. The criterion of utility bears on the requirement that an invention must be useful for the purposes it claims to serve. Surprisingly, the emphasis on plant patents is on *difference* rather than on the superiority of the new plant. Interestingly, the US Senate Report on the subject makes it quite clear that marketing difference is the critical factor, not whether the new plants brings forth an improvement over pre-existing plant varieties.[93] The obvious consequence of this relaxation of a crucial criterion of patentability is the common "proliferation of lines that are genetically different in trivial ways but are marketed as different."[94]

The point is that in recent decades the patent regime has been designed to suit merchandising in plants without any serious pretensions to improvement of plant varieties per se. The direct consequence of this relaxed regime is that, to the detriment of larger social interests, cosmetic changes (which merely create pseudo-varieties whose sole purpose is to separate one commercialized plant from the other) have become standard practice in the seed industry. The trivialization of the criterion of utility for plant patent protection so as to protect pseudo-varieties seems to be the norm. For example, the novelty and utility of Nothrup King Co.'s soybean variety is that it

> is most similar to "Pella," "Cumberland," and "Agripro 25"; however, "S30-31" has *grey pubescence* vs. tawny for "Pella," yellow hila vs. imperfect black for "Cumberland," and white flowers vs. purple for "Agripro 25.[95]

In other words, the so-called S30-31 was given patent-like protection merely because it had a different colour of flower. Commenting on this trend, Klopenburg notes that "it would appear that private breeding work may involve a substantial amount of unproductive effort to achieve uniqueness, and thus protectability, through transfer of non-economic traits such as flower colour."[96] It would be impossible to conceive of patents being granted to mechanical inventions if all that the applicant for a patent could show was that the new invention differed from prior art on grounds of colour (unless colour itself was the defining aspect of the invention).

A direct result of this regime is the appropriation of Southern plants through trivial alterations. According to Meetali Jain, "to date, at least two-thirds of plant genetic resources from India have been patented, through slight alteration in the United States."[97] The inescapable conclusion is that the plant patent system defers to the needs of the seed merchants and traders.

The irony is that, while cosmetic changes in plants are being rewarded with patents in the North, the contributions of traditional farmers, particularly women, continue to be neglected due to deliberate cultural and juridical bias. For example, over the centuries Indian farmers, particularly women, have developed and grown over 30,000 different varieties of rice, none of which received patent protection. Similarly, native Andean farmers have developed hundreds of species of tomatoes, potatoes, maize, and beans. Reputable Western scientists are agreed that "the total genetic changes achieved by farmers over the millennia are far greater than those achieved by the last hundred or two years of more systematic science-based efforts."[98] Yet, in powerful commercial circles, traditionally improved plants are generally construed as raw materials devoid of intellectual credit or property rights.

While the contributions of traditional farmers and breeders may not command the same degree of media attention and circus-like buzz reserved for corporate modification of plants in the North, evidence is overwhelming that they have achieved a far greater degree of plant improvement than have their corporate counterparts in the seed industries, whose efforts are increasingly commerce-driven.[99] For example, in Sierra Leone, local farmers, again particularly women, have developed over seventy varieties of West African indigenous rice. These varieties are based on several useful criteria, such as length to maturity, ease of husking, proportion of husk to grain size, susceptibility to insect attack, behaviour in different soils and moisture levels, cooking time, and so on.[100] The instances of local improvement of plants are legion, and the scientific basis of traditional agricultural practices is no longer in doubt. As a recent report of the Friends of the Earth noted, in order to achieve the phenomenal improvement of plants in the so-called "informal" paradigm, traditional farmers

> employ taxonomic systems, encourage introgression, use selection, make efforts to see that varieties are adopted, multiply seeds, field test, record data [and in fact] ... do what many Northern plant breeders do.[101]

Indeed, even in recent times traditional farmers have often shown better insight into the nature and utility of plants than have the so-called "expert" commercial seed-breeding companies whose loyalties usually lie with the world of stock markets. For example, official government researchers on research stations in the Indian state of Andhra Pradesh rejected a new variety of rice known as "mashuri." Local women rice farmers got hold of this rejected variety of rice, experimented with it in their farms with other rice varieties, and, finding its performance well-suited to local conditions, facilitated its spread to other local farmers. It has been reported in recent times that mashuri is now the third most popular rice variety in the whole of India.[102]

It therefore seems that the continued mischaracterization of traditionally improved plants as raw materials for corporate seed growers and companies in the North is a cultural construct without any objective basis. As notable Canadian activist Pat Mooney has observed, "the argument that intellectual property is only recognizable when performed in laboratories with white lab coats is fundamentally a racist view of scientific development."[103] Arguing in a similar vein, the Crucible Group recently noted that "farmers' fields and forests are laboratories. Farmers and healers are researchers. Every season is an experiment."[104]

It is not only the trivialization of the utility requirement that creates room for the appropriation of plants. Other criteria of patentability, such as repeatability and industrial applicability, have been lowered by judicial and legislative initiatives in order to accommodate the special interests of commercial seed breeders and pharmaceutical corporations.

Plant Biopiracy and the Criteria of Uniform Reproducibility and Industrial Applicability

It is a well known requirement of patent law that, in order for an invention to be patentable, it must be capable of industrial application. Consequently, inventions that cannot be applied in any industry are not patentable. A corollary to the industrial applicability requirement is that the invention be reproducible in such a manner that the copies are indistinguishable from the prototype. All these requirements are clearly tailored towards a mechanized and "conveyor-belt" idea of industrialization and mechanized deployment of capital. Given the linkage to and use of the patent instruments to attract foreign skilled artisans and industry, it is not surprising that this requirement is a crucial aspect of the system of patents.[105] Conversely, inventions that are not known to have any industrial use in the mercantilist context are not legal inventions. Therefore, for indigenous inventions to be patentable, they must satisfy a Western test of industrial applicability.

With reference to plants, industrial applicability and reproducibility require that the processes that led to the first specimen of the new variety should be repeatable. That is to say, a person skilled in the art should, on the basis of the information provided in the disclosure, be able to come to the same result as did the original inventor of the subject.[106] The problem with the application of this principle of law to complex life forms such as plants is that it is not always practicable to reproduce in exact detail the next generations of the original plant life form.

In grappling with this difficulty, the Canadian Federal Court of Appeal noted in *Harvard Oncomouse*[107] that the better approach is to determine the "essential" aspect of the application in question. In other words, the court proposed that a literal construction of the rule of uniform reproducibility should be avoided in determining whether a plant invention clears the hurdle of uniform reproducibility. This minimalist approach to uniform reproducibility recognizes the near impossibility of accurately predicting the idiosyncrasies and variations of genes in complex life forms when people seek to reproduce the original plant invention.[108] However, even with a minimalist approach, a high element of luck may be required in order to reproduce the products of artificial life forms.[109] The point is that, in the absence of special legislation, patents on complex life forms such as plants flout the rule of uniform reproducibility. As the Canadian Court of Appeal noted in *Harvard Oncomouse*, "a complex life form does not fit within the current parameters of the Patent Act without stretching the meanings of the words to the breaking point ... However, if parliament so wishes, it clearly can alter the legislation."[110]

Notwithstanding this judicial restraint on the part of the Canadian courts, some Northern courts have, in the face of legislative inertia, unilaterally expanded the concept of uniform reproducibility.[111] Although the modern patent system on plants has glossed over such problems, the consequence of this relaxed regime constitutes a double standard in relation to non-life, or mechanical, inventions. In addition, it also confers juridical legitimacy on "new" plant "inventions" that fail the test of uniform reproducibility.

Further, the lax plant patent regime often has a tendency to encourage the breach of international agreements on non-patenting of germplasm held in trust in UN agencies such as the International Rice Research Institute, and the CGIAR.[112] To sum up the position on patents on plants and its appropriative function, it is pertinent to note that the US Supreme Court, in *Diamond* v. *Chakrabarty*,[113] articulated the reasons for these radical changes to the law of patents. According to the court, the purpose of the domestic statutes on plant patents was to remove what was considered to be specific "impediments"[114] to the patenting of plants. The court also admitted that the plant patent statutes "relaxed" the enabling requirements for plants. What stands out in this analysis is that the patent system has been used as a deliberate instrument of state policy and that its legal norms may be no more than the congealed political/economic interests of powerful states and commercial/industrial entities. In addition to the appropriation of plants, patent systems also facilitate the appropriation of traditional and indigenous knowledge pertaining to the uses of plants.

How the Concept of Patentability Appropriates Traditional Knowledge of the Uses of Plants

As has been shown, the diverse and rich body of knowledge held and practised by local and traditional farmers, especially women, concerning the uses of plants for food, medicinal, and other purposes is of immense global value and significance.[115] More important, a considerable portion of plants hitherto characterized as "wild" germplasm or "cultivars" have, on closer and less prejudiced examination, turned out to be products of substantial human intervention, even though they may not appear to be commercially useful.[116] Of course, there are still some isolated anachronistic arguments that deny the intellectual merit of traditional farmers and healers.[117] However, the weight of informed and contemporary opinion is that the intellectual contributions of traditional farmers and healers to knowledge about the uses of plants are tremendous. In many cases, the insight and knowledge brought to bear on the varied uses of plants have been nothing short of ingenious. This is the principal reason why modern inquiries into medicinal herbs and plants incorporate and rely on the knowledge of traditional and indigenous peoples.

The statistics on this are overwhelming. For example, 74 percent of the pharmacologically active trees reported by an indigenous group correlated with laboratory tests. In comparison, only 8 percent of random samplings by "formal" scientists showed similar activity.[118] It is estimated that, "absent the aid of indigenous groups, for every commercially-successful drug, at least five thousand species must be tested."[119] In a comparative survey, Michael Balick of the New York Botanical Gardens found that, when formal scientists made use of traditional knowledge, their efficiency of screening plants for medicinal properties increased by more than 400 percent.[120]

It is therefore not surprising that a decisive number of both historical and modern drugs derived from plants have been developed with the insight and experience provided by traditional farmers, healers, and breeders.[121] Scientists have found that 86 percent of the plants used by Samoan healers displayed significant biological activity when tested in the laboratory. Further, crude extracts of plants used by one healer in Belize gave rise to four times as many positive results in laboratory tests for anti-HIV activity than did specimens collected randomly. In another case, between 1956 and 1976 the US National Cancer Institute screened over 35,000 plants for anti-cancer compounds. The program was terminated in 1981 due to poor results. However, a retrospective study conducted on the project concluded that the success rate could have been doubled if medicinal "folk knowledge" had been the only information used to target plant species.

The function of local and traditional knowledge as a "filter" for ordinary guesswork is generally estimated to be five thousand times more effective than random collection. It is for this reason that a considerable number of the giant pharmaceutical and biotechnology companies utilize local healers and breeders as intermediaries when searching for useful plants and TKUP in the South. Currently, there are corporate institutions solely devoted to utilizing local healers and breeders as intermediaries. Some of the well known names in this emerging business include Pharmacognetics of Bethesda, Maxus Petroleum of Dallas, the Carnivore Preservation Trust of the USA, and Shaman Pharmaceutical Inc.[122] Although bioprospecting may not always involve local knowledge of the biochemical properties of plants, it is an undeniable fact that, absent such local knowledge, little would be achieved by the formal scientists in their search for economically useful plants and TKUP.

In contemporary times this body of knowledge on the uses of plants and plant derivatives serves at least four major functions. First, it is a source of direct therapeutic agents.[123] For example, the alkaloid D-Tubercuranine, which is widely used as a muscle-relaxant in surgery, is extracted from a South American "jungle" plant. Modern chemistry is unable to synthesize it in a form comparable to the natural compound, and thus reliance on the natural product continues. Second, TKUP is a starting point for the development of complex semi-synthetic compounds. For example, "saponin"[124] extracts are chemically altered to produce sapogenins, which are necessary for the manufacture of asteroidal drugs. Third, TKUP products often afford a model for new synthetic compounds,[125] and fourth, TKUP serves as a taxonomic marker for the discovery of new compounds.[126]

The huge economic, cultural, and juridical implications of patents has brought to the fore the appropriative role of the patent system in the privatization of TKUP. What is equally significant is the increasing call for a rethinking of the contemporary patent system in its relation to TKUP.[127] Most of these developments have been geared towards highlighting the deficiencies of and inequities in the global patent system, particularly with respect to the absence of any informed participation of local peoples[128] in formulating existing and future policies, legislation, and policy initiatives on the subject of access to and conservation and exploitation of TKUP.[129] While these ideas are worthy of consideration, what seems very important is an appreciation of the mechanisms, especially patents, that appropriate TKUP. The obvious first line of inquiry involves how the patent system has responded to the problem of the so-called "products of nature."

How Patents on "Products of Nature" Appropriate TKUP

In theory, most patent law systems purport to debar the patentability of products of nature. On the basis of this legal principle, mineral products (e.g., iron, gold, aluminum, etc.) are not patentable. Initially, this objection was used as the basis for denying patent protection to derivatives of or extracts from plants.[130] However, as the clout of the pharmaceutical and chemical industries grew, this fundamental postulate of patent law began to crack. Hence, over the years, the "product of nature" exclusion has lost respectability and standing in juridical circles. This process may be said to have started with the product patent in 1910[131] on acetyl salicylic, known medically as aspirin, a quintessential product of TKUP.

In recent times, the distinction between products of nature and artificial inventions has become practically non-existent, especially as patents are routinely granted to subjects pertaining to purified natural products, plant extracts, and DNA sequences. This is quite distinct from the indisputable patentability of a new process even if that process relates to the production of natural products.[132] Of course, patent attorneys and jurists are agreed that process patents are valid regardless of whether these processes deal exclusively with life forms; however, what is becoming increasingly controversial is the patenting of the products themselves.[133] While critics of patents on purified natural biochemical substances have pointed out that such patents should be limited to the process(es) by which the purification or extraction of natural substances is obtained,[134] patents on purified natural substances, plants extracts, and plant DNA sequences tags are becoming routine. Given the preponderance of complex life forms and biochemical compounds in the biodiverse South, the isolation and purification of such compounds becomes biopiracy when the products of these processes are protected under questionable doctrines of patent law. Remarkably, the relaxation of the patent regime has largely remained a revolution executed by the courts. As Thomas Kiley has observed, "there is nothing in the patent statute that says old substances become 'new' when first offered in purified and isolated form. This is law that judges have engrafted on the statute."[135]

It is equally intriguing that law on the patentability of purified biological/biochemical natural substances is inconsistent with regard to purified natural metallic substances. While patent laws across the North have developed a regime of patents on purified biochemical/biological substances[136] that do not ordinarily occur in a pure state in nature, naturally occurring but purified metallic substances have been denied patent protection.[137] Put simply, the patent system makes arbitrary allowance for chemical and pharmaceutical inventions of a biological nature but not for other purified natural products (such as metals). Given that most TKUP is related to the needs of the pharmaceutical and food industries, this bias facilitates the appropriation of traditional knowledge of the uses of plants. Take, for example, US Patent No. 4,673,575 (issued 16 June 1987). This patent, granted to Fox Chase Cancer Center, is on an extract of *Phyllanthus amarus*, a medicinal plant used in India by ayurvedic healers for treating various liver ailments, including jaundice. Researchers at Fox Chase Cancer Center "discovered" that the plant extract was effective against viral hepatitis-B and E.

To further illustrate the bias and illogicality in the law on patenting of purified natural substances, I offer a few additional examples. Two researchers at the University of Wisconsin reportedly received US patents for a protein isolated from a berry from Gabon known as *Pentadiplandra brazzeana*. This berry has been used by communities in Gabon, Central Africa, for centuries because of its incredibly sweet taste. The protein, a natural substance, is 2,000 times sweeter than sugar and does not lose its sweetness when heated.[138] In *Olin Mathieson,* involving product patents for vitamin B12, it was held by a US circuit court that purified vitamin B12 was patentable.

Vitamin B12 occurs naturally but in an impure and unstable state. Strictly speaking, vitamin B12 is not an invention.[139] Yet the court in *Olin Mathieson* held that, given the step between natural and purified vitamin B12, "we think the invention is meritorious ... it did not exist in nature in the form in which the patentees produced it and was produced by them after lengthy experiments."[140]

Furthermore, US Patent No. 5,900,240 was recently granted to Cromak Research Inc., based in New Jersey. The patent was on the diabetic properties of an Indian plant known as "Jamun," whose medicinal properties had been known to Indians for many years.[141] Similarly, in July 1999 an African tree, *Swartzia madagascariensis*, that grows in Zimbabwe and that had long been used by natives for treating fungal infections, was pointed out to two Swiss scientists who "announced" its potential for curing drug-resistant fungal infections, including Athlete's foot, thrush, and some types of eye infections that had no known cure.[142] US Patent No. 5,929,124 (product patent) was granted on derivatives of the plant to the two Swiss scientists, Hostettman and Schalle.

The practice of patenting purified versions of naturally occurring substances is not limited to the United States: it is a common practice in the industrialized countries of the North.[143] In the above-mentioned cases, the underlying theory is that, in spite of the natural origins of the substances in question, because they do not occur in a pure state in nature, purified versions thereof require substantial human intervention and are therefore patentable.

The inconsistency in this theory is that it has never been applied to purified metallic elements that do not occur in a pure state in nature. A case in point is the metal tungsten. Tungsten is an element that is generally used as filament in electric bulbs. It is notorious for being very impure in nature; indeed, its capacity to react with other elements is legendary. It is always found in an oxidized state in nature. In theory, therefore, a purification of tungsten would require substantial human ingenuity, and whoever achieved this feat could be rewarded with a patent as in the above-mentioned cases of aspirin, vitamin B12, and so on. However, in *General Electric Co.* v. *DeForest Radio Co.*, the court of first instance validated the patent on purified tungsten. There is no question that the process for purifying tungsten was patentable. However, what was at issue was the patentability of purified tungsten itself. On appeal, the district court agreed that the purification of tungsten was a "tremendous advance" in technology but, ultimately, it held that the application for a product patent on purified tungsten was not to be countenanced. In a ratio riddled with inconsistencies, the court held that

what [the patentee] produced by his process was natural tungsten in substantially pure form. What he discovered were natural qualities of pure tungsten. Manifestly he did not create pure tungsten, nor did he create its characteristics. They were created by nature.[144]

If tungsten had been a biological compound, it would have had its patent upheld. More important, if the *ratio decidendi* in *Deforest* were applied to all the applications for patents on purified versions of biological substances created or discovered by TKUP, there is no doubt that the modern business of bioprospecting for TKUP would suffer a profound setback.[145] As Michael Davis aptly observes, "the doctrine against the patenting of natural products would appear to be well entrenched in American jurisprudence. However, a completely different and seemingly contradictory rationale has been applied in cases involving bio-molecules."[146]

In addition to vitamin B12, other natural products that have received product patents include purified prostaglandins[147] and adrenalin.[148] Here, Judge Haynsworth sought to rationalize the court's decision with the unhelpful argument that "all of the tangible things which man deals with and for which patent protection is granted are products of nature in the sense that nature provides the basic source material."[149] This "slippery slope" argument had been echoed by Justice Felix Frankfurter in *Diamond* v. *Chakrabarty*. According to Justice Frankfurter, "everything that happens may be deemed the 'work of nature' and any patentable composite exemplifies in its properties 'the laws of nature.' Arguments drawn from such terms for ascertaining patentability could fairly be employed to challenge almost every patent."[150] The sum of this slippery slope argument is, as stated by Michael Davis, that the "product of nature" exclusion in patents is practically dead.

The implications of this juridical revolution have not been lost on scholars. Commenting on the general laxity and double standards of the patent system, John Golden observes:

> The question that remains is how much traction traditional requirements for patentability retain. With regard to the "product of nature," novelty, and enablement doctrines, the answer is practically speaking, "Not much." The "product of nature" doctrine, although still extant, is effectively toothless, because biotechnology by nature involves isolating, and replicating biological materials to produce "unnatural" levels of purity.[151]

In effect, the double standard on the doctrine of purified natural substances, the relaxation of the traditional rules of patentability, and the impact of section 102 of the US *Patent Act,* which excludes oral foreign prior art, combine to facilitate the appropriation of traditional knowledge of the uses of plants.

How the Criterion of Novelty Appropriates Traditional Knowledge
In addition to the appropriative function of the double standard in the

determination of what constitutes patentable subject matter, especially in relation to "products of nature," the concept of novelty in patent law sometimes operates to disregard TKUP when the issue of novelty is addressed by patent examiners in industrialized states. The real problem here is that, under many international treaties and conventions, there is no requirement for patents to adopt a standard of global and absolute novelty. In other words, the international patent regime does not require that patents issued by national patent offices relate to inventions that are absolutely new. The practical implication is that, for example, a TKUP practice may be common knowledge in India yet be considered novel in the United States. With some trivial alteration, such traditional knowledge may don the toga of novelty and be patented in the United States, whose test of novelty is state-centric or WTO-centric. A few examples may suffice to illustrate this point.

The Neem tree (*Azadirachta indica*) is known in several Indian and African communities as a curer of ailments – a "wonder tree." It yields, *inter alia*, a prolific chemical storehouse of pesticides, medicines, cosmetics, dental remedies, and contraceptives. Indian and African communities have known of these useful properties of the Neem tree for centuries. The West was only alerted to the tree's wonderful properties in 1959. However, an American company, W.R. Grace, applied for and obtained US Patent No. 5,124,349 on *Azadirachtin*, a pesticidal chemical ingredient in the plant.[152] Similarly, Suman Dias and Hari Har of the University of Mississippi Medical Center have US Patent no. 5,401,504 on the use of Turmeric (*Curcuma longa*) to make open wounds heal faster. The use of Turmeric for healing sores has been known to Indians for centuries.[153] Although the Turmeric patent was successfully opposed by an agency of the Indian government, it is one of the rare cases in which a dubious patent on TKUP has been reversed through activism and protests.

In another case, British chemist Conrad Gorinsky spent considerable time with the Wapishana Indians of Guyana. During that period the Indians showed him the uses of two plants called *Cunani* and *Tipir*.[154] As Wapishana Spencer notes, "for many days and nights, I was his [Conrad Gorinsky's] guide in the jungle."[155] After Gorinsky left the Wapishana, he applied for and obtained patents on the pertinent chemical properties of both *Cunani* and *Tipir* in the United States and Europe. The first patent granted to Gorinsky covers the Greenheart tree (*Ocotea rodiaei*), which produces *Tipir*. According to his description on the patent application, the active ingredient in the plant is an "efficient antipyretic and useful for treating tumors." The other active ingredient registered by Gorinsky, poly-acetylene, was obtained from the *cunani bush* tree (*Clibadium sylvestre*) and is described as "a powerful stimulant of the central nervous system, as a neuromuscular agent capable of dealing with cases of heart blockage."[156]

In both cases, patents were granted on the ostensible notion that the traditional uses of plants were novelties. While such knowledge of the uses of plants might be novelties in the United States or Europe, certainly it was not a novelty in India or Guyana. If a global test of novelty encompassing written and oral knowledge were in place, then patents based on a trite grasp of traditional knowledge would not be issued as often as they have been in the past thirty years. Without question, section 102 of the US *Patent Act*, which disregards oral prior art, enables the theft of TKUP.

How the Criteria of Utility and Inventive Step Facilitate the Appropriation of Traditional Knowledge

Another aspect of appropriating TKUP through the lax patent system involves a cosmetic rearrangement of the chemical and molecular structure of a biological compound already identified by traditional and indigenous peoples, while retaining the original useful properties of the natural substance. Some researchers are of the view that, by lowering the standards on utility, the research and development systems of many pharmaceutical companies are ultimately wasteful as their main motivation is to tinker with the molecular structure of plant extracts already identified and used by indigenous and traditional peoples for medicinal or therapeutic purposes. The phenomenon of cosmetic rearrangement of molecules is not limited to TKUP. Some "new" drugs involve a trivial rearrangement of the chemical structure of existing patented drugs and hardly constitute a remarkable advancement on existing therapeutic knowledge. According to the US Office of Technology Assessment (OTA), American drug companies (in 1990 dollars) spent $65 million bringing a new drug to market in 1969, $194 million in the 1980s, and $300-350 million in the 1990s. Intriguingly,

> of the 348 drugs introduced by the 25 largest pharmaceutical companies between 1981 and 1988, *only 12 (or 3 percent)* were deemed important therapeutic advances by the FDA [US Food and Administration] ... The vast majority were seen as having little or no potential for advances in treatment.[157]

Whatever justifications the patent office may offer for such patents, the crucial juridical fact is that "supposedly central requirements such as utility and non-obviousness have often merely nibbled at the margins of patentability's broad realm."[158]

In light of this, some scholars have called for a tightening of the regime on patentability. For example, John Golden recently called for "stricter enforcement of the basic hurdles to patentability – novelty, non-obviousness and especially, utility."[159] Similarly, Samuel Oddi has recommended a high standard of patentability that would ensure that only revolutionary inventions and not market-induced detail patents are granted. In Oddi's view the test for patentability should be "whether the claimed invention would be non-obvious to a person skilled in that art *anywhere in the world.*"[160] The emphasis here is on a rejection of parochial limitations on novelty.

What is significant in the above-mentioned cases is that, in addition to an obvious regime of laxity in the patent system, the articulation of what constitutes invention in the significatory language of the dominant narrative of science weaves an implausible novelty into knowledge that is already well known to traditional peoples. Accordingly, the patent system has been turned into a hybridized form of monologic dialogue; that is, while pretending to be speaking the language of plurality, it is in fact speaking a monologue directed towards a narrow epistemological and economic elite.[161] In this strange but familiar world, marginalized epistemological frameworks take on the status of stunned spectators/victims. The exclusiveness of Western scientific jargon as an elite cultural signifier, along with Western juridical formalism itself (which has been globally positioned as the only legitimate and acceptable narrative framework of science), serves to appropriate indigenous knowledge and to marginalize indigenous cultures.

Biopiracy pretends that creativity is an individual effort springing from nothingness – without inspiration and validation from any extant tradition. The reality, however, is that creativity is tied to the inspirational functions of already existing knowledge.[162] Thus the authorial conceit inherent in the patent system and facilitated by a hubristic epistemological regime results in the appropriation of TKUP. To the extent that the international patent system is a narrative that excludes traditional knowledge, it is not only a monologue but also a mimetic discussion among a self-perpetuating band of epistemological and cultural elites.[163] This curious posture and cultural status of the patent system has raised legitimate questions regarding what is to be done to protect plants and TKUP from further appropriation.[164] While some scholars and states ponder this phenomenon, it must not be forgotten that UNCTAD recently noted that

> it is evident that the primary and immediate beneficiaries of the implementation of the TRIPs agreement are likely to be technology and information developers in the industrialized countries. Indeed, the more rapidly and comprehensively the TRIPs agreement is put into place, the greater will be these benefits.[165]

Various suggestions have been made as to how the intellectual contributions of local and traditional farmers, breeders, and healers would be defined, recognized, respected, and, if they so desire, economically rewarded.

International Law and the Challenge of Biopiracy

There is a common notion that the current controversy surrounding biopiracy is attributable to "the advent of biotechnology."[166] This view cannot be wholly correct because the importance of plants and TKUP has always been an abiding part of the political and legal struggle between states and societies. The better view would seem to be that, in an age of globalization, biopiracy has had to confront the growing emancipation of hitherto ridiculed cultures. Until some decades ago it was generally thought that indigenous peoples and cultures were doomed to extinction. However, in the past three decades there has been a renaissance of indigenous worldviews, a greater appreciation of the enormous contribution of indigenous cultures across the globe, and a recognition of the egregious injustices perpetrated against them since the age of colonialism and empire.

Many scholars, activists, NGOs, and international bodies (such as WIPO) have made significant efforts to identify and articulate the indigenous viewpoint within the current globalizing trend of dominant intellectual property regimes. In an era of intense media and academic scrutiny, the discourse on how to locate indigenous peoples problems has been heated. What is striking is the outrage brought about by the intense scholarly attention and political activism directed at the political economics of the patent system and how it appropriates indigenous peoples knowledge. According to Lara Ewens,

> the hypocrisy of western demand for intellectual property protections is twofold: not only do developing countries pay a high price for the patented products that are reintroduced in their countries (yet made from local resources), but developing countries are unable to use the intellectual property framework to protect against the piracy of their own indigenous and local resources and knowledge.[167]

From the analysis so far, a few inferences may be drawn. The first is that current concerns about the overextension of modern patent laws, particularly in the North, are justified.[168] In addition to creating obstacles to future research and innovation, the cluttering of the global intellectual space with liberty-intruding patent rights created by double standards and the relaxation of the patent process not only appropriates and privatizes indigenous peoples knowledge but also stultifies research. Second, the process of globalization seems to consist of the universalization of Western ideologies, which exacerbates the already worldwide distributional inequality.[169] Although modern international law has, in a pontifical manner, sought to address this distributional disequilibrium, some scholars are far from sanguine. As Michael Halewood has argued,

> the international fora within which these issues have evolved so far are not likely to accommodate what would otherwise be the next most predictable

and potentially desirable development in this area of law, i.e. international legal obligations on states to provide specially tailored or *sui generis* intellectual property protections for indigenous knowledge.[170]

Contemporary international law has been criticized for not offering legally enforceable multicultural rights in matters pertaining to intellectual property rights,[171] particularly in a way that would help the South.[172] In light of this, it has attempted to respond to biopiracy in a number of ways.[173]

International Law of State Responsibility and the Challenge of Biopiracy

Generally speaking, it seems that contemporary international law has not fully appreciated the appropriative functions of the global web of patent laws and systems. Indeed, the very question of whether biopiracy is a reality or a piece of Third World rhetoric is a controversy that often pits the North against the South.[174] In examining the role of international law in combating biopiracy, the first issue to be determined is its role vis-à-vis state responsibility on matters pertaining to appropriation of property. Since patents are generally construed as property rights, international law on state responsibility for appropriation of tangible property would apply.[175]

Surprisingly, while international law has well known legal norms of redress for injuries suffered as a result of national appropriation of tangible property of aliens, it has not yet formulated any reliable remedy for state or individual victims of biopiracy. It is arguable that, on the basis of the Lockean theory of a moral-cum-legal right to intellectual property, plant varieties created or modified by indigenous peoples, and indigenous knowledge itself, are as much a manifestation of property as is tangible property.[176] Although Reichman's argument on this issue is with respect to the interest of the North, its logic applies with equal force to traditionally modified plants and TKUP. According to Reichman:

> The notion that intellectual property depends entirely on the place where protection is sought continues to preclude the formation of private international remedies for entrepreneurs who invest in intellectual goods that are reproduced and commercially exploited on foreign soil. At the same time, gaps in the public international law of state responsibility that protects alien property in general permit states to seize the form of alien property that has become the key to economic growth through technological innovation.[177]

In light of the contemporary convergence of the patent system with the regimes on plants and TKUP it seems that various state initiatives have been preoccupied with the question of instituting a regime on access to plants and TKUP, as though physical access is the fundamental mode of effecting appropriation.[178] The reality is that appropriation is a juridical as well as an epistemological process. It is the exclusive legal right over appropriated plants and TKUP that offers the incentive and legitimacy for biopiracy. Accordingly, effective legal measures against biopiracy must address the juridical processes that facilitate the process of biopiracy.

Combating Biopiracy: International Law and Access to Plants and TKUP

Within the convergent confines of the provisions of Article 27 of the TRIPs agreement and the provisions of the CBD, which reaffirm the sovereignty of states over plant life forms within their respective territories, it seems that the prevalent response of the South to the problem of biopiracy has been to enact domestic laws restricting access to plants and TKUP. Such domestic laws have tended to incorporate rules of ethics in research, such as obtaining the prior informed consent (PIC) of the affected local community that owns the plant or TKUP at issue. For example, the Republic of the Philippines, by virtue of its Executive Order No. 247 dated 18 May 1995, has taken some steps towards mitigating the perceived inadequacies in the present regime pertaining to access to plant resources. Section 2 of the order prescribes that

> prospecting of biological and genetic resources shall be allowed within the ancestral lands and domains of indigenous cultural communities only with the prior informed consent of such communities, obtained in accordance with the customary laws of the concerned community.[179]

However, it is instructive that the CBD,[180] which reaffirms a regime of national sovereignty on plants, leaves open the question of national sovereignty or lack thereof over intellectual property that may be attached to traditionally improved plants and TKUP. Although Article 8 (j) empowers states to make laws or create initiatives that would protect useful traditional practices, there is no precise category of legal rights recognized under international law for plants modified or created under traditional knowledge frameworks. This is quite remarkable but not unusual because, at the end of the day, it is for the domestic legislator, acting within the confines of international minimum standards on intellectual property, to determine which inventions and knowledge are capable of being accorded legal protection. The forms of legal regimes needed to protect traditional knowledge in the domestic domain are within the jurisdiction of the domestic legislator. In this context legal mechanisms for the protection of knowledge are almost always products of the cultural contexts and political realities within which knowledge is protected.

More significantly, the CBD does not take a categorical stand as to whether or not the patent concept is helpful to the ideals of conservation and equitable use of plants and TKUP. In this regard, Articles 11 and 16(5) of the CBD oblige contracting parties to "as far as possible and as appropriate, adopt economically and socially sound measures that act as incentives for the conservation and sustainable use of components of biological diversity."[181] It is arguable that the patent system may be one of those economic "measures" that may play a role in the process of conservation and sustainable use of plants and TKUP. But this is debatable. The CBD does not state clearly whether the patent system plays a "sound" or positive role in the conservation and sustainable use of plants and TKUP. Even if the patent system were to be construed as a sound economic and social measure for the promotion of conservation and sustainable use of plants and TKUP, it would seem obvious that such a determination must be made within the ambits of the objectives of the CBD itself and not in detraction thereof. Article 16 of the CBD lends further weight to this interpretation. Article 16 (5) provides as follows:

The Contracting Parties, recognizing that patents and other intellectual property rights may have an influence on the implementation of this Convention, shall cooperate in this regard subject to national legislation and international law in order to ensure that such rights are supportive of and do not run counter to its objectives.[182]

It is equally arguable that the placement of the above-mentioned paragraph in Article 16 (which deals mainly with issues of transfer of technology) implies that the role of patents in the regime on plants and TKUP is pertinent only with reference to transfer of technology *simpliciter*. However, these speculations hardly answer the question of what the response of international law has been to the phenomenon of appropriation of plants and TKUP through the patent system. The crux of the matter is that states and leading academics are not agreed on the proper role of the patent system in respect of life forms, particularly plants.

The distrust that some states, particularly from the South, have for the modern patent system would have remained a purely domestic matter were it not for the globalization of the patent concept as a fundamental component of the global trade machinery designed by the industrial world. With the WTO/TRIPs arrangement practically swallowing the idealistic legal norms established by the CBD, it would appear that the sphere of domestic legislative competence on patents on plants and TKUP is limited to what the WTO/TRIPs agreement permits. In effect, in the absence of any globally binding instrument identifying and/or outlawing appropriation of plants and TKUP, state response to the phenomenon must be a limited distillate of jural responses contrived within the narrow ambits of the exceptions in Artcle 27 of the TRIPs and the relevant provisions of the CBD. This is a narrow path and a task suited for the boldest and most innovative of domestic legislators.

It must be admitted that the dominant methods of legal protection of knowledge are incompatible with indigenous worldviews and values. A careful study of the evolution and development of the dominant models of intellectual property law – especially patents, trademarks, copyrights, and industrial designs – clearly shows that those models evolved in ways that put them at odds with indigenous epistemic narratives. More important, the indigenous peoples' quest for equality implicates their demand for cultural integrity. Hence, in diverse international instruments dealing with economic and cultural rights, international law has sought to reconceive the question of biopiracy as a violation of human rights of indigenous peoples, especially economic and cultural rights. Through a remarkable array of conferences, declarations, workshops, and resolutions, indigenous groups, activist-scholars, and various organizations have articulated norms of behaviour in respect of indigenous knowledge on the uses of plants.

This approach is a reflection of the social justice and economic consequences of biopiracy.[183] For example, Article 15 of the International Covenant on Economic, Social and Cultural Rights provides for the "right to benefit from the protection of the moral and material interests resulting from any scientific, literary or artistic production."[184] Although such conventions and other UN declarations[185] affirming the existence of cultural and economic rights have significant moral, normative, and psychological virtues and benefits, care must be taken not to deploy them absolutely in the language of human rights. This is because the patent system is not founded solely on the concept of human rights. There is no state in the world that has a purely human rights-based theory of the patent system. Even the United States, which operates what may be described as the world's most aggressive and liberal patent regime, ultimately recognizes that patents are governmental privileges dictated by perceived economic, social, and political interests.[186]

No inventor, whether in a traditional or a Western framework, has a "human right" to his/her invention. Within the bounds of international law the state determines what gets to be patented, the duration of the patent, the procedures for obtaining the patent, and, where the need arises (as determined by the state), further restrictions or conditions may be imposed on the patent regardless of the interests of the patent holder. Accordingly, intellectual property activists and domestic legislators who deal with the scourge of biopiracy should be careful when (re)conceiving juridical rules or norms. Of course, appropriation of plants and TKUP has multiple consequences that have human rights dimensions. These consequences are the legitimate subjects of the human rights regime, but they should not lead one to presume that there is a human rights basis to the patent system. In the absence of clear, specific, and obligatory international law on appropriation of plants and TKUP, and in the absence of innovative ideas from the various states, it seems that many have taken solace in the world of "soft law"[187] and other non-binding but persuasive norms of behaviour.

Biopiracy and the Soft Law Approach in International Law

Although international law may have survived the tedious and fruitless debate concerning whether, in the absence of a coercive supranational authority, its juridical efficacy as law qua law continues to be valid,[188] there remains an abiding tradition of distinction between specific and precise legal obligations and exhortatory "soft-law recommendations"[189] that may elicit little or no state compliance.[190] The former is considered to be positive international law while the latter is considered to be soft law.[191] Generally speaking, there is no definition of soft law under international law. Declarations, resolutions, and other non-binding public statements of states and international bodies that encourage certain modes of behaviour may be categorized as soft law.[192] In other words, notwithstanding the retreat of Austinian positivism in the interrogation of what constitutes the "legal" nature of international law, positivist clarity and obligatoriness is still a defining characteristic of international law and marks out legal norms from those non-binding norms that are, at best, precatory notions of behaviour at the twilight of legal obligation.

This is not to say that soft laws lack normative persuasiveness; rather, international law theorists of the positive school would like to argue that those "embryonic" and exhortatory rules constitute an adjunct to emerging international law.[193] On the other hand, liberal scholars would opine that soft laws often spawn binding treaties, expand and invigorate existing legal regimes, and normativize general international law.[194] In addition, this school of thought has argued that soft law may harden into customary international law or become part of treaties. For example, Birnie argues that resolutions of the UN General Assembly, although not generally binding per se, "may become so on the basis of state acceptance of them as such, either when adopted or in their subsequent practice in relation to them."[195] In other words, declarations, resolutions, and the like, while sometimes non-binding, may be likened to signposts that point the way towards future treaties and customary international law. As Saraf has argued:

> In the absence of an effective law execution machinery, it is difficult to separate resolutions and declarations of international organisations which create binding norms from those which are merely recommendatory. In

any case these resolutions are legal data for the purpose of determining the state of law at a particular time.[196]

In effect, this school of thought would argue that soft law plays an important role in developing, concretizing, and sustaining norms of international law.[197] This complex existence of soft law presents a living paradox in the sense that soft law both *is* and *may become* international law. This may be the case where the norm in question has been the subject of repeated reaffirmation by various states or international entities.

Although acts of repeated declaration on a certain norm do not, by themselves, amount to a source of or evidence of international law (it is what states do that matters), such repeated declarations may provide an indication of what states or other subjects of international law consider to be the law. As the International Court of Justice (ICJ) held in *Nicaragua:*

> The effect of consent to the text of such resolutions cannot be understood as merely that of a "reiteration or elucidation" of the treaty commitment undertaken in the Charter. On the contrary it may be understood as an acceptance of the validity of the rule or set of rules declared by the resolution by themselves.[198]

In this sense, the distinction between positive international law and soft law seems blurred.[199]

However, there is another school of thought that casts considerable doubt on the existence, utility, and desirability of soft law.[200] A consideration of the arguments of this latter school yields the impression that the rejection of the soft law theory of international law is a direct function of a binary conception of law; to wit, that legal order should not be watered down by norms that have an indeterminate and uncertain quality of bindingness. That is to say, law is either binding or it is not, and thus to speak of soft law is to celebrate the redundancy[201] of bindingness arrived at through dubious gradations of normativity.[202] There is substantial logic in this argument, especially in environmental law and global social justice issues where states prefer to evade clear obligations and direct issues by seeking refuge in platitudinous and exhortatory declarations that lack binding force.[203]

In recognition of the various loopholes in modern international patent law, which facilitate and legitimize biopiracy, it is fair to conclude that the various declarations, resolutions, and reports of fact-finding missions on the subject constitute an emerging international soft law against biopiracy. The first of such initiatives signalling an emerging international norm on biopiracy is the Mataatua Declaration on the Cultural and Intellectual Property Rights of Indigenous Peoples. Convened by the nine indigenous nations of Mataatua in the Bay of Plenty Region of Aotearoa (New Zealand) in

June 1993, the declaration recognizes that dominant regimes of intellectual property protection are inadequate to protect the intellectual property of indigenous and traditional peoples. The second initiative is the 1994 International Consultation on Intellectual Property Rights and Biodiversity. The Consultation was at the instance of the Coordinating Body for the Indigenous Organizations of the Amazon Basin (COICA). At the COICA Summit the Mataatua Declaration was endorsed. Indeed, the COICA statement is in many respects a radical document. For example, according COICA's statement of 30 September 1994:

> Prevailing intellectual property systems reflect a conception and practice that is:
> colonialist, in that the instruments of the developed countries are imposed in order to appropriate the resources of indigenous peoples;
> racist, in that it belittle and minimizes the value of our knowledge systems;
> usurpatory, in that it is essentially a practice of theft.[204]

In 1995 a final statement issued at the end of the South Pacific UNDP Regional Consultation on Indigenous Peoples and Intellectual Property Rights echoed COICA values. The following year, the International Congress of Ethnobiology issued the Declaration of Belem, recommending, *inter alia*, that

> nobody should be allowed to present research findings in its future conferences if she or he has not shared the findings in an easily comprehensible manner with the providers of the knowledge in local language; taken their concurrence for sharing their knowledge with others; informed them of the possible commercial interests of their own or other third parties, sourced or cited the providers just as they would cite a fellow researcher.[205]

Such an instrument, which is representative of most recent declarations by public-spirited NGOs and civil society agents, seeks to ethicize access to traditional knowledge on the uses of plants. In the same year, in Seattle, indigenous peoples issued the Indigenous Peoples Seattle Declaration on the Occasion of the Third Ministerial Meeting of the WTO.

The recommendations and code of ethics of the International Society of Ethnobiology and similar declarations have been popular amongst activists, scholars, and some states, particularly in the South. It has been suggested that these ethical prescriptions should form part of a new protocol to the CBD or at least form part of the corpus of domestic law on access to and exploitation of plants. These principles may be enumerated as follows: the entitlement of indigenous groups to intellectual property rights regime of their own; the principle of prior rights, which recognizes that local and traditional peoples have proprietary rights over their plant resources, whether wild or cultivated; and the principle of local self-determination for indigenous peoples over their resources;[206] full disclosure; active participation; prior informed consent and veto; confidentiality; respect for indigenous worldviews; active protection of indigenous peoples with regard to intellectual property laws and access to indigenous knowledge; precaution, compensation, and equitable sharing of the benefits of TKUP; support for indigenous research; and restitution for appropriated plants and TKUP.

Some of the current domestic juridical responses to the phenomenon of biopiracy have embodied the foregoing principles.[207] Similarly, there have been several fora of traditional and indigenous peoples where the appropriative function of the patent system has been denounced. The psychological value of these types of public statements is not in doubt. More important, a number of developments at policy level in the international arena have imbibed the norms advocated by indigenous groups, activist-scholars, and various interested parties. Indeed, international organizations such as UNCTAD, UNESCO, the FAO, the UNDP, the IFF, the ILO, and WHO have opened preliminary discussions on integrating indigenous-knowledge-protection protocols into the intellectual property debate at the global level.

It is therefore correct to say that there is an emerging soft law in international law on biopiracy. While it is true that "declarations" made by persons and entities that international law has not yet recognized as having the capacity to generate soft law are the softest of soft law, given the recent trend in international law to grant standing to some of these non-state entities, they ought not to be summarily dismissed as ineffectual.[208]

Interestingly, similar declarations and statements have emanated from international organizations such as WIPO and the CBD. These statements, *ex facie,* constitute part of the emerging *lex ferenda* on biopiracy. In particular, statements by the Conference of the Parties (COP) of the CBD on Article 8 (j) of the CBD must be taken seriously. Article 8 (j) requires parties to "respect, preserve, and maintain knowledge, innovations, and practices of indigenous and local communities embodying traditional lifestyles ... and promote their wider application with the approval and involvement of such knowledge." In 1994 the COP opened discussion over the implementation of Article 8 (j).[209] The CBD has, through its Working Group (WG) on Art. 8 (j) and Related Provisions, provided legal advice with respect to evolving mechanisms (legal and normative) for the protection of indigenous knowledge. It is remarkable that the group gives strong legal consideration to the subnational and customary traditions for the protection of knowledge. The WG is also in working communication with the UN Permanent Forum on Indigenous Issues (PFII). The PFII is a subsidiary organ of the Economic and Social Council of the United Nations. More important, it serves as an advisory body on indigenous issues in the context of the council's mandate relating to economic and social development, culture and environment, and education and human rights. Significantly, the PFII has recommended the "establishment of an international ethical code on bioprospecting in order to avoid biopiracy and ensure the respect for indigenous cultural and intellectual heritage."[210]

One of the most important points to emerge from the initiatives of COP and PFII is the recognition that, as part of their customary law, indigenous and traditional peoples have their own systems for the protection and transmission of traditional knowledge. Consequently, it can be said that soft international law is increasingly recognizing that indigenous peoples are the appropriate authorities in matters relating to the protection of indigenous knowledge.[211] In this regard, the activities of WIPO deserve mention. WIPO established the Global Intellectual Property Issues Division, which conducted a global fact-finding mission that evaluated the intellectual property challenges facing indigenous groups around the world. The report detailed three different indigenous protocols for the protection of indigenous peoples knowledge: (1) trade regimes over traditional songs, designs, and dances; (2) ritual regimes for the protection of traditional medicinal knowledge; and (3) customary laws on the use of traditional images and artistic works. All these protocols have the capacity to deter unlawful appropriation of indigenous knowledge. WIPO has also undertaken studies on the applicability of customary laws and protocols to the protection of indigenous peoples knowledge. Similarly, WIPO has the Intergovernmental Committee on Genetic Resources, Traditional Knowledge, and Folklore, where member states deliberate on issues related to the protection of traditional knowledge.

The point here is that there is a gradual shift in international intellectual property law towards legal plurality, especially on matters pertaining to indigenous peoples knowledge. This shift is clearly in recognition of the inherent inability of dominant intellectual property rights regimes to protect indigenous peoples knowledge. More important, it can no longer be disputed that even the best-intentioned applications of dominant intellectual property protocols inevitably expose indigenous peoples knowledge to theft and biopiracy.

From the points noted above, there is no question that the protection of traditional knowledge has gradually moved from the peripheries to centre stage in global discussions concerning intellectual property rights and trade. Further to the declarations and statements of NGOs, activist groups, indigenous groups, international groups, and so on, it is significant that the Doha Ministerial Declaration of 14 November 2001 expressly mandated the Council for TRIPs to look into ways of protecting traditional knowledge. Paragraph 19 of the Doha Declaration reads in part as follows:

We instruct the Council for TRIPS, in pursuing its work programme including under the review of Article 27 (3) (b), the review of the implementation of the TRIPS Agreement under Article 71 (1) and the work foreseen pursuant to paragraph 12 of this Declaration, to examine, *inter alia*, the relationship between the TRIPS Agreement and the Convention on Biological Diversity, the protection of traditional knowledge and folklore, and other relevant new developments raised by Members pursuant to Article 71 (1).[212]

Despite the notable achievements of the bodies mentioned above, it would be overly enthusiastic to categorize the principles, norms, and sentiments that have emanated from these diverse bodies as part of international law on biopiracy; they are not. The bottom line, therefore, is that international law, especially in binding protocols, has not yet fully risen to the challenge of biopiracy. Indeed, to the extent that appropriation of plants and TKUP keeps some industries in the North afloat, it would be naive to expect a sea change in the international legal regime. According to Lara Ewens, "because of the immense investment western corporations have made in plant genetic resources and plant genetic research, and of the important potential biotechnology offers ... modification of the [patent] system is likely to come from within if [it comes] at all."[213]

Conversely, if the position between the North and the South were reversed, it would not be out of place to expect a new and effective legal regime to deal with appropriation of plants and TKUP. For example, when it became obvious that the existing patent system could not be contorted to suit the interests of computer chip makers in the North, the 1989 Washington Treaty on Intellectual Property in Respect of Integrated Circuits was quickly negotiated and concluded. Yet, with respect to protection of TKUP, which is predominantly a resource from weak Southern states, the contrast is clear. As Drahos argues:

> In contrast, the issue of protection for indigenous knowledge has remained just that, an issue. Proposals and models have been put forward but little in the way of concrete, binding law has emerged ... Basically international norms for the protection of indigenous knowledge have thus far taken the form of model laws and declarations by NGOs – in other words, the softest of soft law.[214]

Without discounting the important policy gestures at the international level, prudent and circumspect states are better off devoting greater attention to the domestic forum for ways and means of dealing with the phenomenon of biopiracy and privatization of plants and TKUP. Such measures may be characterized as *sui generic* in the context of Article 27 (3) of TRIPs.

One such measure designed to defeat biopiratical patents involves the registration of traditional knowledge as a way of avoiding the argument that such knowledge has not been published for the purposes of determining patentability.

Biopiracy and the Option of Registration of Traditional Knowledge
Apart from the emerging ferment of legislative activities and initiatives across the South on the issue of regulating access to plants and TKUP, another initiative being considered by state and regional authorities is the registration of TKUP for the purposes of regulating contracts for benefit sharing with so-called end-users. The concept of registration of the occurrence, practices, propagation, and varied uses of TKUP has attained considerable international juridical recognition.[215] The idea is to compile and keep a secret register of uses of TKUP, which will form the basis of any contract between local communities and "end users" of TKUP.[216] This concept has found legislative support in countries such as India,[217] Uganda, Peru, and South Africa.[218]

Two principal reasons have been postulated as the rationale for this concept. The first is that most of the world's patent systems do not have any respect or protection for oral knowledge pertaining to TKUP; thus, in order to protect against its appropriation on the grounds of non-publication in writing, it is argued that such knowledge should be inscribed in written forms. The other reason is that traditional knowledge frameworks are facing extinction in light of cultural homogenization through Westernization. Therefore, it would seem prudent to record knowledge of TKUP before it is too late.[219]

The register of uses of TKUP is an interesting idea, but certain issues need to be addressed. First, unless states appreciate the appropriative function of the patent system and are willing to invalidate those patents on TKUP obtained in a manner inconsistent with the intention of the CBD,[220] there may be problems with achieving the objectives behind the collection of databases. In other words, it must be ascertained that databases maintain the requisite degree of confidentiality necessary to frustrate unscrupulous commercialization or appropriation of the TKUP database in question.

Thus, in the absence of either procedural safeguards, or what Crucible Group I has termed a "convincing Global Morality,"[221] it may be naive to expect that mere documentation would derail the appropriation of plants and TKUP through the patent system. Patent law requires that, for a published prior art to defeat the novelty of an invention, the former must be specific, direct, and unambiguous. This is a very stringent test, and it can be used to avoid liability. In any event, it has already been shown that the requirement of printing and publication of knowledge, especially in the United States and Japan, may sometimes yield unexpected results.

In addition, it should be noted that the drive towards documentation of TKUP might imply that TKUP is invariably oral and that the practitioners are unlettered and primitive people. This is a reckless generalization. A careful perusal of the history of traditional peoples shows that, in addition to oral narrative frameworks, some of them have some of the world's oldest writing cultures.[222] Furthermore, the secrecy regimes that govern medicinal (and spiritual) practices of local healers and breeders question the assumption that all TKUP is in the so-called public domain.[223] As the WIPO Report reaffirmed, there are traditional protocols of intellectual property protection that rely on secrecy.

The concept of registry of uses may perpetuate the myth that all TKUP and traditionally improved plants are "raw materials," without sufficient intellectual input to warrant property rights unless and until they are "improved" in Western laboratories. It is not possible to argue that only Northern multinational corporations and public research institutions are engaged in plant improvement and the generation of knowledge on the uses of plants. This is not only implausible but also inconsistent with modern international law, which recognizes the experimental, observational, and cumulative nature of traditional farming and healing practices.[224]

More important, the characterization of traditionally improved plants and TKUP as raw materials reduces the claims and struggles of traditional and indigenous peoples to a grumble for monetary profits. This position cannot be correct: the claims of traditional and indigenous peoples include a legitimate recognition of their intellectual contributions to the mosaic called human civilization.[225] As the WIPO Report has noted, "they [TKUP practitioners] do not wish to be confined to the role of mere purveyors of resources and know-how for the benefit of commercial interests in which they have no participation."[226] Attention should also be paid to the immense "bargaining" advantages possessed by influential and powerful commercial bioprospectors.

This is not to say that registration databases are not useful. The point is that the concept of registration of traditional knowledge should seriously consider the sophistication of institutions and legal concepts that facilitate biopiracy. Unless adequately regulated, the register of uses may function as a catalyst for predatory commercial bioprospecting and biopiracy. Another *sui generic* option involves modifying the patent system in order to create defensive patents for indigenous peoples.

A Framework for Modifying the Patent System

Ideally, the solution to biopiracy would be the institutionalization of indigenous or traditional intellectual property regimes at both national and transnational jurisdictions. It is significant that both the WIPO and the

CBD have been working on such models. However, given the relative powerlessness of traditional communities to effect such changes in good time, another option could be to modify the patent system to create defensive patents. It is ironic that, notwithstanding the classical patent system's theoretical and cultural incompatibility with traditional peoples, this suggestion may mitigate the phenomenon of biopiracy.[227] The favoured option would be a regime of communal patents. Certain reasons compel this audacious proposal. The first is that, although the patent system has global repercussions, it is fundamentally a state privilege, and any domestic legislator with enough skill can conform to global standards on patents while effectively pursuing a domestic anti-biopiracy agenda. The surprising thing here is that states in the South that complain about appropriation have not adopted the same instrumentalist approach to patents as have their counterparts in the North. While the latter perceives the patent system to be a malleable tool, the former perceives it to be a sacrosanct foreign institution – writ in stone and not amenable to manipulation at the local level. The patent system is an instrument of state policy, and this fundamental maxim seems to have been lost on states of the South.

Second, in the absence of a property right over traditionally improved plants and TKUP, the bargaining power and position of traditional societies is fundamentally compromised. As a matter of juridical strategy, property rights fundamentally alter the bargaining process for access to and equitable sharing of the benefits of plants and TKUP. As R.H. Coase sagely observed, "in any negotiations between two parties over property rights the initial distribution of the property rights will determine who has to pay and who does not."[228] Thus, property rights allow the holder of the right to appropriate the value of the resource to which the right relates and thus set the tone of the bargain. More important, they also determine both the direction of wealth transfer and define *ab initio* whose intellectual input is being transferred, to whom, and the terms of that transaction.

If there are no property rights in traditionally improved plants and in TKUP, then certain actors will simply construe and appropriate such intellectual contributions as a "free input of production."[229] Accordingly, in the absence of Western restraint in vesting property rights in appropriated plants and TKUP, it is of the utmost importance that the South place a form of property rights on traditionally improved plants and TKUP. The best method for doing this would involve such states looking inward and recognizing the traditional legal mechanisms governing ownership of such intellectual-cum-cultural practices.

With respect to the issue of modifying the patent system itself at the domestic level, the arguments of "cultural purists," who are determined to keep traditional societies away from the "contamination" of the patent system, must be anticipated, confronted, and resolved. In this regard, certain misconceptions must be dispelled. It cannot be denied that patents represent the Western liberal ideology of capitalist profits and commodification. It would therefore be tempting to conclude that an introduction of the patent concept into the traditional cultural framework will contaminate the values of traditional peoples. However, this notion is nonsensical and intellectually lazy.

The patent system can be used defensively. As Peter Drahos has already noted, allowing Western free market patents on TKUP will not necessarily pollute the cultural practices of traditional peoples. In his words, the argument on cultural purity

> tends to oversimplify a more complex picture. Property rights in indigenous knowledge do not necessarily have to lead to open trade. They can be used to prohibit or regulate such a trade. It is a mistake to think that property only has an appropriation function. It also functions as a means of self defence or survival ... The crucial issue is not whether trade in culture will corrupt those engaged in it, but rather whether we can devise regulatory forms that will allow indigenous people to pursue their own economic interests in the use of their cultures in ways that are consistent with their aspirations for the preservation and evolution of those cultures. Devising such regulatory forms seems best left to indigenous people and the positive law of individual states.[230]

Thus, moving from the paternalistic and implicitly condescending treatment of traditional knowledge as an anthropological exoticism that is to be kept in a gilded cage, we need to "extend the existing system rather than to create a new system to provide the same protection."[231] This response to the patent debacle is grounded in common sense and pragmatism. At the moment, local communities and the plant-diverse/culturally diverse countries of the South simply lack the political clout necessary to institutionalize their autochthonous regimes of intellectual property rights and to make such indigenous forms of intellectual property binding in international law.

In effect, patents can be used defensively as a means of protecting against biopiracy. Recently, this view has gained ground among scholars of the phenomenon of appropriation. According to Naomi Roht-Arriaza:

> One obvious response to the appropriation of indigenous and traditional knowledge and its fruits is to modify existing national and international intellectual property protection regimes to encompass the informal innovations of indigenous and traditional communities.[232]

Thomas Cottier has also articulated this idea:

The very concept of IPRs, however, does not prevent states from introducing feasible systems which are cheap and simple. They are often called petty patents and complement the more sophisticated patent system which is designed for industrial purposes. Switzerland for example, relies upon a national patent system which is cheap and simple mainly by the fact that novelty is not examined at the stage of examination and registration. Such questions are left to courts only in case those patents are being challenged. Interestingly, these patents do not fare worse than patents had been examined for novelty at registration.[233]

Thus, the case for a simple, inexpensive, and essentially defensive communal patent system has been gaining attention in modern scholarship. As I have argued elsewhere, it seems that what is needed is a creative adaptation of the patent system and its jural concepts.[234] Thomas Cottier agrees:

IPR systems could adapt to such needs by way of introducing new forms of ownership. Nothing prevents us from introducing novel communitarian titles; the system need not be limited to individual ownership. In sum, modern IPRs, appropriately adjusted can make a considerable contribution to valuation and dissemination of grassroots innovation and building of sustainable technologies.[235]

It would therefore seem that, given the instrumentalist and malleable nature of the patent system, another experimentation would not be out of place, especially where such innovation serves a social and equitable purpose. Defensive patents have also become part of the juridical response by the CGIAR, which controls the IARCs. For example, as Halewood has noted:

The CGIAR centres have recently endorsed a system-wide scheme whereby individual centres may engage in *defensive patenting of their own innovations simply to prevent profit driven companies from appropriating their work and commercializing it.* Simultaneously, the centres adopt the policy that they will not seek to commercialize these intellectual property-protected innovations.[236]

The point is that the patent system is an instrumentalist mechanism, and thus communal patents, as instruments designed to achieve a measure of distributive justice, are an attractive option. In effect, in spite of the existing problems of integrity and instrumentalism, states of the South have neither fully explored nor adequately adapted the patent system to suit the interests of their traditional communities.

Given that powerful states use and exploit the patent system to advance their national interests, it would be naive of the South to proceed as if the system was beyond counter-manipulation. However, in proceeding to sketch the outlines of what may be described as a proposed regime of a communal patent system, certain issues must be resolved. These include:

1. whether TKUP and traditionally modified plants rise to the level of patentable subject matter[237]
2. the supposed communal and collective nature of traditional inventions as opposed to the much celebrated individuality/originality of Western invention
3. the so-called "public" and stale nature of innovations and inventions in the traditional domain.

The first issue operates on the assumption that an invention must rise above the level of prior art before it can be protected with a patent. While this assumption is taken as a given, the reality of the matter is that the determination of what constitutes a patentable novelty is subjective.[238] In any case, it is common knowledge that bioprospectors rely on the insight of traditional farmers and healers in their search for better crops and medicines. TKUP and traditionally modified plants are not inherently unpatentable merely because of the cultural paradigm within which they are developed.

The second issue relates to the argument that the ostensibly communal nature of life in traditional societies is inconsistent with the purported individualism of the inventive process.[239] The contention of this school of thought is that the patent system is predicated on the idea of the inventor as an individual qua individual and of the inventive process itself as an exercise in solitary experimentation and lonely mental agony.[240] The image that is created and propagated in the Western world concerning inventorship is one of individualism, and the resultant invention is seen as a product of the inventor's individual genius. The idea is that the patent system is designed to compensate the individual qua the inventor.

None of these assumptions withstands scrutiny. The myth of inventorship as a solitary and heroic quest is arrogant and ludicrous. Every invention is part of the mosaic of human culture and rests on pre-existing ideas and inventions. The culture of individualism in Western liberal democracies involves a misapprehension of the social structure as well as the process of inventiveness in that it supposes that inventions today are predominantly carried out by the likes of Benjamin Franklin, James Watt, or Filippo Brunelleschi working in conditions of near solitary confinement. Apart from the fact that inventorship is incremental in nature, in modern times inventorship involves a multitude people working away on specific projects, building upon existing ideas and knowledge, exchanging ideas and information, and generally operating as a community.[241] As Kuhn's path-breaking research observes:

> The transformation of technology and of economic society during the last century negates completely the patent law assumption as to the nature of the inventive process ... In the modern research laboratories, tens, hundreds of men [sic] focus upon single, often minute problems; inventions become increasingly inevitable.[242]

David Safran concurs with Alfred Kuhn. In his words:

> In this age, most inventions result from corporate research efforts ... a growing number of these inventions are the result of the work of several research and development teams that are located in different countries.[243]

In light of this template shift in the inventive process, it is not a coincidence that patent jurisprudence inevitably created the legal fiction of corporate/employer's rights to inventions that were made by employees.[244] The inescapable conclusion is that "collective invention is a common and determinant force in both local economies and the world economy."[245] Interestingly, it has not been suggested that the communality of inventions in "formal" scientific frameworks is a bar to their patentability. Yet traditional knowledge frameworks have been singled out, and the inventions obtained within have been maligned on the grounds that they are communal in nature. It is equally intriguing that, in other areas of collective intellectual input, the theory of individual authorship does not follow its own rules. For example, the non-individualistic authorship of films has been largely obviated by nominating a person, usually the producer, as the "author" of the work. Needless to say, several hundreds of hands work to produce a movie. It is therefore clear that, according to basic logic, Western knowledge frameworks are as mosaical, accretional, and communal as are traditional and non-Western frameworks.

Furthermore, it would seem that the alleged hermetic boundary between individualism and collective creativity involves a conflation of cultural communalism and collective inventiveness. It is probable that, in some cases involving inventiveness in a communal society, an individual, deriving inspiration from surrounding and pre-existing knowledge, may create new inventions or reinterpret existing knowledge to create something "of intricate detail and complexity, reflecting great skill and originality."[246]

The third argument on the patentability of TKUP and traditionally improved plants is that they are matters of common knowledge and reside in the so-called public domain. There are several cultural biases, along with factual flaws, in this theory. In the first place, the direct implication of this argument is that TKUP is merely a form of heritage frozen in time. This culturally prejudiced notion is simply absurd. Neither international law[247] nor common sense supports the notion of a stagnant and ossified concept of traditional knowledge. Like all other forms of knowledge framework, TKUP is incremental and dynamic. In the second place, there is no principle or rule of international law on patents that prescribes any maximum number of years in which a particular invention that has been made public must be patented before it loses its patentability on the grounds of lack of novelty. For sound public policy reasons the Paris Convention sets a lower limit of only one year. In other words, an invention may be in common use for any number of years (prescribed by the local legislator) without losing its novelty and thus becoming patentable in spite of being in common use. Interestingly, this important aspect of patent law has escaped the attention of most commentators and scholars. The issue here is that the ease with which information is publicized in various states or societies varies, and the local legislature is entitled to take this factor in consideration.

For example, the Kuwaiti patent law limits novelty to twenty years after the first public use of the invention.[248] The Libyan patent law of 1959 provided for a period of fifty years of open use preceding the date of submission of the patent application.[249] The point is that there is no upper limit on public use prior to patenting an invention. Indeed, some international organizations have strongly suggested that special allowance be made when dealing with patents for TKUP for traditional peoples. For example, the recent WIPO Report on the needs of traditional knowledge practitioners notes that "an extended grace period for traditional knowledge holders ... would give informal innovators additional time to research possibilities of commercialization."[250]

In the third place, it is not true that all TKUP is in the public domain. As John Frow notes, "indigenous cultural systems are not built upon a principle of open access but are highly regulated and restricted; they are built upon secrecy as much as upon openness."[251] Native healers are notorious for the secrecy regimes surrounding their knowledge of the uses of herbs. The secrecy surrounding their immense knowledge of the medicinal and pharmacological uses of plants ensures their continued power and influence in traditional societies. The elaborate ritual, magic, and spirituality that often surrounds traditional medical practice is a crucial aspect of the secrecy regimes imposed on TKUP and traditionally modified plants by traditional herbalists.[252] Each product of TKUP should be critically examined on its merits to determine whether or not it is part of the public domain.

Moreover, assuming that a particular product of TKUP is in the public domain, it is well to note that placement of such knowledge without consent does not *ipso facto* extinguish a right of claim to intellectual property. In addition to the legal validation of traditional methods of ownership, dissemination, and transfer of traditional knowledge, a regime of defensive community patents on TKUP and traditionally modified plants is quite feasible. It is significant that some states, such as Brazil, are instituting similar proposals.[253] As Eugenio de Silva has observed, "the [Brazilian] law also adopts a new concept for the application and patentability of indigenous industrial property rights, when it establishes a principle for the protectability of indigenous traditional knowledge."[254]

Outlines of a Community Patent System[255]
In formulating any broad proposal for community patents, the first juridical issue to be addressed is the question of legal persons who may apply for patents on TKUP and traditionally modified plants. The Eurocentric conception of the patent system appears to limit the category of such entities to natural persons and corporations. However, it is significant that, even in the defunct socialist countries, which were the crudest European approximation of collective and communal existence, inventions were routinely granted to "socialist organizations" or other such multiplicities of individual human persons. For example, communist Romanian law on patents[256] provided for such groups and "collectives" to apply for patents. Similar provisions were made in the communist patent laws of the defunct Republic of Yugoslavia.[257] Even capitalist and individualistic societies like the United States provide for joint inventorship.[258]

The point is that countries and cultures where collective inventorship is prevalent need not confine themselves within the confines of Western jurisprudence on legal personality. It is the undoubted prerogative of states to create, grant, or deny legal personality to artificial persons. The creation of various types of moral or artificial persons is a function of the values, cultures, and ideology of each domestic legal system. Accordingly, the suggested solution lies in a juridical consummation of the already existing forms of legal personhood in traditional and non-Western societies. The relevant persons may include families, villages, bands, clans, kindreds, and/or any other legal persons recognized as such by and in the customary law of the cultures or peoples concerned.

The problem here is that dominant cultures and jurisprudence have insisted on defining for others, especially colonized, marginalized, and disempowered peoples and cultures, who or what constitutes a legal person. Under this colonial regime, legal persons such as "families," "kindreds," "clans," "age-grades," "the spirit of the unborn," "ancestral spirits," and other forms and categories of legal personality have been reduced to exotic curiosities for modern anthropologists, sociologists, and jurisprudes. It is remarkable that some countries, including the Philippines[259] and Brazil, have taken steps towards reaffirming that categories and types of legal persons are not limited to those acceptable to Eurocentric biases.

In Brazil, for instance, pursuant to Bill PL N. 2.057 of 23 October 1991,[260] indigenous peoples in their communal aggregation have been recognized as possessing legal personality. And this recognition is not a function of any requirements for registration or of the magnanimity of the government. Under the proposed legislation, "indigenous communities, or any of their members, have the right to apply for a patent of invention, utility model or industrial design which has been developed utilizing their traditional collective knowledge."[261] In the event of any overlap of communities, patents granted in the name of one community at the expense of another may be administratively rectified. Similarly, in the event of disputes relating to priority, it may be suggested that traditional arbitration methods be adopted.

It is also interesting that many traditional societies are leery of having their sacred plants and TKUP "commodified" through the patent process. In the absence of real alternatives capable of binding the global community, this attitude may need to be replaced with a more astute and pragmatic response to the dilemma. The concern over commodification is legitimate. However, an astute reading of the patent law system and the well-meaning but misguided attempts by some scholars to present traditional peoples as "folks" desirous of living in a culturally "uncontaminated" world shows that such romantic (mis)representations of traditional and indigenous peoples have been both exaggerated and misconceived. As Thomas Cottier and Peter Drahos have already noted, the communal patent system is primarily a defensive concept.

In any event, the discontent with commodification may arise from a misunderstanding of modern international patent law. Under Article 27 of the TRIPs agreement, which constitutes the global minimum threshold for patent law, there is no strict obligation to use an invention. In fact, there are thousands of patented inventions that have never been used and may never be used. Even among those inventions that have managed to be commercialized, a host spent decades on the shelf. The most well known include the television and the fax machine, which spent over forty and seventy years on the shelf, respectively. In short, given the global trend towards the abolition of compulsory use of the patents, the fears of commoditization of sacred TKUP and plants may be exaggerated.

In addition, it bears repeating that patent terms have never remained constant; rather, they have always reflected an instrumentalist function in the hands of states.[262] There is no universally prescribed or binding upper limit to the duration of a patent grant. Article 27 of the TRIPs agreement only sets the minimum number of years for patent grants. Therefore, the local legislature in the affected states may protect the spiritual and religious significance of some TKUP for traditional and indigenous populations by significantly lengthening the duration of and/or renewing such communal patents at the request of the affected traditional community.

As regards the question of costs of setting up such a patent system to implement the communal regime of patents, it is a well known fact that a substantial part of the costs of running a patent system is spent on the maintenance of patent lawyers and examiners. It is suggested that poor countries should take a leaf from European countries that do not maintain a large army of patent examiners, especially in matters pertaining to "petty patents." The prudent thing would be to dispense with the examination of communal patent applications. This is already the practice in some European patent law jurisdictions (e.g., France and Switzerland). In such cases the burden of proof shifts to the person who asserts that the patent grant is not supportable. Interestingly, Article 34 of the TRIPs agreement shifts the burden of proving non-infringement of a patented invention to the defendant.

Furthermore, gene-rich states that have complained of having their plants and TKUP appropriated should consider a twin regime of non-recognition of such patents within their own domestic jurisdiction. They may also use the good offices of their states to launch both diplomatic and organized public protests against such patents. Patents on TKUP and traditionally improved plants should not be recognized in states that have legitimate grievances with such patents, particularly where they were obtained without the prior informed consent of the affected traditional community or indigenous people. It is interesting to note that, during the negotiation of the CBD, the Government of Ethiopia made a similar suggestion:

> We express dissatisfaction with the provisions protecting patents and other intellectual property rights without commensurate regard for informal innovations, especially in Article 16, paragraph 2, which opens the way for use by countries with the technological know-how of genetic resources and innovations from countries without the know-how in patents and other intellectual property rights and for taking them out of the reach of even those countries which created the very genetic resources and innovations.[263]

The Government of Ethiopia thus suggested that, at a later date, the following paragraph be added to Article 16 of the CBD:

> Where a technology, organism or genetic material which is patented or legally protected in any other way as an intellectual property has incorporated an organism or organisms, a genetic material or materials, a technology or technologies or any other traditional practice or practices originating in another country or countries, the patent or other intellectual property shall not be valid in the country or countries of origin of any of its component parts; and the benefits accruing from the application of the patent or

other intellectual property right in other countries shall be equitably shared between the holder or holders of the protected right and the country or countries of origin.[264]

Although the Ethiopian suggestion is worthy of serious consideration, it suffers from the erroneous assumption that states may not *suo motu* enact laws in their domestic jurisdictions invalidating or rendering unenforceable those patents that make a mockery of the obligations created by the CBD and accepted by contracting parties.[265] It cannot be denied that some of these issues deal with the unresolved question of liability in international law pertaining to contradictory obligations of states. At the moment, it would seem that states tend to resolve this dilemma by leaning towards the obligation that most converges with their perceived national interests. In the process, the opposing obligation may fall into disuse.

On the issue of publicized protests, it is arguable that such practices drain legitimacy from international practices such as biopiracy. It can hardly be disputed that the recent public demonstrations against institutions of globalization and their supporting legal framework have questioned the efficacy and legitimacy of these institutions and extorted a measure of change. Another way of resolving the problem of biopiracy would be for states to clarify the law on this issue by adding protocols to both the CBD and the TRIPs agreement. In sum, it seems that the global community cannot escape the imperative of rationalizing and clarifying the conflicts embedded within such international instruments as the CBD and the TRIPs agreement.

There are some binding multilateral and regional arrangements that follow the reasoning of the suggestions made above. For example, the Andean Community, in its Decision 486 on Biological, Genetic, and Traditional Knowledge, provides that:

> The Member Countries shall ensure that the protection granted to intellectual property elements shall be accorded while safeguarding and respecting their biological and genetic heritage, together with the traditional knowledge of their indigenous, African-American, or local communities. As a result, the granting of patents on inventions that have been developed on the basis of material obtained from that heritage or that knowledge shall be subordinated to the acquisition of that material in accordance with international, Andean Community, and national law.[266]

In addition, Article 3 of Decision 486 provides that "Member Countries recognize the right and authority of indigenous, African-American and local communities to decide on their collective knowledge."[267] Further, Article 26 of Decision 486 provides that the application for a patent for an invention must contain:

(h) a copy of the contract for access, if the products or processes for which a patent application is being filed were obtained or developed from genetic resources or byproducts originating in one of the member countries.

Paragraph 26 (i) further specifies that the application must include

[a] copy of the document that certifies the license or authorization to use the traditional knowledge of indigenous, African-American, or local communities in the Member Countries where the products or processes whose protection is being requested were obtained or developed on the basis of the knowledge originating in any one of the Member Countries, pursuant to the provisions of Decision 391 and its effective amendments and regulations.[268]

Interestingly, under Article 75, the competent authority could nullify a patent in cases where the relevant copy of the contract evidencing access was not attached to the application for the patent. While these legal instruments and provisions may seem draconian, they appear to have been conceived in the spirit of outrage against the inequity in the world's appropriative major patent systems.

Another option worth pursuing is the adaptation of the common law rule of *parens patriae;* that is, the state should be obliged to protect and take into custody the rights and privileges of its citizens for discharging its obligations.[269] Given that multinational corporations have immense clout, far in excess of what aggrieved traditional communities and indigenous peoples could dream of mustering in protesting cases of biopiracy, it would be worthwhile for states desirous of protecting traditional knowledge to consider enacting laws that would enable them to pursue legal proceedings on behalf of traditional communities in matters related to biopiracy. Given that it is one of the duties of the state to secure for its citizens the enjoyment and protection of economic, cultural, and social rights, it is arguable that, in cases where those citizens are unable to assert those rights by themselves, the state ought to step in, especially given that such knowledge and products are ultimately national assets. In other words, the state itself has an independent interest in the consequences of the appropriation it seeks to protest. Therefore, there is no known reason why states cannot assume this function, especially in cases of egregious appropriation of TKUP.[270] Even if such lawsuits are lost in the courtrooms, they may still have some normative value and make biopirates uncomfortable.

Certain conclusions are due for consideration. First, the international intellectual property rights community should give legal effect to indigenous protocols on the protection of indigenous peoples knowledge. Second, the doctrines and principles of the world's major patent systems should be re-evaluated, especially within the context of those legal principles that facilitate the appropriation of indigenous peoples knowledge. Third, the international patent law regime should be modified to enhance the exchange of information between the North and the South and also to accommodate critical analyses of the ramifications of the patent system. Fourth, the possibility of an alternative regime to the patent system on ownership of plant resources should be vigorously pursued.[271] An option in this regard involves the issue of elevating the concept of farmers rights (now recognized in international law by virtue of the FAO Treaty on Plant Genetic Resources) into a community-based alternative to "the private property framework of traditional IPRs."[272] At present there are three bills at the Congress of Philippines providing for the establishment of such community intellectual rights protection.[273] These bills seek the incorporation of the ideas already discussed here in respect of traditional knowledge on the uses of plants.

Apart from such measures detailed above, states should also pursue a tighter regime on patent disclosure. As Professor Oddi has argued, "the statutory requirements for enabling disclosure could be quite specific in requiring specifications, blueprints, dimensions, chemical compositions, exact temperatures, pressures, bill of materials, equipment requirements, etc."[274] In other words, applicants for patents involving products of TKUP should fully and completely disclose the elements highlighted by Oddi. This would afford patent examiners the material to make determinations as to whether a purported invention is real or is merely a pseudo-invention.

With regard to the issue of publication of TKUP, there is a need for extra caution as this may function as a catalyst for appropriation. Indeed, studies by Kate and Laird show that:

> Eighty percent of all companies that use ethnobotanical knowledge (only half of those interviewed) rely solely on literature and databases as their primary source for information. This fact has significant implications for benefit-sharing and suggests that academic publications and transmission of knowledge into databases – rather than field collections on behalf of companies – are the most common route by which traditional knowledge travels from a community to the commercial laboratory. Companies therefore have access to knowledge in ways that do not trigger benefit-sharing.[275]

The obvious point is that states should modify their patent laws and laws on physical access to TKUP to deal with this reality. Further, gene-rich states should wean themselves of the idea that, in order for a binding international instrument on biopiracy to be global, it needs the financial support and membership of all other states in the world. A considerable number of binding international law instruments started off with minimal state ratifications. Waiting on a universal convention is simply unrealistic.

It is also suggested that, towards a regime of absolute global novelty, an international bureau similar to the International Patents Bureau[276] could be established to provide the various patents offices with "reasoned opinions regarding the novelty of inventions in respect of which applications for patents have been filed with the respective national industrial property services."[277] This bureau would also have the duty of cross-checking various national offices for claims of novelty. The basis for determining what constitutes "state of the art" or "prior art" should recognize the validity of oral information. An additional measure would involve passing laws elevating oral knowledge to the same status as written knowledge. In effect, no patent would be recognized anywhere in the world if its claims had been anticipated by oral knowledge. Indeed, this is one of the legislative initiatives adopted by the Government of India in December 1999.[278]

In the event that some states do not want to be part of this proposed international bureau on novelty of inventions, it could be established only by those states to whom it would be of benefit. If this regime is instituted, then a major normative benefit would be that patents on TKUP or plants that were not certified by the bureau would lack legitimacy. In addition to a documentary search for what constitutes "prior art," the bureau should also engage in a search of oral tradition of the claimed invention. The patent system should be held up to its own theoretical best standards. The bureau should also serve as a bridge between the WTO TRIPs and the CBD, particularly on the vexed issue of the relationship between IPRs and the CBD. The member states of this proposed bureau would undertake to make the domestic validity of their patents conditional on compliance with the bureau's requirements pertaining to a global standard of novelty.

Finally, given the increasing relevance of PBRs, as a *sui generic* model for the protection of intellectual property on improved plants, it is necessary to examine how PBRs may implicate the debate on biopiracy.

Biopiracy and Plant Breeders Rights

The TRIPs agreement provides for the protection of plant varieties "either by patents of by an effective *sui generis* system or by any combination thereof."[279] One of the more well known *sui generis* property regimes is the concept of PBRs. The origins of this genre of rights have been traced to the efforts of six Western European states. In the United States, although the *Plant Patent Act of 1930*[280] laid the juridical basis for patents on plants, PBRs were created by the *Plant Variety Protection Act of 1970* (PVPA).[281] This latter legislation extended legal protection to sexually reproduced (i.e., seed) plants. The significance here is that most important food crops in the world (e.g., wheat, maize, rice, etc.) are sexually reproduced. Thus, control over them has deep economic and political consequences. Until the TRIPs agreement, attempts by the American agribusiness industry to introduce PBR legislation in Australia, Canada, and Ireland were rebuffed by various interest groups, including organized labour, farm, church, and environmental groups.[282]

The reasons for this opposition to PBRs even in the North are not difficult to discern. PBRs were created for the benefit of commercial plant breeding companies. The potential for such corporations to have an iron grip on global food supply has remained hotly debated. However, by "tying-in" other aspects of economic activities as part of the enlarged rubric of "global trade," those who opposed PBRs were defeated.[283] Today, PBRs are practically entrenched at the international level courtesy of the UPOV Conventions, which have been promoted as a *sui generis* option to patents on plants.[284] The UPOV Convention of 1978 was replaced by UPOV 1991.

Thus, at the time the TRIPs negotiations were in progress, the UPOV Convention of 1978 was in force. Unlike UPOV 1978, UPOV 1991 mandates the protection of all plant varieties and disallows the exchange or sale of seeds by farmers. It provides for the exclusive right of reproduction of the protected variety. Given that these stringent rights are created with a lowered threshold for plant variety protection, it is arguable that UPOV 1991 may facilitate the appropriation of plant varieties by the powerful seed merchants and plant breeders of industrialized states.

Furthermore, UPOV 1991's limits on farmers rights and breeders exemptions lead to the view that states from the Third World may find UPOV 1978 more tolerable.[285] Despite the accession of many Third World countries to UPOV 1978 (in compliance with WTO rules), the usefulness of PBRs to Third World states is still not clear. There have been case studies in Argentina, Chile, and Uruguay, the results of which remain inconclusive.[286] Further, it has been asked whether PBRs have the capacity to reward community-based innovation in the plant regime.

In addition, the appropriative function of these conventions is a matter of legitimate concern. Commenting on the US PBR's legislation, the *PVPA*, Kloppenburg comments: "The PVPA is but the most recent of a variety of juridical strategies taken by private enterprise to extend the reach of the commodity-form to encompass plant germ plasm."[287] The fundamental problem with PBRs is the absence of any regulatory framework to ensure that only superior varieties of plants, or plants that represent an improvement on existing varieties, are accorded legal protection. In effect, proprietary rights over plants are obtainable under PBRs regimes upon fulfillment of the conditions of novelty, uniformity, and stability (consistent phenotypic reproducibility). For example, Article 5 of the International Convention for the Protection of New Varieties of Plants, 2 December 1962, which sets the international standard for granting of PBRs, is instructive. It provides that "the breeder's right shall be granted where the variety is new, distinct, uniform and stable."[288] There is no consideration for improved quality or improved utility of the plant.

In addition, it bears noting that, like patents, the UPOV conventions have no global or absolute standard for determining novelty. The implication is that mere "discovery" of new plant species or varieties could give rise to legal protection under the PBRs regimes established by either of the UPOV Conventions. Given the preponderance of plant genetic and varietal diversity in the dense rainforests of Third World countries, the potential for mischief has not been taken into consideration in setting the standard for novelty in PBRs regimes. There are both geographical and commercial contingencies for ascertaining novelty in PBRs regimes. Similarly, the criterion of distinctness has relevance only in a marketing, as opposed to a utilitarian, sense. As Kloppenburg notes, "the PVPA was pursued by the seed industry primarily as a mechanism for permitting the *differentiation* of its products"[289] rather than as a reward for useful improvement of plants. Similarly, the criteria of uniformity and stability are specially tailored to comport with the vagaries of plant genetics and commercialization.

The irony is that, in the face of low standards for granting proprietary rights under PBRs, the weight and scope of PBRs under UPOV 1991 is much stronger than under UPOV 1978 and is almost patent-like in its rigour. For example, in addition to prescribing a minimum duration of twenty years for PBRs, the UPOV Convention of 1991 outlines the scope of the PBRs thus:

Subject to Articles 15 and 16, the following acts in respect of the propagating material of the protected variety shall require the authorization of the breeder:

(i) production or reproduction (multiplication),

(ii) conditioning for the purposes of propagation,

(iii) offering for sale,

(iv) selling or marketing,

(v) exporting,

(vi) importing,

(vii) stocking for any of the purposes mentioned in (i) to (vi) above

(b) The breeder may make his authorization subject to conditions and limitations.[290]

For the aforesaid reasons, some commentators have argued that "the scope of the right foreseen by the UPOV Act of 1991 goes far beyond that required by the UPOV 1978 Act. In fact, the 1991 revision has brought the Convention more in line with patent law."[291] As Philippe Cullet has observed, "the latest revision of the Convention adopted in 1991 has further strengthened the rights of commercial plant breeders. This includes the obligation for member states to provide protection to all plant genera and species ... Overall, in the 1991 version, plant breeders rights have become akin to weakened patents and the conceptual distinction between the two is now blurred."[292] The cumulative impact of the relaxation of the criteria for patentability of plants and TKUP in an age of tighter PBRs regimes under UPOV 1991 is, in effect, a virtual convergence of patents and PBRs with regard to the plant regime.[293]

6
Conclusion

In the preceding chapters I examined the origins and development of the patent system, its globalization, and, ultimately, its political economy vis-à-vis indigenous peoples knowledge and plant diversity. My analysis focuses on the North-South debate on biopiracy, and I discuss the issues raised within the context of loss of plant and cultural diversity. I argue that international agricultural research centres sometimes facilitate the appropriation of plant life forms and TKUP. This contentious aspect of my analysis is conducted within the context of North-South relations and the inappropriate use of the concept of common heritage of mankind in matters pertaining to plant germplasm.

I indict patents on indigenous knowledge on the uses of plants and plant genetic patents as mechanisms for the appropriation of Third World resources. I demonstrate that the contemporary patent system of powerful states has been significantly manipulated and retrofitted to suit the interests of seed merchants and pharmaceutical companies. These factors operate within a social culture of prejudice and disrespect for non-Western forms of epistemology. Consequently, indigenous peoples and local communities have been exploited and impoverished both economically and culturally.

I now look at some of the consequences of the erosion and appropriation of plant life forms and TKUP by both international institutions (especially the international agricultural research centres) and the patent systems of powerful states. First, it is becoming increasingly clear that a handful of multinational corporations with primary interests in the stock markets control the global supply of seeds and related agricultural inputs. More significantly, agricultural inputs are being "tied in" with the global food supply to create an oligopoly. The implications of this trend for global food supply and agricultural security are matters of extant debate and concern. Second, there are also concerns about the implications of the emerging dispensation for human health and the safety and integrity of the environment. As I have argued elsewhere, since the lowering of the threshold for patentability of

TKUP and the emergence of genetic patents, there is increasing evidence of patents on subjects that constitute "novelty without innovation," especially in the pharmaceutical industry.[1]

Other consequences include the increasing concentration of power in the hands of a few multinational seed breeders and the erosion of plant genetic diversity. Related to this is the fact that the patent system has largely encouraged the precipitate and premature release into the environment of genetically modified plant life forms without adequate assessment of human safety issues and compatibility with the various interacting niches of the ecosystem. The role of the principle of precaution in this dispensation is a matter of deep concern. It seems, however, that this principle is emerging as a powerful tool for sober consideration of the wider implications of the role of patents in environmental safety and health. These issues, in themselves and by mutual interaction, raise various questions of law and human rights. They also compel a reappraisal of the concept and practice of development. It is difficult to theorize about these complex, multidimensional issues, and there are no quick-fix solutions. However, we must address these questions if only to provoke further debate.

Patents and Plants: The Question of Global Food Security

A major argument put forward by proponents of patents on plants and TKUP is that the patent system offers an economic incentive for promoting the sustenance of plant genetic diversity and encouraging scientific improvement of plants. Although there may be some element of validity to this free-market argument for patents on plants, there is compelling evidence showing that the industrialization of agriculture is one of the greatest causes of erosion of plant genetic and species diversity.[2]

The reasons for this are not hard to find. The commercially successful industrialization of agriculture thrives on uniformity of crops. Uniformity means ease of cultivation and harvest, which translates into greater profit. Given that the patent system plays a significant role in the commercialization of inventions, it follows that it undermines the essence of plant diversity by focusing undue attention on so-called economically useful plants.[3] The historical trend is that plants that offer little or no appeal to the industrialist machinery are often ignored, and in some cases the market drives such plants into the peripheries (if not extinction). For example, in the United States a survey of seed banks showed that chufas, martynia, and rampion have been lost entirely.[4] It is estimated that 8,000 to 10,000 different varieties of apples have been named and recorded throughout the history of the world. Today, only about twelve to fifteen varieties account for 95 percent of all commercially grown apples.

In other words, plant species, varieties, and even entire crops that hold little commercial appeal (at least as determined by "market forces") are jettisoned for commercially exploitable varieties. Ironically, plant genetic material offers security to both the local and the industrial farmer.[5] Plant breeding activities are dependent upon landraces. This is not to say that industrial or mechanized agriculture has not significantly boosted supply of food and agricultural produce; however, the unnecessary tragedy is that patents on plants may constitute a perverse incentive and thus undermine the basis of sustainable agriculture. Indeed, agronomists are agreed that the world today produces more food per inhabitant than ever before. This remarkable revolution may be partly attributed to the mechanization of agriculture and the large-scale addition of agricultural inputs.[6]

One consequence of the erosion of plant genetic diversity is that the capacity of the economically preferred plants to resist pests and diseases is compromised. The marketability of plant produce is not necessarily coterminous with the inherent superior quality of the plants to be marketed or selected for mono-cropping. Given the potential utility of plants that market forces may erroneously dismiss as economically useless, the short-sighted depletion of the plant genetic pool can be both costly and dramatic.[7]

In addition to the danger in the erosion of the plant genetic base, patents on plants seem to create a regime that enables the emergent biotechnology[8] industry to integrate the corporate food chain with agribusiness and the products of chemical multinational corporations. This may be inferred from the increasing trend towards breeding or genetically engineering crops to suit the needs of the corporations that manufacture pesticides and agricultural additives. In other words, the same corporations that are acknowledged as the "seed giants" also own subsidiary chemical, fertilizer, and pesticide corporations.[9] For example, multinational corporations such as Hoescht, ICI, Sandoz, and Shell are all involved in the genetic modification of life forms.[10]

The vertical integration of these interests may explain why contemporary plant crops require large amounts of pesticides, herbicides, and fertilizers. It is therefore not coincidental that research and development on the part of biotechnology conglomerates is now aimed at "fostering, rather than reducing, dependence on company products. In addition to engineering plants to resist natural hazards, corporations engineer plants to resist certain chemicals as well."[11] For the above-stated reasons, critics have largely construed patents on plants as reflective of the struggle for economic and political control of the global food system by multinationals.[12] The role of the patent system as a catalyst in the commercialization of agriculture facilitates the domination of the global food and agricultural market by a handful of seed corporations. For example, at the end of 1995 the Hoechst group held 86,000 patents and patent applications. In 1997 Novartis held more than 40,000 patents worldwide. At the end of 1995 approximately 3.84 million patents were in force worldwide. Indeed, "the world's food

supply is primarily controlled by three dominant food chains – Cargill/
Pharmacia, ConAgra, and Novartis/ADM – which all hold shares of the 'gene
to dinner table.'"[13]

Monsanto, DuPont, Aventis, Novartis, and AstraZeneca control nearly all
of the world trade in genetically modified crops. The top ten global seed
companies control an estimated one-quarter to one-third of the $30 billion
annual commercial seed trade, and the world's top ten agrochemical corporations account for 91 percent of the $31 billion agrochemical market world-
wide. In addition, the top five vegetable seed companies control 75 percent
of the global vegetable seed market, and four companies control 69 percent
of the North American maize seed market. By the end of 1998 a single company controlled 71 percent of the US cotton seed market.[14]

Another aspect of this emerging trend is the granting of the so-called
"sweeping patents," or broad-spectrum patents, to corporations. For example,
Agracetus, now owned by Monsanto, was issued US patent Nos. 5,004,863
and 5,159,135 for its claimed innovative particle bombardment (biolistic)
method of transforming the soya bean. The patent right in question was a
very broad-spectrum patent awarded on all existing genetically modified
soya bean varieties and for all other genetically modified crops where the
same particle bombardment method was used. Though opposed by other
corporations, such as DeKalb and Syngenta, the biolistic method is well established in laboratories across the world. The European patent office vide
European patent No. 301,749 issued a similar patent.[15] Similar patents on
cotton seed granted by the US Patent Office to Monsanto (US Patent Nos.
4,940,835; 5,352,605; 5,530,196) have been successfully opposed by various
NGOs and activist scholars.[16] The consequences of this emerging trend towards global food security and global access to food are of increasing concern, especially given that the contemporary state of global access is already
bleak. The Brandt Report, written in the 1980s, states that one-fifth or more
of all the people in the Southern half of the world suffer from hunger and
malnutrition. The World Food Council's 1977 report stated that 500 million
people in the South suffer extreme deprivations of food, and current estimates show that over 800 million people are chronically undernourished.[17]

Another aspect of the implications of patents on plants and the biopiracy
of indigenous biocultural knowledge is the set of legal barriers on access to
plant genetic resources. These barriers are erected by states who wish to
protect their plant genetic resources from appropriation. Unless properly
instituted, however, excessive property rights on plant resources will affect
the way research is conducted and germplasm is exchanged.[18] This pattern
of domestic activity will inevitably affect how food is grown, processed,
and sold across the globe. As the seed giants use their patents to acquire
monopoly rights over plant genetic resources, there seems to be a juridical
backlash from the South. The direct implication is that not only is the era of

free access to plant genetic diversity over but also the debates on the legal ownership of plant germplasm stored in the IARCs gene banks are and will continue to be acrimonious.

In addition to the backlash from the South on restriction of access to plant germplasm, it seems that lax patenting and commercialization of "terminator technology"[19] poses a significant threat to the security and sustenance of plant genetic diversity. This technology prohibits farmers from growing second-generation crops from the same seed; every farming season they must go to the seed merchants for new seeds. The seed business has long complained that, although plant genes from the South have been tremendous in improving crop yields and resisting plant diseases, the reproducibility of seeds meant that farmers would not return yearly for fresh seeds. This would of course damage the profits of the seed giants. Terminator technology offers a new way of retaining economic control over plant genetic resources. This technology, otherwise known as genetic use restriction technology, refers to the use of an external chemical inducer to turn on or off a plant's genetic traits (in this case, to induce seed sterility in plants). The prototype is US Patent No. 5,723,765 granted to the USDA and the Delta Pine Land Company. It is a technology that, after one season, blocks genetically altered seeds from germinating. As I have argued elsewhere, the implications of this technology for farmers across the world are enormous.[20]

Considering that at least 1.4 billion people rely on farm-saved seed for their annual crop and farming activities, the implications of the terminator technology are devastating and irreversible. For example, unsuspecting farmers whose farms are near farms planted with terminator technology plants may have their crops ruined by escaping genes from the patented seeds. In other words, the impact may not be limited to farmers who purchase artificially sterilized seeds. As *Schmeiser*[21] (a case tried in Saskatoon, Saskatchewan) shows, the probability that genetically engineered seed may escape into undesired parts of the environment cannot be foreclosed.[22]

Patents and Biodiversity: Issues of Health and Environmental Integrity

Patents on plants may encourage the premature commercialization of genetically modified food crops without sufficient empirical assurances regarding their safety, juridical implications, and environmental hazards. In the debate on the social utility of patents on genetically modified plant life forms, it bears noting that genetic engineering is not a natural extension of traditional crossbreeding methods. Under traditional crossbreeding (or hybridization) nature allows only the mixing of genes from the same or closely related species.[23] In contrast, genetic engineering allows scientists to completely defy the natural reproductive boundaries established over billions of years of evolution and human intervention. Thus, "through invasive virus or instruments like gene guns, DNA from completely unrelated organisms like fish and strawberries, bacteria and soybeans, or humans and pigs, for example, can be intermingled."[24]

Biotechnology involves a template shift with huge potential for good, but it also has huge potential for the opposite. Proponents of genetic engineering of plants have made impressive arguments that, notwithstanding the troubling concerns about the safety of their products for humans and the environment, the benefits of these products are profound and outweigh any risks. Benefits of genetic modification include longer shelf life for many agricultural products, adding colour to natural fibres before harvesting, conferring resistance to pests and fungi, and facilitating the use of herbicides on harmful weeds.[25] Another notable argument for genetically engineered plants[26] is that "absent biotechnology developments, one might doubt that traditional plant-breeding techniques could increase the world's food supply enough to feed the estimated global population of 9.4 billion in the year 2050."[27] Impressive as these seem, there is still need for caution. The undoubted promise of biotechnology must not becloud its danger.

On the question of food supply, it has been conclusively shown by Nobel Laureate Amartya Sen[28] and others that, even without the promises of biotechnology, the world today produces more food per inhabitant than ever before. Food experts agree that there is enough food to provide 4.3 pounds for every person every day: 2.5 pounds of grain, beans, and nuts; about a pound of meat, milk, and eggs; and a pound of fruits and vegetables.[29] Accordingly, the problem of global hunger is one of poverty rather than lack of food.[30] Indeed, the excess food produced in the word is often destroyed to buoy up the price of commodities at the global food market. Therefore, the argument that genetic modification of food is necessary in order to feed the starving billions may not be correct.

In addition, the risks, profits, and dangers of biotechnology are not evenly spread or shared. While the seed and industrial corporate giants reap the economic benefits, a large proportion of the environmental risks and dangers are borne by the poor. According to Graziano:

> Agriculture world wide is at risk [and] Third World countries are the most threatened, because they have more diverse varieties of crops, they have to feed more people, and they house much of the world's biodiversity. They are also more susceptible to change, because most biogenetically engineered agriculture is produced in Third World countries. Biotech companies commonly disregard the vulnerability of Third World ecosystems and expose them to possible catastrophic environmental damage.[31]

Although the agribiotechnology industry has been quite consistent in arguing that fears of the potential negative effects of genetic modification of plant life forms are exaggerated, if not unfounded, concerns remain as to the human and environmental safety of genetically modified plants. The possible negative impacts of genetically modified plants stretch from the soil to the human food chain and to the human immune system. For example, even though a genetically modified food is chemically similar to its natural counterpart, this may not be adequate evidence that it is safe for human consumption. Other factors that should be considered but that have been ignored include the biological, toxicological, and immunological consequences of genetically engineered plant life forms.

Interestingly, tests conducted by British scientist Arzad Pusztai showed that, when rats were fed with genetically modified potatoes, there was damage to their immune system and internal organs.[32] In addition, there are reported cases of transgenic super-weeds and pesticide-resistant plants arising from genetically engineered plants.[33] In 1994 a genetically engineered bacterium developed to aid in the production of ethanol produced residues that rendered the land infertile. New crops "planted on this soil grew three inches and fell over dead."[34] Similarly, in 1996 it was reported that scientists discovered that ladybugs that had eaten aphids that had eaten genetically engineered potatoes died. Some of these reports may be alarmist, but the importance of the interaction of plants with the wider environment cannot be denied.[35]

As Richard Strohlman, emeritus professor at the Department of Molecular and Cell Biology, University of California at Berkeley, has emphasized:

> Genes exist in networks, interactive networks which have a logic of their own. The technology point of view does not deal with these networks. It simply addresses genes in isolation. But genes do not exist in isolation. And the fact that the industry folks don't deal with these networks is what makes their science incomplete and dangerous.[36]

An independent and reputable science review panel commissioned by the British government issued a report in 2003. It found no scientific case for ruling out all genetically modified crops and their products; nor, however, did it give them blanket approval. It argues that, worldwide, after seven years of consumption of genetically modified products by humans and livestock, there are no verifiable ill effects. The report emphasizes that genetic modification is not a single homogenous technology and that its application needs to be considered on a case-by-case basis, bearing in mind the emerging principle of precaution in public international law.[37]

Biopiracy and the Precautionary Principle in International Law
It stands to reason that, in designing policies regulating the granting of patents on genetically engineered plants, legislators should consider questions

relating to the environmental and human safety of those products.[38] Patents are instruments of national policy and must reflect an informed balance of the general interests of society.[39] Therefore, in the issuance of patents or in designing policies for the patent systems of the world, it would be prudent for national patent systems and the emerging global order on patents to take into consideration the environmental implications of patented genetically modified plants. However, there is considerable doubt whether states would embrace this suggestion. Moreover, as James Buchanan has convincingly argued, the history of biotechnology in the global law-making process is a tale of the elimination of real caution in treaty making.[40] For example, since 1992, when biotechnology was slated to be the subject of one of the conventions to be concluded under UN auspices, pressure from the United States, Japan, and Germany ensured that it was downgraded from the status of a convention to that of "an issue" connected with biodiversity.

Hence, although the question of regulation of genetic modification of plants is remarkably complex and controversial,[41] a major stumbling block to a rational appreciation of its ramifications is the huge amount of capital outlay and investment involved in the biotechnology and agribusiness industry.[42] However, international law seems to have grasped the peculiar risks of genetically modified plant life forms – hence, the Cartagena Protocol and other juridical initiatives. The Cartagena Protocol is a follow-up to Article 19.3 of the CBD, which obliges parties to the convention to consider the need for, and modalities of, a protocol on the safe transfer, handling, and use of living modified organisms (LMOs) that may have an adverse effect on biodiversity. The Cartagena Protocol constitutes a juridical recognition of both the potentials and perils of genetic modification of plants. More significantly, it entrenches the principle of precaution in relation to genetically modified plants.

The Protocol adopts a three-pronged principled approach; namely, advanced prior informed consent,[43] the precautionary approach,[44] and the establishment or maintenance of national measures.[45] The essence of this instrument is to institute a regime for the safe handling and use of genetically modified plants and other life forms. Given its wide ambit, its normative import may extend to other areas where the law on biological diversity converges with emergent human rights, such as the "right to safe food" and a safe environment.[46] Our expectations of the normative implications of the Cartagena Protocol should be sober and modest, especially since it contains a "savings" clause emphasizing that it does not preempt rights and obligations provided for in other international agreements and organizations, particularly those dealing with international trade. One such international agreement is the Agreement on the Application of Sanitary and Phytosanitary Measures (SPS).[47]

For example, preambular recitals 9 to 11 emphasize that the obligations created by other international agreements have not been diminished or compromised by the commitments made to the Cartagena Protocol. In other words, given that the normative link between the Cartagena Protocol and other international agreements on trade and the environment has not been clarified, one cannot be sanguine about the prospects of it creating a progressive and stringent regime of liability for potentially dangerous genetically modified plant life forms and their handling, particularly those with patent protection.[48] The protocol encompasses the so-called LMOs-FFPs (modified organisms that are intended for use as food or feed or for processing). This class covers such widely modified plants as corn, soy, wheat, canola, tomatoes, seeds, and so on, which may or may not be introduced into the environment. In effect, genetically modified plants that may have an impact on human health as well as on the environment as a whole are included under the juridical ambit of the Cartagena Protocol.

The protocol marks a significant milestone in the evolutionary process of the principle of precaution in international environmental law. It can therefore be argued that the norm of precaution[49] in international environmental law has, for good reasons, become one of the most powerful legal norms in recent times for the regulation not only of genetically modified plants but also of other human activities that pose potential but unproven dangers to human health and to the environment. In simple terms, the norm of precaution prescribes that, in some cases where the costs of action are low and the risks of inaction are high, preventive action should be taken, even without scientific certainty about the problem being addressed.[50]

The principle of precaution is an evolving norm with manifold applications and, in practice, it gives governments and environmental regulatory agencies a useful amount of discretion in setting environmental policy, even if there is no definitive scientific basis for the precautionary measures taken.[51] Thus, governments and environmental regulatory agencies must decide, in the face of empirical uncertainty and paucity of scientific data, how high the risks are likely to be (beyond monetary terms) and what types of action, if any, should be called for. Of course, whether states can resist counter-pressures from vested interests in genetically modified plants is another question.

Still, in a commonsensical way, the precautionary principle ensures that an activity that poses a threat to the environment may be prevented. It emphasizes avoidance. Notwithstanding problems of definition[52] and conceptual ambiguities, it may be argued that the precautionary principle has emerged as a full-blown principle of international environmental law.[53] Operationally, key elements of the principle of precaution may include an evidentiary threshold, a burden of proof, a duty owed to the international society as a whole, and a policy for action in the face of scientific uncertainty or ignorance. Historically, the precautionary principle is an improvement upon the erstwhile "assimilation approach"[54] to regulations on the environment. The assimilationist policy operated under the assumption that the environment can assimilate a certain amount of pollution without a collapse of the ecosystem. However, the complexity of the ecosystem makes it impossible to predict with certainty the maximum capacity of the environment to absorb pollution.

The problem with the precautionary approach is that, although it would be ideal to have every ecosystem conserved, the realities of life, the immense global pressure that champions a particular model of economic development, and the imperative needs of peoples (whether political, economic, or cultural) compel otherwise. Inevitably, hard choices have to be made, and the "safety first" slogan does not really help in such situations.[55] Even without the complexities of genetically engineered plants, no one really knows how the various species in their natural state interact in the ecosystem, and thus the precautionary principle may not shed much light on the complicated workings of nature. However, the case for precaution cannot be denied.

The Human Rights Dimension of Biopiracy

Generally speaking, the unwillingness of the contemporary international patent law framework to address the needs of and injustices suffered by marginalized peoples and cultures has led to a shift towards the adoption of the human rights paradigm as a normative platform for addressing its perceived inadequacies.[56] Although intellectual property rights have thus been recognized as having impacts on human rights,[57] there is near-conclusive juridical and scholarly consensus that patents are not founded solely on a theory of human rights. However, it would be equally erroneous to argue that erosion and appropriation of plant life forms and TKUP do not raise serious issues with repercussions in the human rights paradigm.

Article 15 of the Covenant on Economic, Social and Cultural Rights institutes a juridical threshold for some of the human rights implications of the erosion and appropriation of plants and TKUP.[58] It provides thus:

> The State Parties to the present Covenant recognize the right of everyone:
> (a) To take part in cultural life;
> (b) To enjoy the benefits of scientific progress and its applications;
> (c) To benefit from the protection of the moral and material interests resulting from any scientific, literary or artistic production of which he is the author.[59]

In cases where traditional use of plants pertains to the culture of a people, it seems beyond doubt that biopiracy constitutes both an individual and collective violation of an internationally recognized and protected right to culture. Even though economic, social, and cultural rights have traditionally been marginalized in the human rights discourse and praxis, there is no doubt among scholars that they are human rights in the full sense of the term, with all the legal obligations attendant thereto.[60] In effect, the inability or unwillingness of states to address and redress the human rights implications of biopiracy and the erosion of plants is a political weakness rather than a function of a purported absence of juridical basis for the protection of such economic, social, and cultural rights. Indeed, international human rights law provides such techniques as the duty to report on the progress made in the domestic terrain for the protection of human rights.[61] While this technique and related methods may have some normative virtues, if states really want to redress the human rights implications of the appropriation and erosion of plants and TKUP, then they must reconfigure and rethink at the domestic level the social justifications for patents on plants and TKUP.

Perhaps the reporting technique of the CESCR (Committee on Economic, Social, and Cultural Rights) should be used to encourage the creation of domestic legal regimes that protect and reward indigenous and traditional contributions to biodiversity innovation in a manner consistent with their respective cultural values and aspirations.[62] This approach will of course raise questions about empowering marginalized cultures and peoples in multicultural and multiethnic states. These question may not be adequately addressed without reference to the claims for the cultural and political self-determination of peoples.[63] The point is that states remain central in both the human rights arena and in the implementation of international patent law norms, and when both regimes converge it is the responsibility of states to chart a socially balanced course. Although some scholars argue that the scope of domestic manoeuvrability has been foreclosed by the TRIPs agreement, states still have a remarkable discretion in redressing the human rights violations constituted by biopiratical activities and systems.

However, while it is apparently easy to identify global forces responsible for the appropriation and erosion of plants and TKUP, it would be erroneous to proceed as if domestic political and economic factors were blameless in the continuing appropriation of these phenomena.[64] Many local communities in gene-rich states operate under conditions of medieval serfdom, with local potentates literally living off the penury of traditional and indigenous peoples. A major consequence of this domestic oppression is that the protection of plants and TKUP suffers. As domestic elites and privileged classes or races align their interests with commercial appropriators of plants and TKUP, there is little hope that domestic legislative initiatives designed to promote the interests of poor peasant farmers will come to fruition.[65] In order to begin ameliorating this problem, concerned states must rethink their land ownership policies and laws. For example, the Brazilian government, in conjunction with the Brazilian Landless Workers Movement (a civil society), has in twelve years redistributed over twenty-one million hectares of previously unused private land to more than 300,000 landless peasants. This measure removes some of the immense pressure on overpopulated spaces.

Biopiracy and the Crisis of Development in the Third World

The pertinent question here is the extent to which predatory patents on plants and appropriation of TKUP affect the normative function of the concept of the right to development.[66] It can hardly be denied that the economic and cultural losses of indigenous peoples arising from appropriation of plants and TKUP constitutes a severe violation of the right to development, especially of marginalized peoples. As noted in *Guinea/Guinea Bissau Maritime Delimitation Case*, the "legitimate claims of the parties as developing countries and the right of the peoples involved to a level of economic and social development which fully preserves their dignity"[67] is a right protected by international law.

The concept of a right to development surfaced at the international level in Strasbourg in 1972, when Senegalese jurist Keba Mbaye, in his capacity as the president of the International Commission of Jurists, delivered a lecture entitled "The Right to Development as a Human Right."[68] Since then there has been a fierce and largely polemical debate among scholars concerning whether there is a right to development, at least in the normative sense.[69] In a positivist sense the crux of the juridical ferment is that, since rights correlate to juridical obligations, to whom is the right to development owed and who is obliged to ensure development?

It seems that this debate, framed in the Hohfeldian paradigm, is arid, excessively positivist, and, in any case, ignores the normative dimension of the conception of development as a human right. As Zalaquette has argued:

> Development and human rights differ in many respects, but both relate to those questions of survival and justice which have shown the most potential for conceptual expansion. It is no surprise that attempts to endow development proposals with normative strength tend to converge with efforts to enlarge human rights.[70]

It is probably in recognition of the normative utility of the concept of a right to development that it has been eloquently postulated that the essence of normative legality is not so much "the form of the norm-bearing instrument as ... whether ... it is a reflection of mutual expectations that establish a self-sustaining reciprocal equilibrium in the behaviour of states."[71] The concept of a right to development thus would encompass economic growth, self-reliance, satisfaction of all basic needs, and the fulfillment of all human rights in the pursuit of social justice within and among states.

Although it may be argued that there is no precise formulation of the right to development, it seems that the approximation of the above-mentioned sentiments has been encapsulated by the Commission on Human Rights and the UN General Assembly, which articulate the concept of development as the "equality of opportunity for development."[72] In other words, the notion of a right to development encompasses the material and non-material needs of all human beings, respect for human rights, the opportunity for full participation, recognition of the principles of equality, and a degree of individual and collective self-reliance.[73]

Largely propelled by the South and some sympathetic countries of the North, the concept of a right to development has achieved a measure of internal energy and definition.[74] As Okafor argues, "the right to development is in essence (but not exclusively) a claim to the means of development: financial capital, human capital and technology."[75] Waart adds that "the right to development ... embodies the entitlement of individuals and peoples to an international order which provides for a just and adequate realization of the universally recognized human rights."[76] Whether or not the right to development is a human right in the narrow legalistic sense is of little moment.[77] What is important is that it is at least a potentially useful concept,[78] especially in the attempt to institutionalize a normative global regime dedicated to international co-responsibility in addressing the fundamental need for all human beings to have a decent existence.[79]

The right to development is a human right because humanity cannot exist without development. Karel Vasek's separation of political and civil rights from economic and social rights is virtually dead in learned circles:[80] all human rights are interrelated and indivisible.[81] This means that, given the comprehensive economic, social, cultural, and political aspects of the concept of human development,[82] the right to development is an expansive and organic concept affirming the interdependence of rights.

Long before the 1986 UN Declaration on the Right to Development, the African Charter on Human and Peoples' Rights had recognized vide its Article 22 that

> all peoples shall have the right to their economic, social, cultural development with due regard to their freedom and identity and in the equal enjoyment of the common heritage of mankind ... States shall have the duty, individually or collectively, to ensure the exercise of the right to development.[83]

This norm also reiterates the individual and collective nature of the right to development and affirms its juridical character.[84] It is probable that, because the right to development has been expounded in the context bridging the economic gap between "poor" and "rich" states,[85] international law, in its bias towards powerful states,[86] has been insensitive to the cause of development.[87] Hence, the North has persistently opposed the existence of a right to development. It is a rule of customary international law that obligations do not bind states that have persistently objected to the validity of the obligation in question.[88] Although, in principle, the North agrees to support the quest for development in the South, this is subject to access to the latter's coveted raw materials. This means free access to plant life forms and TKUP as well as the continued existence of the institutions and juridical mechanisms that facilitate biopiracy.[89] As Professor Ignaz Seidl-Hohenveldern has noted, "there appears to be some merit in complaints that economic colonialism by the former metropolitan power has not yet come to an end, but has been replaced by neo-colonialism."[90]

Biopiracy: Some Thoughts for Action
This book has examined the emerging relationship between patents, plants, and TKUP. The concept of traditional knowledge of the uses of plants as it affects the patentability of such knowledge is a complex and controversial issue. More important, there is a need to rethink global attitudes and regimes on the patentability of plants and TKUP. The mechanisms for rewarding and protecting innovations and inventions must have a transparent and consistent set of guiding principles. Since the patent system represents the most ubiquitous and established institutional and juridical mechanism for protecting and rewarding innovations and inventions, it follows that it must also eschew cultural prejudices and biases. Here, the instrumentalist nature of the patent system comes into play. Questions remain as to what purpose it should be made to serve.

It lies in the hands of both states and international organizations, particularly regional organizations, to articulate what they want to achieve through the patent system and other juridical concepts. Of course, this cannot be done without rethinking the shrinking ambit of domestic competence in matters of intellectual property legislation.[91] At the moment, the instrumentalist nature of the patent system largely panders to global entities primarily interested in creating larger economic profits for a few powerful industrial interests/states.[92]

There are clearly losers in the age of globalization.[93] Whether there should be any losers is a question of morality, ethics, social justice, and human survival.[94] Perhaps the problem is not that all human cultures are "globalizing" but, rather, that one culture[95] – backed with tremendous economic, political, and juridical might – is globalizing.[96] And, as a consequence, the cultural and juridical mosaic of the globe is shrinking to a monochrome.

Given that, fundamentally, the patent system is a socioeconomic instrument, the question is how it can be put to the best service of the larger society, particularly with respect to plants and TKUP. This would require a clear and rigorous overhaul of the patent systems of the world. Contrary to the WIPO Report, which indicates that the cases of appropriation of plants are mere instances of "bad patents" and not an indictment of the patent system as a whole,[97] the reality is that the problem of erosion and appropriation of plants and TKUP is systemic both juridically and institutionally. Accordingly, the patent system needs critical re-evaluation, reordering, and redirection.

First, the appropriateness of patents on plants and TKUP should be reconsidered. It would be good to create a special incentive regime for innovations in plants and products of TKUP. In devising this mechanism, particular attention should be paid to the peculiarities of traditional societies and cultures. It is the innovations of local farmers and traditional communities that are most important with regard to utilizing and conserving plant life forms and TKUP. Although these people have been able to achieve this without the aid of the patent system, it would be unrealistic for the world to continue to pretend that they do not need to be legally protected from the predatory practices of appropriators. Presently, the application of the patent concept to plants and TKUP serves interests that are at odds with the real and substantial innovators of plant life forms and TKUP.

Second, it is of the utmost importance that the basic concepts of the patent system be critically re-evaluated and redefined. The criteria for patentability must be clarified and provided with a consistency that is presently lacking. One of the major consequences of this regime of permissiveness and inconsistency is that not only is the patent system tarnished but also its integrity is open to serious doubts. It is absurd that the patent system, which asserts that it was established to reward inventors for their socially beneficial inventions, has no clear definition of invention. Even national patent laws that purport to define what constitutes invention provide little more than clues to enable the courts to figure out what might pass as an invention. Regrettably, "neither the TRIPs agreement, nor the Paris Convention gives any definitions of what an invention should be. The EPC does not define the term 'invention' while US patent law only gives a definition of what may be patented."[98]

Similarly, the lowered criterion for patentability of inventions constricts the global intellectual space and extorts fees and rent from the general public, particularly in the fields of pharmacy and biotechnology. With a deluge of dubious patents, the distinction, if any, between inventions and discoveries has become blurred. This has severe erosive and appropriative repercussions on the regime of plant genetic diversity. According to Leskien and Flitner:

> In the field of biotechnology, defining the precise line between unpatentable discoveries and patentable inventions may lead to specific problems that have yet to be resolved. This is because most products of biotechnology are or are based on genes or cells that have been taken from nature or isolated from pre-existing living micro-organisms, plants, animals or humans ... It is in fact a very thin line that separates invention from discovery under both US law and EPC.[99]

Indeed, no line of demarcation, thin or otherwise, exists between inventions and discoveries in relation to plant patents.[100] Neither Article 27 of the TRIPs agreement[101] nor any other juridical instrument contains any specific boundary between discoveries and inventions in relation to plant forms. This regime is unacceptable.

Further, the criteria of reproducibility, utility, specification, and non-obviousness have been significantly watered down for the purposes of the pharmaceutical and biotechnology industries. In the absence of clear legislative provisions, the double standard involved does no credit to the patent system. The patent system needs to be restricted and its ambiguities need to be cleared up. At the very least, "inventions" and "innovations" in plants and TKUP that do not meet the criteria for patentability for mechanical inventions should not be patented. This would require a global and absolute standard of novelty, a strict enforcement of the classical requirement of specification, and a substantial enhancement of the tests of industrial applicability and utility. Unless this happens, the patent system, especially in its application to plants and TKUP, will remain an engine of mischief and deception and society will continue to pay rent for undeserving "inventions."[102] A regime that privileges a set of weak standards for pharmaceutical and biotechnological products while leaning hard on mechanical inventions is an engine of mischief and a disservice to the public.

The least the patent system can do is to ensure that inventions that it seeks to reward are inventions in the full sense of the word. In the absence of a global standard of absolute novelty and dichotomy between *de jure* and *de facto* novelty, the emerging global patent regime lacks credibility and integrity. If a patent grant should be upheld beyond the borders of the state where it was issued, the beneficiary of that privilege must show that the invention in question is truly novel anywhere in the world. Otherwise, such inventions should enjoy only domestic respect. In an age of rapid dissemination of information, there is no compelling reason why geographical limits should loom large in the determination of novelty.

In addition, we should be serious about the test of utility of inventions. Those who intend to reap the benefits of patents on plants and TKUP must at least show the usefulness of their inventions and "improvements." Thinking particularly about PBRs, it is counter-productive to grant awesome proprietary powers of control of plants to entities or persons whose innovations serve no useful societal ends. Given that, practically speaking, PBRs have become another name for patents, the element of utility should form part of the requirements for granting PBRs.

The same rigorous approach should be applied to the requirements of specification and industrial applicability. Where applicants for plant patents are unable to fully disclose the nature of their inventions as required by the relevant patent legislation, the courts should, in strict compliance with patent legislation, reject such applications. The corollary is that the spate of judicial activism in the field of patents on plants and other life forms should be revisited or stopped. As *Diamond* v. *Chakrabarty* has shown, the courts, particularly in the United States, have often seized the initiative from the legislature in "creating" new laws on patents. The enormity of the public consequences of private control of life forms and the complexity of the issues make an irrefutable case for legislative preeminence in this regard. Courts should be careful in pre-empting the legislature. Legislative inactivity is no excuse for judicial usurpation of the task of legislation, especially in areas of high social and economic sensitivity such as ownership of plants life forms and TKUP. Where there are no special laws dealing with such highly sensitive issues, courts of law should not write new legislation while pretending to interpret "existing" law.

However, in appreciation of the political and economic realities that often characterize the struggle for control of plant genetic resources across the globe, it would be good for gene-rich but politically/economically weak states to pursue the suggestions made above within the ambits of regional frameworks. It would be practically impossible for weak states to create a new global legal mechanism for the protection of plants and TKUP. They lack both the political clout and economic muscle to do this. Their best option, therefore, is to regionally modify the existing structure in order to reflect their particular concerns, priorities, and values. The importance of regional patent systems, which reflect the peculiar attributes and concerns of a group of states with similar economic and cultural backgrounds, cannot be over-emphasized. One of the most effective and viable patent regimes is that of the European Union. Given the member states' similar historical and cultural backgrounds, the relative coherence and efficiency of the EU patent framework suggests that international patent regimes work better when the parties share values and priorities. Applying this approach to countries and patent systems of Africa, Asia, and South America would be a first step in the process of combating plant genetic erosion and biopiracy.

Therefore, groups of states may consider adopting, at a regional level, defensive measures designed to curb the phenomenon of erosion and appropriation of plants and TKUP. In this regard, I would suggest that gene-rich states should explore the option of restricting access to plant genetic material by those states with notoriously prejudicial and appropriating patent systems. This approach may work best when gene-rich states act like cartels under a continental or regional framework. For example, when the United States refused to sign the CBD in 1992, "Venezuela stopped signing new agreements for scientific collaboration with United States companies that wished to study genetic resources."[103] This singular act helped to galvanize support for the CBD within the US biotechnology industry and led to the consequent ratification of the CBD by the US government.[104]

To realize these proposals, I suggest that states wean themselves of the notion that any effective international or multilateral treaty must have the ratification of all the powerful and economically strong states. Gene-rich states need not wait upon the powerful states before they assert themselves in this regard. Their sheer number already offers leverage and the potential for the creation of customary international law on the question of appropriation of plants and TKUP. Strategically, any such treaty or convention should be effective with twenty or thirty ratifications. For example, the UPOV Convention, which today constitutes the baseline for PBRs, went into effect with only six ratifications.[105] At present there are forty-four ratifications of the UPOV Convention, and most of the ratifiers are from the North.

Countries in the regions of Asia, Africa, and South America have more than enough states to bring such treaties into effect. Thereafter, any number of states may join if they so wish. This is not unusual. The South should learn to play the politics of norm creation by using its sheer numbers to generate codes of conduct and normative rules that promote the sustenance of plant genetic diversity and TKUP. A notable initiative in this regard is the *Andean Common Code on Intellectual Property*.[106]

This agreement incorporates some of the suggestions made in this book. For example, on the question of oral knowledge constituting part of prior art in the determination of novelty, Article 2 thereof provides that "prior art is all that has been accessible to the public by written or *oral description*."[107] It is arguable that this novel step would not have been possible at a global level because of the enormous impact it would have on many patent applications (which would fail the test of novelty if oral knowledge were made part of prior art). The problem, however, seems to be that most Southern states construe the problem of patents and plant genetic diversity through the lens of financial help from the powerful Northern states. Without discounting the relevance of money, it seems that the emphasis on fiscal support from the North is exaggerated.

No single state is wholly independent and self-sufficient when it comes to plant life forms. This compels international cooperation, good faith, and, more important, a reconciliation of the parallel and discordant legal regimes on plant life forms and TKUP instituted by the CBD, the FAO, and the WTO. It would be unhelpful to maintain a gladiatorial state of affairs between the regimes created by these international institutions. The problem seems to be that, in most cases, the international law-making process, particularly in this age of increasing corporate power, has tended to place trade and profit above conservation and sustainability.

This attitude is both wrong and short-sighted. Trade is essential, but it must be understood that it does not exist for itself and that it does not exist outside of the environment. Only the living can pursue trade, and they can do so only if there is something to buy, sell, or barter within a sustainable environment. The collapse of the cod fishery in Newfoundland, Canada, and anchovy in Chile are grim testaments to the inability of the environment to recover from rapacious exploitation. Of course, the environment is magnificent in its elasticity, but it is equally true that humankind knows little about the dynamics of environmental stability and regenerative capacity.

Access to and sustainable use of plant life forms need not assume the nationalistic chauvinism that has somewhat become the South's response to the acquisitive and predatory practices of multinational agribusiness. Co-operation between states on plant germplasm is unavoidable – indeed, it is imperative. Cooperation should not be confused with duplication of regimes. Presently, there is a duplication of efforts by the CBD and the FAO, particularly with regard to legal ownership of plant life forms stored in *ex situ* gene banks. It is expected that the newly adopted FAO International Treaty on Plant Genetic Resources will bridge the gap and harmonize both regimes.

Although previous FAO instruments are non-binding, the new treaty may have achieved normative clarity and consistency. In addition to the compensation mechanisms currently created by the new treaty with regard to germplasm stored in IARC gene banks, greater emphasis should be laid on free access to that germplasm within a strict regime of non-patentability. In addition, international researchers and commentators should de-ethnicize the language relating to TKUP as such language validates the social context within which biopiracy takes place. It must be realized that international scholars are influential when it comes to shaping ideas and validating or invalidating attitudes and perceptions. This is a responsibility that scholars and researchers would do well to exercise with even-handedness and integrity. It would be unseemly if scholars, as moulders and moderators of opinion (which often congeals into law or government policy), interposed their prejudices in the discourse leading towards norm creation.

Accordingly, with respect of the issue of patents and plants, use of terms such as "ethnobiological knowledge,"[108] which imply that non-Western biological knowledge has no scientific basis and is valid only in an "ethnic" setting, should be eschewed. Apart from the fact that these so called "culture-bound" innovations supply more than one-quarter of the world's medicines and over 90 percent of the world's food supply, there is no valid basis for the aspersions connoted by such terms as "ethnicity." Scholars should desist from propagating their cultural prejudices as though they were irrefutable truths. Traditional knowledge is an empirical form of knowledge,[109] and scholars who deal with the problematic issue of biopiracy would do well to use language that is respectful of other cultures.[110]

It is imperative to recognize the intellectual worth and global value of knowledge, regardless of its cultural context. As the Crucible Group observed:

As conventional science has adapted its language and approach to appreciate the experimental research of farming communities, it has also had to reconsider its understanding of what is "known" and "unknown," "wild" and "undomesticated." The discovery that a "jungle" in West Africa was, in fact, an intentionally-developed agro-forestry system spurred the reappraisal of long-held assumptions. Many indigenous and farming communities eschew the term "wild," for example, arguing that the term testifies to the limitations in the information available to conventional science. A so-called wild plant may be protected and nurtured if not actively bred. It is very often used and planted.[111]

It is impossible to conserve the world's plant diversity without respecting and conserving cultural diversity. The loss of traditional farm communities, languages, and indigenous cultures has direct impacts on plant diversity. It is not a coincidence that the highest levels of plant diversity occur at the sites of the world's richest linguistic life. Studies show that "ten out of 12 'mega-diversity' countries identified by the International Union for Conservation of Nature and Natural Resources rank among the top ten countries in endemic languages."[112] As languages are increasingly lost, the global community loses the knowledge associated with the uses of those plants nurtured by the affected communities. This is nothing less than a phenomenal loss of human intellectual capital. And this is an area where current domestic state policies on cultural diversity should be seriously reappraised.

Another vital point involves the question of domestic environmental self-determination. While it may be fashionable to highlight the inequities of the global legal and economic order, particularly in the context of the North-South divide, it is equally true that domestic factors have wreaked as much, if not more, havoc on marginalized peoples and cultures. The oppression and domination of indigenous peoples and other marginalized cultures seems far more acute in their domestic jurisdictions, in both North and South, than any conceivable hardships they may experience within the contexts of the notorious North-South divide.

Local elites and dominant cultures that subjugate local and traditional peoples need to rethink the domestic juridical and institutional factors that lead to the erosion and appropriation of plants and TKUP. This would have to take place under the juridical rubric of various international law instruments[113] and institutions that recognize the importance of minority rights and the entrenched human rights of traditional and indigenous peoples to own property individually and/or collectively, their right to maintain and enjoy their culture, and their right to equal protection under the law. The potential of the human rights initiative in addressing and redressing these issues seems promising. As a corollary, I suggest that pressure be brought to bear on states that are egregious abusers of minority rights and deniers of domestic environmental self-determination.

A few progressive changes in this regard may be mentioned. For example, in October 1997 the Philippine government passed the *Indigenous Peoples Rights Act*. This law "effectively bestows ownership of resources within ancestral domains ... to indigenous peoples."[114] The law, enacted in the face of "opposition from many influential groups whose interests would be diminished by returning ancestral rights to [our] indigenous communities,"[115] empowers the 20 percent of Filipinos who are indigenous by granting them the right to control their own lands.[116]

The law creates a mechanism whereby a Certificate of Ancestral Domain Title (CADT) is awarded by the newly created National Commission on Indigenous Peoples to indigenous cultural communities (ICCs). Once a CADT has been awarded, a mining company or other extractive industry must secure the ICC's consent before logging permits can be granted. Of course, this gives some leverage to the indigenous community in negotiating and dealing with agencies whose activities have the potential to damage the ecosystem or to appropriate plant life forms and TKUP. In the absence of serious respect for and protection of traditional and indigenous peoples and cultures, and in the absence of protection of knowledge at the domestic level, international posturing will remain a mere sideshow to the relentless erosion and appropriation of plants and TKUP. The inescapable fact is that the concept of land and legal norms and institutions pertaining to its ownership cannot be divorced from the rhetoric and legalese of biopiracy.[117]

Furthermore, it is by a process of environmental and local self-determination that the concept of national sovereignty over plant life forms within the boundaries of states will come to have real significance. National and regional initiatives on access to plant life forms and equitable sharing of the benefits from biological diversity will be meaningless unless the local and traditional farmers who are in daily contact with plants have an effective role in the formulation and execution of policies designed to institute this new legal regime.

It should be mentioned that states like Thailand, Ecuador, Peru, Bolivia, and Venezuela have adopted various laws that empower local and traditional communities in matters pertaining to access to and equitable sharing of the benefits of plant life forms. States should safeguard their environment and its resources as this is what constitutes both the base and the superstructure of trade. In this regard, the need for caution should not be equated with alarmism. It follows that technologies that promote the protection of the environment should be rewarded. There is a role for the patent system in environmental protection, particularly with regard to the need to incorporate environmental security as a criterion for the patentability of genetically modified plants. Given the importance of plants and the environment to human development, states should be very careful in promoting or rewarding technologies whose impact on them is questionable.

Finally, regarding the problem of development, it seems clear that the lineal conception of development and the emphasis on high rates of consumption of resources is flawed. The continued use of urbanization, industrialization, social mobility, occupational differentiation, free enterprise, and conspicuous consumption of goods and services as the parameters for measuring development is anathema to the values of cultural and plant genetic diversity. The argument that plant biodiversity is eroding quickly because there are no economic incentives to aid its conservation is, at the very least, exaggerated. The granting of IPRs for products of modern plant breeding is certainly not the only way of conserving biological diversity. In strictly "ecological terms, incentives to speed the spread of industrial-style agriculture are rather counterproductive and should be properly balanced with other legal or economic measures limiting their destructive effects."[118]

Biopiracy is a function of institutional and juridical structures operating within a complex milieu of notions of cultural superiority/inferiority. Therefore, if the rhetoric on protecting plant genetic diversity and global cultural diversity is to congeal into a course of action[119] that will enhance environmental and cultural sustenance, then the epistemological and cultural divide between the North and the South must be bridged. We must create avenues for a conversation between cultures, and this is not to occur solely at the marketplace.[120] This involves a multiple engagement necessitating a reconsideration of our social and ethical values. It is both an individual and a collective enterprise, the purpose of which is to maintain our well-being if not our survival.[121]

Notes

Chapter 1: Introduction

1 Catherine Hoppers, ed., *Indigenous Knowledge and the Integration of Knowledge Systems: Towards a Philosophy of Articulation* (Capetown, South Africa: New Africa Books, 2003).

2 D.M. Warren et al., eds., *The Cultural Dimensions of Development: Indigenous Knowledge Systems* (London: Intermediate Technology Publications, 1995).

3 John Frow, "Public Domain and Public Rights in Culture" (1998) 13 Intellectual Property Journal 39.

4 The concept of public domain is an ill-defined space: it is what remains after all other rights covered by intellectual property regimes have been unitized, privatized, and commercialized. See David Lange, "Recognizing the Public Domain" (1981) 44 Law and Contemporary Problems 4; Jessica Litman, "The Public Domain" (1990) 39 Emory Law Journal 968.

5 Peter Drahos, "Indigenous Knowledge and the Duties of Intellectual Property Owners" (1997) 11 Intellectual Property Journal 179 at 180. Notwithstanding contemporary principles of private international law concerning the question of extraterritorial adjudication and enforcement of patents, it seems possible, if not probable, that the globalization of the patent regime would provide for the extraterritorial judicial enforcement of patent suit injunctions. For a further examination of this issue, see John Adams, "Litigation beyond the Technological Frontier: Comparative Approaches to Multinational Patent Enforcement" (1996) 27 Law and Policy in International Business 277. However, there is a renewed interest in the World Intellectual Property Organization (WIPO) negotiating and concluding a treaty on settlement of disputes between states on matters related to intellectual property rights (IPRs), particularly patents. See *Draft Treaty on the Settlement of Disputes in the Field of Intellectual Property*, International Bureau of WIPO, WIPO Doc. SD/CE/V/2 (8 April 1993); Thomas Vanaskie, "The European Patent Conventions: State Sovereignty Surrendered to Establish a Supranational Patent" (1977) 1 Association of Student International Law Societies International Law Journal 73.

6 Robert Samuel Summers, *Instrumentalism and American Legal Theory* (Ithaca: Cornell University Press, 1982) at 20.

7 Drahos, *supra* note 5 at 190.

8 Peter Drahos, *A Philosophy of Intellectual Property* (Dartmouth: Aldershot, 1996) at xi. See also Alex Geisinger, "Sustainable Development and the Domination of Nature: Spreading the Seed of the Western Ideology of Nature" (1999) 27 Environmental Affairs 43.

9 L.C. Becker, *Property Rights: Philosophic Foundations* (London: Boston, 1977); L. Lindhal, *Property and Change* (Dordretch: Martinus Nijhoff Publishers, 1977); A.M. Honore, "Ownership" in A.G. Guest, ed., *Oxford Essays in Jurisprudence* (London: Oxford University Press, 1961).

10 Stephen Munzer, *A Theory of Property* (New York: Cambridge University Press, 1990).

11 *Ibid.*

12 Drahos, *supra* note 5 at 209. See also R. Eisenberg, "Patenting the Human Genome" (1990) 39 Emory Law Journal 721; T. Roberts, "Broad Claims for Biotechnological Inventions" (1994) 9 European Intellectual Property Review 373.
13 Konstantinos Adamantopoulos, ed., *An Anatomy of the World Trade Organization* (The Hague: Kluwer Law International, 1997) at 1.
14 Terry Nardin and David Mapel, eds., *Traditions of International Ethics* (Cambridge: Cambridge University Press, 1992) at 10.
15 M. Kenney, *Biotechnology: The University-Industrial Complex* (New Haven: Yale University Press, 1986).
16 Jules Coleman, *Markets, Morals and the Law* (Cambridge: Cambridge University Press, 1988) at 114.
17 R. Rajagopalan, "Common Heritage: The Ecovillage Approach" in R. Rajagopalan, ed., *Common Heritage and the 21st Century* (Malta: International Ocean Institute, 1998) at 159.
18 Kristin Shrader-Frechette, "Environmental Ethics and Global Imperatives" in Robert Repetto, ed., *The Global Possible: Resources, Development, and the New Century* (New Haven: Yale University Press, 1985) at 97.
19 Anthony Stenson and Tim Gray, *The Politics of Genetic Resource Control* (London: Macmillan Press, 1999); Klaus Bosselman, "Plants and Politics: The International Legal Regime Concerning Biotechnology and Biodiversity" (1995) 7 Colorado Journal of International Environmental Law and Policy 111; Inamul Haq, "The Problem of Global Economic Inequity: Legal Structures and Some Thoughts on the Next 40 Years" (1979) 9 Georgia Journal of International and Comparative Law 507.
20 Sevine Ercmann, "Linking Human Rights, Rights of Indigenous People and the Environment" (2000) 7 Buffalo Environmental Law Journal 15.
21 For a recent discussion on policy implications in patent law, see Benjamin Enerson, "Protecting Society from Patently Offensive Inventions: The Risk of Reviving the Moral Utility Doctrine" (2004) 89 Cornell Law Review 685.
22 Tom Greaves, ed., *Intellectual Property Rights for Indigenous Peoples: A Source Book* (Oklahoma: Society for Applied Anthropology, 1994); Stephen Brush and Doreen Stabinsky, eds., *Valuing Local Knowledge: Indigenous People and Intellectual Property Rights* (Washington/Covelo: Island Press, 1996); Graham Dutfield, "Implementing the Biodiversity Convention," <http://www.users.ox.ac.uk/~wgtrr/impcbd2.htm>; Graham Dutfield, "Report on the Fourth International Congress of Ethnobiology, Lucknow, India, November 1994," <http://users.ox.ac.uk/~wgtrr/congeth.htm>; Graham Dutfield, "Report on the Second Conference on Cooperation of European Support Groups in the UN Decade of Indigenous Peoples, Almen, Netherlands, 3-5 May 1996," <http://users.ox.ac.uk/~wgtrr/almen.htm>; Graham Dutfield, "The Costa Rica Biodiversity Law: A Brief Summary," <http://users.ox.ac.uk/~wgtrr/crley.htm>; The Crucible Group, *People, Plants, and Patents: The Impact of Intellectual Property on Trade, Plant Biodiversity, and Rural Society* (Ottawa: IDRC, 1994) and *Seeding Solutions: People, Plants, and Patents Revisited* (Ottawa: IDRC, 2000); "Biopiracy Update: A Global Pandemic" RAFI Communiqué, September/October 1995; "Controversy Still Steaming over 'Counterfeiting' Basmati: Indian Government Prepares to Challenge Basmati Patent in US," RAFI Geno-Type, 4 January 2000; Bob Dillen and Maura Leen, "Biopatenting and the Threat to Food Security: A Christian and Development Perspective," CIDSE Press Release, Brussels, 10 February 2000; "Mexican Bean Biopiracy: US-Mexico Legal Battle Erupts over Patented 'Enola' Bean, Plant Breeders' Wrongs Continue...," RAFI Geno-Types, 17 January 2000; Vandana Shiva, "Biopiracy: Need to Change Western IPR Systems," The Hindu, Wednesday, 28 July 1999, 3.
23 Pat Mooney, *The Seeds of the Earth: A Private or Public Resource?* (Ottawa: Inter Pares, 1979); Pat Mooney and Cary Fowler, *Shattering: Food, Politics and the Loss of Genetic Diversity* (Tucson: The University of Arizona Press, 1990).
24 Thomas Franck and Steven Hawkins, "Justice in the International System" (1989) 10 Michigan Journal of International Law 127.
25 Michael Goldman, ed., *Privatizing Nature: Political Struggles for the Global Commons* (London: Pluto Press, 1998).

Chapter 2: Patents, Indigenous and Traditional Knowledge, and Biopiracy

1 Jack Kloppenburg Jr., *First the Seed: The Political Economy of Plant Biotechnology, 1492-2000* (Cambridge: Cambridge University Press, 1988) at xii and 49. Plants may be defined as those members of the taxonomic kingdom distinguished by their multicellular status, the ability to produce their own food from inorganic matter by the process of photosynthesis, and the possession of rigid cell walls containing cellulose. See *Webster's Unabridged Dictionary* (New York: Random House, 1997) at 1481.

2 Graham Dutfield, "TRIPs-Related Aspects of Traditional Knowledge" (2001) 33 Case Western Reserve Journal of International Law 233; Daniel Gervais, "Spiritual But Not Intellectual? The Protection of Sacred Intangible Traditional Knowledge" (2003) 11 Cardozo Journal of International and Comparative Law 467.

3 Chidi Oguamanam, "Between Reality and Rhetoric: The Epistemic Schism in the Recognition of Traditional Medicine in International Law" (2003) 16 St. Thomas Law Review 59-108.

4 Chris Cunneen and Terry Libesman, *Indigenous People and the Law in Australia* (Sydney: Butterworths, 1995).

5 Michael Balick and Paul Alan Cox, *Plants, People and Culture: The Science of Ethnobotany* (New York: Freeman and Company, 1996); Rajiv Sinha, *Ethno-Botany: The Renaissance of Traditional Herbal Medicine* (Jaipur: Ina Shree Publishers, 1996); Erwin Ackerknecht, *Medicine and Ethnology* (Baltimore: Johns Hopkins Press, 1991).

6 Kerry Ten Kate and Sarah A. Laird, *The Commercial Use of Biodiversity: Access to Genetic Resources and Benefit Sharing* (London: Earthscan, 1999) at 29.

7 George Meyer, ed., *Folk Medicine and Herbal Healing* (Springfield, IL: Charles Thomas Publisher, 1981).

8 George Foster, *Traditional Societies and Technologies Changes* (Delhi: Allied Publishers, 1973); Martha Johnson, ed., *Lore-Capturing Traditional Environmental Knowledge* (Ottawa: Dene Cultural Institute and the International Development Research Centre, 1992); Peter Morley and Roy Wallis, eds., *Culture and Curing: Anthropological Perspectives on Traditional Medical Beliefs and Practices* (London: P. Owen, 1978).

9 Since the dismantling and delegitimization of the moral and legalistic justifications for colonization of peoples by European powers, international human rights discourse and practice have shifted focus to the consequences of postcolonialism. Here, the historical and contemporary injustices afflicted on indigenous peoples have assumed prominence. This trend was given greater impetus by the Martinez Cobo Report of the late 1980s. See Martinez Cobo, *Study of the Problem of Discrimination against Indigenous Populations*, UN Doc. E/CN.4/Sub.2/1986/7 Add. 4, UN Sales No. E.86.XIV.3.3; *The United Nations Draft Declaration on the Rights of Indigenous Peoples*, UN ESCOR, Commission on Human Rights, 11th Sess., Annex 1, UN Doc. E/CN.4/Sub.2 (1993); Heather Archer, "Effect of United Nations Draft Declaration on Indigenous Rights on Current Policies of Member States" (1999) 5 Journal of International Legal Studies 205; James Anaya, "Environmentalism, Human Rights and Indigenous Peoples: A Tale of Converging and Diverging Interests" (2000) 7 Buffalo Environmental Law Journal 1; Lakshman Guruswamy, Jason Roberts, and Catina Drywater, "Protecting the Cultural and Natural Heritage" (2000) 7 Buffalo Environmental Law Journal 47; Rainer Grote, "The Status and Rights of Indigenous Peoples in Latin America" (1999) 59 ZaorRv-Heidelberg Journal of International Law 497.

10 Cobo Report, *ibid.*

11 International Labour Organization Convention 169 Concerning Indigenous and Tribal Peoples in Independent Countries, 7 June 1989, reprinted in (1989) 28 International Legal Materials 1382; Dudmundur Alfredsson, "The Rights of Indigenous Peoples with a Focus on the National Performance of the Nordic Countries" (1999) 59 Heidelberg Journal of International Law 529. There are other definitions of indigenous people in international law. See for example, "Indigenous and Tribal Peoples," <http://ecocouncil.ac.cr/indig>. See also, Cobo Report, *supra* note 9 at 29.

12 The terrible conditions of First Nations peoples are a matter of common knowledge. See, for example, Colin Samson et al., eds., *Canada's Tibet: The Killing of the Innu* (London:

Survival, 1999) at 4. The Innu in Canada are thirteen times more likely to commit suicide than is the average Caucasian Canadian. See *Choosing Life: Special Report on Suicide among Aboriginal Peoples* (Ottawa: Minister of Supply and Services, 1995). It is well documented that indigenous peoples suffer from alcoholism and social/family dysfunction, which are attributed to colonization and domination.

13 Michael Halewood, "Indigenous and Local Knowledge in International Law: A Preface to Sui Generis Intellectual Property Protection" (1999) 44 McGill Law Journal 952; Mark Hannig, "An Examination of the Possibility to Secure Intellectual Property Rights for Plant Genetic Resources Developed by Indigenous Peoples of NAFTA States: Domestic Legislation under the International Convention for Plant Varieties" (1996) 13 Arizona Journal of International and Comparative Law 175; Janet McDonnell, *The Dispossession of the American Indian, 1887-1934* (Bloomington: Indiana University Press, 1991); Helen Venne, *Our Elders Understand Our Rights: Evolving International Law Regarding Indigenous Rights* (Penticton, BC: Theytus Books, 1998); James Crawford, ed., *The Rights of Peoples* (Oxford: Oxford University Press, 1988); Catherine Brolman et al., eds., *Peoples and Minorities in International Law* (Dordrecht: Martinus Nijhoff, 1993); Sarah Pritchard, ed., *Indigenous Peoples, the United Nations and Human Rights* (London: Zed Books, 1998); Mary Ellen Turpel, "Indigenous People's Rights of Political Participation and Self-Determination: Recent International Legal Development and the Continuing Struggle for Recognition" (1992) 25 Cornell International Law Journal 579; Maivan Clech Lam, "Making Room for Peoples at the United Nations: Thoughts Provoked by Indigenous Claims to Self-Determination" (1992) 25 Cornell International Law Journal 603; Raidza Torres, "The Rights of Indigenous Populations: The Emerging International Norm" (1991) 16 Yale Journal of International Law 127. But see R.H. Barnes et al., eds., *Indigenous Peoples of Asia* (Ann Arbour, MI: Association for Asian Studies, 1995), showing that the concept of indigeneity may also apply to Asia (e.g., the Ainu of Japan).

14 The Salt Water, or Blue Water, theory posits that colonialism applies exclusively to the relationship between European colonial powers and their "overseas" territories. This seems to carry a vestigial bias for the notion that colonial empires were established by sea-powers, the implication being that expansion into contiguous land masses was not colonialist. Another disturbing element of this theory is that territories in which the colonialists persisted in their settlement (e.g., Canada, the United States, and Australia) are treated differently than are colonized territories that the colonizers ultimately vacated (e.g., Nigeria). See L.C. Bucheit, *Secession and the Legitimacy of Self-Determination* (New Haven: Yale University Press, 1978).

15 *Draft Report of the World Intellectual Property Organization (WIPO) Fact-Finding Missions on Intellectual Property and Traditional Knowledge (1998-1999)*, Geneva, Switzerland. Copies of this report are available from WIPO headquarters at, WIPO, 34 Chemin des Colombettes, 20, Geneva, 1211 Switzerland. Fax number +41-22-338-8120 or by email at: ffm-report-comments@listbox.wipo.int.

16 WIPO Report, *supra* note 15 at 28.

17 The *Illustrated Oxford Dictionary* defines a "tribe" as a "group of primitive families or communities, linked by social, economic, religious, and blood ties having a common culture and dialect, or a recognized leader." Ironically, some of the so-called "tribes," like the Yorubas and Igbos of Nigeria, number over 40 million, respectively, and occupy territories larger than that occupied by a long list of European countries. Needless to say, it would be considered insulting to describe less populous peoples, such as the Irish, Welsh, English, Scottish, and Dutch, as tribes.

18 The *Oxford English Dictionary* 2nd ed. Vol. 28 (1987) at 503 defines "tribe" as "a race of people, frequently applied to a group of primitive people, especially, a primary aggregation under a chief or headman."

19 The Convention Concerning the Protection and Integration of Indigenous and Other Tribal and Semi-Tribal Populations in Independent Countries (otherwise known as ILO 107), 26 June 1957, reprinted in 328 U.N.T.S. 247, is perhaps the worst international law instrument. It castigates non-Western cultures and peoples as devoid of culture and desperately in need of "development." This convention is explicitly based on the notion that non-Western cultures, knowledge, and practices are backward and primitive. Widespread rejection of this

racist convention led to its revision in ILO 169. See Richard Guest, "Intellectual Property Rights and Native American Tribes" (1995-6) 20 American Indian Law Review 111.

20 Art. 8 (j) CBD (1992) 31 International Legal Materials 813. See also Daniel Gervais, *The TRIPs Agreement: Drafting History and Analysis* (London: Sweet and Maxwell, 2003) at 58.

21 Graham Dutfield, "The Public and Private Domains: Intellectual Property Rights in Traditional Ecological Knowledge" *Oxford Electronic Journal of Intellectual Property Rights*, <http// users.ox.ac.uk/~mast>; Russell Barsh, "Forests, Indigenous Peoples and Biodiversity: Contribution of the Four Directions Council" Submission to the Secretariat of the Convention on Biological Diversity (1996).

22 For example, Article 8(j) of the CBD provides that "each contracting party shall, as far as possible and as appropriate subject to national legislation, respect, preserve, and maintain knowledge, *innovations* and practices of indigenous and local communities."

23 The Crucible Group, *People, Plants and Patents: The Impact of Intellectual Property in Trade, Plant Biodiversity and Rural Society* (Ottawa: IDRC, 1994); Roy Ellen, Peter Parkes, and Alan Bicker, eds., *Indigenous Environmental Knowledge and Its Transformations: Critical Anthropological Perspectives* (Amsterdam: Harwood Academic Publishers, 2000).

24 Sumathi Subbiah, "Reaping What They Sow: The Basmati Rice Controversy and Strategies for Protecting Traditional Knowledge" (2004) 27 Boston College International and Comparative Law Review 529.

25 According to a scholar of the patent system, John Jewkes: "It is easy enough to perceive the weaknesses, even the absurdities, of the patent system and the reasons why conflicting opinions as to its value are to be found. Its very principles are paradoxical ... it is a crude and inconsistent system. The standards of patentability, the patent period, the conditions attached to the patent have varied greatly from time to time in the same country and many as between different countries. The patent system lacks logic. It postulates something called 'invention' but in fact no satisfactory definition of 'invention' has ever appeared, and the courts, in their search for guiding rules, have produced an almost incredible tangle of conflicting doctrines ... The system, too, is wasteful. It gives protection for 16 years (or thereabouts) whilst in fact over nine-tenths of the patents do not remain active for the whole of this period ... It is almost impossible to conceive of any existing social institution so faulty in so many ways. It survives only because there seems to be nothing better." See John Jewkes, D. Sawers, and S. Stillerman, *The Sources of Invention* (London: Macmillan, 1969).

26 Carlos Primo Braga, "The Economics of Intellectual Property Rights and the GATT: A View from the South" (1989) 22 Vanderbilt Journal of Transnational Law 243; Ian Brownlie, "Legal Status of Natural Resources in International Law" (1979) Recueil de Cour 1; Jonathan Carlson, "Strengthening the Property-Rights Regime for Plant Genetic Resources: The Role of the World Bank" (1996) 6 Transnational Law and Contemporary Problems 91.

27 Naomi Roht-Arriaza, "Of Seeds and Shamans: The Appropriation of the Scientific and Technical Knowledge of Indigenous and Local Communities" (1996) 17 Michigan Journal of International Law 919.

28 See, for example, Vandana Shiva, *Biopiracy: The Plunder of Nature and Knowledge* (Cambridge, MA: South End Press 1997).

29 Paul Heald, "The Rhetoric of Biopiracy" (2003) Cardozo Journal of International and Comparative Law 519.

30 Dutfield, *supra* note 2 at 235; Graham Dutfield, "What Is Biopiracy?" International Expert Workshop on Access to Genetic Resources and Benefit Sharing, 2004.

31 Chidi Oguamanam, "Localizing Intellectual Property in the Globalization Epoch: The Integration of Indigenous Knowledge" (2004) 11:2 Indiana Journal of Global Legal Studies 135-169.

32 International Chamber of Commerce: The World Business Organization, "Policy Statement: TRIPs and the Biodiversity Convention: What Conflict?" Commission on Intellectual and Industrial Property, 28 June 1999, <http://www.iccwbo.org/home/statement/1999/trips-and-biodiversity.asp>.

33 Remigius Nwabueze, "Ethnopharmacology, Patents and the Politics of Plant Genetic Resources" (2003) 11 Cardozo Journal of International and Comparative Law 585.

34 Ikechi Mgbeoji, "Beyond Rhetoric: State Sovereignty, Common Concern, and the Inapplicability of the Common Heritage Concept to Plant Genetic Resources" (2003) 16 Leiden Journal of International Law 821-37.

35 Dutfield, *supra* note 30 at 4.

36 Rosemary Coombe, "The Recognition of Indigenous Peoples' and Community Traditional Knowledge in International Law" (2001) 14 St. Thomas Law Review 285.

37 James Buchanan, "Between Advocacy and Responsibility: The Challenge of Biotechnology for International Law" (1994) 1 Buffalo Journal of International Law 221.

38 William Lesser, *Institutional Mechanisms Supporting Trade in Genetic Materials: Issues under the Biodiversity Convention and GATT/TRIPs* (Geneva: UNEP, 1994) at 41.

39 For a judicial pronouncement on the appropriative impact of modern patent law, see Lord Hoffman in *Merrell Dow Pharmaceuticals Inc. and Another v. HN Norton and Co. Ltd.*, [1996] 33 Intellectual Property Reports 10; [1996] Report of Patent Cases 76. See also Steven Rothschild and Thomas White, "Printed Publication: What Is It Now?" (1988) 70 Journal of Patents and Trademark Office Society 42.

40 See, "Bolivian Farmers Demand Researchers Drop Patent Andean Food Crop" RAFI Press Release, 18 June 1997.

41 In Sanskrit the word "basmati" means "Queen of Fragrance" or "fragrant earth." This species of rice is known for its slender, aromatic long grain and unusually delicate texture. See RAFI Geno-Types, 1 April 1998.

42 *Oxford Dictionary and Thesaurus* 111 (Oxford: Oxford University Press, 1996).

43 Subha Ghosh, "Globalization, Patents and Traditional Knowledge" (2003) 17 Columbia Journal of Asian Law 73 at 74.

44 Stephen Ladas, *Patents, Trademarks, and Related Rights: National and International Protection* (Harvard, MA: Harvard University Press, 1975) at 6. For the argument on the British origin of patents, see J. Gordon, "Patent Law Reform" (1906) 55 Journal of the Society of Arts 26. For an alleged German origin of patents, see Harold Wegner, "TRIPs Boomerang-Obligations for Domestic Reform" (1996) 29 Vanderbilt Journal of Transnational Law 535 at 538; compare with Mladen Vukmir, "The Roots of Anglo-American Intellectual Property Law in Roman Law" (1991) 32 IDEA: The Journal of Research and Technology 123.

45 Owen Lippert, "One Trip to the Dentist Is Enough" in Owen Lippert, ed., *Competitive Strategies for the Protection of Intellectual Properties* (Vancouver: The Fraser Institute, 1999) at 131.

46 According to Maximillian Frumkin, the Brunelleschi patent "was a real invention patent, as good in subject matter as any of those dealt with in 1947 by the British Patent Office." See Ulf Anderfelt, *International Patent Legislation and Developing Countries* (The Hague: Martinus Nijhoff, 1971) at 4; Bruce William Bugbee, *The Early American Law of Intellectual Property: The Historical Foundations of the United States Patent and Copyright Systems* (unpublished doctoral thesis) at 70.

47 Moureen Coulter, *Property in Ideas: The Patent Question in Mid-Victorian Britain* (Missouri: Thomas Jefferson University Press, 1991) at 7.

48 The Venetian Statute of 1474 was enacted with a large majority (116 for, ten against, and three abstentions). Although it was written in old Venetian dialect, which is difficult to translate, Professor Luigi Sordelli's version (reproduced below) is the most widely accepted. The overt masculinization of inventions is also obvious. Owing to its seminal status, the Act is reproduced in extenso:

> There are in this city, and also there come temporarily by reason of its greatness and goodness, *men* from different places and most clever minds, capable of devising and inventing all manner of ingenious contrivances. And should it be provided, that the works and contrivances invented by them, others having seen them could not make them and take their honor, *men* of such kind would exert their minds, invent and make things which would be of no small utility and benefit to our State. Therefore, decision will be passed that, by authority of this Council, each person who will make in this city any new ingenious contrivance, not made heretofore in our dominion, as soon as it is reduced to perfection, so that it can be used and exercised, shall give notice of the same to the office of our Provisioners of Common. It being forbidden to any other in any territory and place of ours to make any other contriv-

ance in the form and resemblance thereof, without the consent and license of the author up to ten years. And, however, should anybody make it, the aforesaid author and inventor will have the liberty to cite him before any office of this city, by which office the aforesaid who shall infringe be forced to pay him the sum of one hundred ducates and the contrivance immediately destroyed. Being then in liberty of our Government at his will to take and use in his need any of the said contrivances and instruments, with this condition, however, that no others than the authors shall exercise them. [Emphasis added]

See Ladas, *supra* note 44 at 6-7. It should be noted that, in the Western paradigm, invention was primarily a masculine affair. Women were thought to be incapable of having "scientific" abilities. For a fuller account of female contributions to Western science and inventiveness, see Autumn Stanley, *Mothers and Daughters of Invention* (Piscataway, NJ: Rutgers University Press, 1993).

49 By taking the grant of patents outside the wide ambit of royal favour, the Venetian Statute was far ahead of the Stuart Statute of Monopolies, which it antedated by over a hundred years. On this basis, the patent granted Jacopo Acontio in 1565 for an actual invention is probably the first English patent (rather than the earlier grants made for the introduction or importation of already existing skill, trade, or industry). See Lynn White Jr., "Jacopo Acontio as an Engineer" (1967) 72 American Historical Review at 432.

50 Christine Macleod, *Inventing the Industrial Revolution: The English Patent System, 1660-1800* (Cambridge: Cambridge University Press, 1988) at 11. Galileo Galilei received a patent under that statute in 1594.

51 Ikechi Mgbeoji, "The Juridical Origins of the International Patent System: Towards a Historiography of the Role of Patents in Industrialization" (2003) 5 Journal of the History of International Law 403-22.

52 William Aldous et al., eds., *Terrell on the Law of Patents* (London: Sweet and Maxwell, 1982) at 1.

53 Fritz Machlup, *An Economic Review of the Patent System* (Study of the Subcommittee on Patents, Trademarks, and Copyrights of the Committee on the Judiciary, United States Senate, 85th Congress, Second Session. Study No. 15) at 1. An example is patents of nobility.

54 Laurinda Hicks and James Holbein, "Convergence of National Intellectual Property Norms in International Trading Agreements" (1997) 12 American University Journal of International Law and Policy 769.

55 Machlup, *supra* note 53 at 7; J.M. Laine, "Infringement of Patents by Intention" (1901) 17 The Law Quarterly Review 201; Dana Rohrabacher and Paul Crilly, "The Case for a Strong Patent System" (1995) 8 Harvard Journal of Law and Technology 263.

56 Machlup, *supra* note 53 at 2.

57 The concept of property has dominated legal thought since the hazy origins of jurisprudence. The literature is enormous and beyond the scope of this book. However, the following texts would reward readers: Kenneth Vandevelde, "The New Property of the Nineteenth Century: The Development of the Modern Concept of Property" (1980) 29 Buffalo Law Review 325; Emily Sherwin, "Two-and Three-Dimensional Property Rights" (1997) 29 Arizona State Law Journal 1075; Joseph Singer and Jack Beerman, "The Social Origins of Property" (1993) 6:2 Canadian Journal of Law and Jurisprudence 217; R.S. Bhalla, "The Basis of the Right of Property" (1982) 11 Anglo-American Law Review 57; R.S. Bhalla, "Legal Analysis of the Right to Property" (1981) 10 Anglo-American Law Review 180.

58 Adam Smith, *The Wealth of Nations*, ed. Edwin Canaan (London: Grant Richards, 1904) at 103.

59 J. Locke, *Two Treatises of Government*, ed. P. Laslett (Cambridge: Cambridge University Press, 1967), section 27.

60 Kenneth Swan, "Patent Rights in an Employee's Invention" (1959) 75 The Law Quarterly Review 77.

61 Otherwise known as commodification, this is the process and ideology of transforming otherwise interhuman and human-nature relationships into corporeal articles of trade with a ghostly objectivity and life of their own. A worker's labour, for instance, thus becomes a

commodity. The rationalizing ideology behind this notion is capitalism, which is as ideological as any other socioeconomic theory. Under capitalism, cooperative and communal interests are transformed into articles of trade and are subordinated to self-interest in the "free-market." This is probably why the contemporary narrative framework of the patent system is incapable of adequately addressing some of the ethical and moral concerns raised by its expansion into various non-mechanical manifestations of life. For an excellent analysis and evaluation of the commodification of human labour and the environment, see Norman Spaulding III, "Commodification and Its Discontents: Environmentalism and the Promise of Market Incentives" (1997) 16 Stanford Environmental Law Journal 293. See also L. Bently and B. Sherman, "The Ethics of Patenting: Towards a Transgenic Patent System" (1995) 3 Medical Law Review 275.

62 Historically, "individualism" is "a nineteenth-century word." See Will Kymlicka, "Individual and Community Rights" in Judith Baker, ed., *Group Rights* (Toronto: University of Toronto Press, 1994) at 17. The term was first used in its French form, "individualisme," and grew out of the general reaction to the French Revolution. See Alan Gilbert, *Democratic Individuality* (Cambridge: Cambridge University Press, 1990); Theodore M. Benditt, *Rights* (New Jersey: Rowman and Littlefield, 1982); Steven Lukes, *Individualism* (Oxford: Basil Blackwell, 1973).

63 "Man" here refers to the gender-specific term of the word as the early patent institution, like most other juridical institutions, was largely a masculine affair. See Rosemary Coombe, "Challenging Paternity: Histories of Copyright" (1994) 6 Yale Journal of Law and the Humanities 407.

64 Walter Ullmann, *The Individual and Society in the Middle Ages* (Baltimore: Johns Hopkins Press, 1966) at 45-72.

65 Oscar Wilde, *The Soul of Man under Socialism* (1891) in *Oscar Wilde's Plays, Prose Writings and Poems* (London: Everyman's Library) at 258. But see Edmund Burke in Kymlicka, *supra* note 62 at 120 at 1. However, it is a subject of debate as to whether the romantic notion of the individual inventor accords with contemporary realities or, at any rate, whether this paradigm tells the whole story of creativity. See Angela Riley, "Recovering Collectivity: Group Rights to Intellectual Property in Indigenous Communities" (2000) 18 Cardozo Arts and Entertainment Law Journal 175.

66 J. Morrell and A. Thackeray, *Gentlemen of Science: Early Years of the British Association for the Advancement of Science* (Oxford: Clarendon Press, 1981) at 3-12.

67 Bently and Sherman, *supra* note 61 at 110.

68 R.H. Tawney, *Religion and the Rise of Capitalism* (Baltimore: Penguin Books, 1947).

69 William Seagle, *Men of Law: From Hammurabi to Holmes* (New York: Macmillan, 1947) at 254.

70 Pierre Trudeau, "The Values of a Just Society" in Thomas Axworthy, ed., *Towards a Just Society* (Toronto: Viking Press, 1990) at 363-4; Brian Morris, *Western Conceptions of the Individual* (New York: St. Martin's Press, 1991); Shlomo Avineri and Anver de Shalit, eds., *Communitarianism and Individualism* (Oxford: Oxford University Press); Brian Lee Crowley, *The Self, the Individual and the Community* (Oxford: Clarendon Press, 1987).

71 Bugbee, *supra* note 46 at ii (emphasis added).

72 Frederick Abbott, Thomas Cothier, and Francis Curry, eds., *The Making of the International Intellectual Property System* (The Hague: Kluwer Publications, 1999) at 228.

73 As Abbott notes, "societies, for centuries, evolved on the basis of informal transfers of knowledge and technological advances in know-how, from masters to students, from fathers to sons, from mothers to daughters." See *ibid.*

74 Peter Drahos, *A Philosophy of Intellectual Property* (Dartmouth: Aldershot, 1996) at 61.

75 *Ibid.* at 15. See R.H. Lowie, *Primitive Society* (New York: Routledge, 1920) at 235-43.

76 Drahos, *ibid.* See Byron Good, *Medicine, Rationality and Experience: An Anthropological Perspective* (Cambridge: Cambridge University Press, 1994) at 5-20; Ann McElroy and Patricia Townsend, *Medical Anthropology in Ecological Perspective* (Boulder, CO: Westview Press, 1989).

77 Kronstein and Till, "A Re-Evaluation of the International Patent Convention" (1947) 12 Law and Contemporary Problems 765.

78 G.W.F. Hegel, *Philosophy of Right,* trans. T.M. Knox (Oxford: Oxford University Press, 1967) at 151.
79 Michael Gadbaw and Timothy Richards, eds., *Intellectual Property Rights: Global Consensus, Global Conflict?* (Boulder, CO: Westview Press, 1988) at 18.
80 Susan Tiefenbrun, "Piracy of Intellectual Property in China and the Former Soviet Union and Its Effects upon International Trade: A Comparison" (1998) 46 Buffalo Law Review 1; Tao-Tai Hsia and Kathryn Haun, "Laws of the People's Republic of China on Industrial and Industrial Property" (1973-74) 38 Law and Contemporary Problems 274.
81 See, for example, Joseph Singer, *The Edges of the Field: Lessons on the Obligations of Ownership* (Boston: Beacon Press, 2000); Gregory Alexander, *Commodity and Propriety: Competing Visions of Property in American Legal Thought, 1776-1970* (Chicago, IL: The University of Chicago Press: 1997).
82 For a succinct analysis of the problematic nature of theories on patents, see Samuel Oddi, "Un-Unified Theories of Patents: The Not-Quite-Holy Grail" (1996) 71 Notre Dame Law Review 267.
83 The French patent law of 1791 reads thus: "Every novel idea whose realization or development can be useful to society belongs primarily to him who conceived it, and it would be a violation of the rights of man in their very essence if an industrial invention were not regarded as the property of its creator." Interestingly, de Boufflers, the French jurist who drafted the French patent law of 1791, admitted that the natural rights theory of the French patent law was sheer propaganda. See Anderfelt, *supra* note 46 at 16.
84 Edith Penrose, *The Economics of the International Patent System* (Connecticut: Princeton University Press, 1974), at 89.
85 *The Role of Patents in the Transfer of Technology to Developing Countries,* Report of the Secretary-General, United Nations (New York: Martinus Nijhoff, 1964) at 9.
86 UN Report, *ibid.* Adding greater force to this observation, P.J. Michel noted that "patent systems are not created in the interest of the inventor but in the interest of the national economy. The rules and regulations of the patent system are not governed by civil or common law but by political economy." As quoted in, S. Vedaram, "The New Indian Patents Law" (1972) 3 International Review of Industrial Property and Copyright Law 39 at 41.
87 M. Bruce Harper, "TRIPs Article 27.2: An Argument for Caution" (1997) 21 William and Mary Environmental Law and Policy Review 381.
88 The *Atomic Energy Act of 1988* 42 U.S.C. subsection 2181 (a) provides that "no patent shall hereafter be granted for any invention or discovery which is useful solely in the utilization of special nuclear material or atomic energy in an atomic weapon." See also *National Aeronautics and Space Act of 1958,* 42 U.S.C. 2457; Virginia Geoffrey, "Do the Atomic Energy Act and the NASA Act Promote Adequate Advancement?" (1961) 43 Journal of Patent Office and Society 624.
89 Section 181 of the United States *Patent Act* provides, *inter alia:*

Whenever the publication or disclosure of an invention by the granting of a patent, in which the Government does not have a property interest might, in the opinion of the Commissioner, be detrimental to the national security, he shall make the application for patent in which such invention is disclosed available for inspection to the Atomic Energy Commission, the Secretary of Defense, and the chief officer of any other department or agency of the Government designated by the President as a defense agency of the United States ... If, in the opinion of the Atomic Energy Commission, the Secretary of a Defense Department ... publication or disclosure of the invention by the granting of a patent therefore would be detrimental to the national security, the Atomic Energy Commission, the Secretary of a Defense Department ... shall notify the Commissioner and *the Commissioner shall order that the invention be kept secret and shall withhold the grant of a patent* for such a period as the national interest requires. [Emphasis added]

90 Eric Schiff, *Industrialization without National Patents – The Netherlands, 1869-1912, Switzerland, 1850-1907* (New Jersey: Princeton University Press, 1971), at 3.
91 Penrose, *supra* note 84 at 32.

92 Melvin Kranzberg, "The Technical Elements in International Technology Transfer: Historical Perspectives" in John McIntyre and Daniel Papp, eds., *The Political Economy on International Technology Transfer* (New York: Quorum Books, 1986) at 31; Samuel Oddi, "The International Patent System and Third World Development: Reality or Myth?" (1987) 63 Duke Law Journal 831; Samuel Oddi, "TRIPs: Natural Rights and a Polite Form of Economic Imperialism" (1996) 29 Vanderbilt Journal of Transnational Law 415.

93 Samuel Oddi, "Beyond Obviousness: Invention Protection in the Twenty-First Century" (1989) 38 American University Law Review 1097. Other examples include the automatic transmission, which was invented in 1904 but was commercialized only in 1937; the cotton-picker (1850-1942); magnetic recording (1898-1939); penicillin (1928-44); radar (1904-35); silicon (1904-48); television (1905-40); and xerography (1937-50). For an analogy of the patent grant to a hunting licence, see Allan Topol, "Patents and Hunting Licenses: Some Iconoclastic Comments and an Irreverent Solution" (1968) 17 American University Law Review 424; B.V. Hindley, *The Economic Theory of Patents, Copyrights, and Registered Industrial Designs* [Background Study to the Report on Intellectual and Industrial Property] (Ottawa: Economic Council of Canada, 1971).

94 Robert Sherwood, Vanda Scartezini, and Peter Dirk Siemsen, "Promotion of Inventiveness in Developing Countries through a More Advanced Patent Administration" (1999) 39 Journal of Law and Technology 473; Robert Sherwood, "Human Creativity for Economic Development: Patents Propel Technology" (2000) 33 Akron Law Review 1; Donald Gregory, Charles Saber, and Jon Grossman, *Introduction to Intellectual Property Law* (Washington, DC: BNA Books, 1994); William Kingston, *Innovation, Creativity and the Law* (Dordrecht: Kluwer Academic Publishers, 1990).

95 H. Stafford Hatfield, *Inventions and Their Use in Science Today* (London: Pelican Books, 1939) as quoted in Eric Schiff, *supra* note 90; Harvey Bale, "Patent Protection and Pharmaceutical Innovation" (1996-97) 29 International Law and Politics 95. Another concurring voice enthused that "it was not ... by accident that the patent system had its origin in England, nor that the industrial revolution was the inevitable consequence." See H.G. Fox, as quoted in Schiff at 90 at 9. Fox continues, "the dross of abuse and impropriety in the monopoly system had to be refined in the furnace of experience before the gold of the present patent system emerged to take its place as the greatest contributory factor to modern industrial progress" (*ibid.*).

96 Quoted in Schiff, *supra* note 90 at 12. These emotional tributes to the patent system are not limited to the British or continental Europeans. Across the Atlantic, a United States patent attorney vowed, "the strongest evidence of the value of the American patent system is our industrial economy, which has been built largely upon groundwork of patented inventions" (*ibid.*). Adding a flavor of democratic heroism to this panegyric to patents, another American patent attorney declared "the defence of the democratic world depends largely on American industry, which owes its present strength in large part to traditional American patent policy." See Karl Lutz, "A Proper Public Policy on Patents: Are We Adopting the Soviet View?" (1951) 37 American Bar Association Journal 943. According to the American Patent Society, "the patent system is the foundation of American enterprise. It has ... contributed to the achievement of the highest standard of living that any nation has ever enjoyed" (Schiff, *ibid.*).

97 Robert Merges, "Battle of Lateralisms: Intellectual Property and Trade" (1990) 8 Boston University International Law Journal 239; Robert Sherwood, *Intellectual Property and Economic Development* (Boulder, CO: Westview Press, 1990) at 2. But see Robert Sherwood, "Intellectual Property Systems and Investment Stimulation: The Rating of Systems in Eighteen Developing Countries" (1996-7) 37 IDEA: The Journal of Research and Technology 261.

98 Abbott, *supra* note 72.

99 William Lesser, *Sustainable Use of Genetic Resources under the Convention on Biological Diversity: Exploring Access and Benefit Sharing Issues* (CAB International, Oxford, 1997) at 167.

100 *Ibid.* See also, *The Impact of Intellectual Property Rights Systems on the Conservation and Sustainable of Biological Diversity and on the Equitable Sharing of the Benefits from Its Use*, UNEP/CBD/COP/3/22, 1996. But see Barkev Sanders, "The Economic Impact of Patents" (1958) 2 Patents, Trademark, Copyright Journal of Research and Education 340.

101 Abbott, *supra* note 72 at 223; Richard Carr, "Our Patent System Works: A Reply to the Melman Report" (1960) 4 Patent, Trademark, Copyright Journal of Research and Education 5. But see Siegfried Greif, "Patents and Economic Growth" (1987) 18 International Review of Industrial Property and Copyright Law 191.

102 Anderfelt, *supra* note 46 at 28.

103 Ikechi Mgbeoji, "Patents and Traditional Knowledge of the Uses of Plants: Is a Communal Patent Regime Part of the Solution to the Scourge of Biopiracy?" (2001) 9 Indiana Journal of Global Legal Studies 163. For example, as late as 1975 Cesar Milstein and Georges Kohler decided not to patent their path-breaking and commercially valuable invention of mono-clonal antibody-producing hybridum cells. See Bhupinder Chimni, "Hard Patent Regime Completely Unjustifiable" in Subrata Chowdhry et al., eds., *The Right to Development in International Law* (Dordrecht: Martinus Nijhoff, 1992) at 315; Andrew Currier, "To Publish or to Patent, That Is the Question" (2000) 16 Canadian Intellectual Property Review 337. See also Robert Golden, "Biotechnology, Technology Policy, and Patentability: Natural Prod-ucts and Invention in the American System" (2001) 50 Emory Law Journal 155.

104 The "serendipity phenomenon" is the faculty of making unexpected discoveries or inven-tions by accident. The word "serendipity" was coined by Horace Walpole after the title of the fairy tale "The Three Princes of Serendip," in which the heroes were always making accidental discoveries. Serendip was an ancient name for Sri Lanka (Ceylon).

105 *Time Magazine* [Canadian edition], 4 December 2000, "The Best Inventions of the Year" at 21. Serendipitous inventions include Roentgen's X-Ray, Fleming's penicillin, Goodyear's vulcanization of rubber, Edison's phonograph, Bessemer's steel-making, Maybach's carbu-retor, Macintosh's raincoat, Daguerre's photographic process, and Nobel's dynamite and plywood.

106 *Ibid.* Other famous accidental inventions include Post-it-Notes, Coca-Cola, Scotchguard, Teflon, Gore-Tex, cornflakes, chemotherapy, and Slinky. In addition, there is the related phenomenon of spin-off inventions. These are collateral products. Inorganic paint, the walking wheelchair, maintenance-free lubricated bearings, and sight-controlled switches are all spin-off inventions from the United States National Aeronautical and Space Agency (NASA).

107 The average cost of obtaining a patent in mid-eighteenth-century England was the princely sum of 350 pounds. Recent studies indicate that, excluding maintenance fees, it costs an average sum of US$40,000 to secure a patent in Japan, US$17,265 in Germany, US$15,785 in Norway, US$14,000 in Finland, US$14,625 in Austria, US$8,335 in France, and US$7,090 in the United Kingdom.

108 Anthony Bourget, "Protecting Inventions in the Former Soviet Union" (1991-22) 10 Wis-consin International Law Journal 1; Adolf Dietz, "Trends Toward Patent Rights in Social-ist Countries?" (1971) 2 International Review of Industrial Property and Copyright Law 155; M. Hoseh, "The USSR Patent System" (1960) 4 Patent, Trademark, Copyright Journal of Research and Education 220; S.J. Soltysinski, "New Forms of Protection for Intellectual Property in the Soviet Union and Czechoslovakia" (1969) 32 Modern Law Review 408.

109 W.H. Price, *English Patents of Monopoly* (London: H. Milford, Oxford University Press, 1906) at 62.

110 T.S. Ashton, *The Industrial Revolution, 1760-1830* (Oxford: Oxford University Press, 1948).

111 Phyllis Deane, *The First Industrial Revolution* (Cambridge: Cambridge University Press, 1965).

112 Peter Mathias, *The First Industrial Nation: An Economic History of Britain* (New York: Charles Scribner's Son, 1969).

113 Coulter, *supra* note 47 at 3. According to Coulter, "the question of whether patents for invention "served as a stimulus to or a drag upon inventive activity during the industrial revolution will probably remain open" (Coulter, *ibid.* 23). But see Gerald Mossinghoff, "The Importance of Intellectual Property Protection in International Trade" (1984) 7 Bos-ton College International and Comparative Law Review 235.

114 Ashton, *supra* note 110 at 10. The example of James Watts's quarter-century domination of steam engine construction in Britain comes in handy in Ashton's conclusions. But see David Landes, *The Unbound Prometheus: Technological Change and Industrial Development in Western Europe from 1750 to the Present* (Cambridge: Cambridge University Press, 1969) at

199; H.I. Dutton, *The Patent System and Inventive Activity during the Industrial Revolution* (Manchester: Manchester University Press, 1984) at 104.

115 Coulter, *supra* note 47 at 23.

116 Schiff, *supra* note 90 at 5. For a modern examination and rebuttal of the "patents-propel-inventiveness thesis," see Samuel Oddi, "Reality or Myth?" *supra* note 92 at 40.

117 Macleod, *supra* note 50 at 16.

118 Schiff, *supra* note 90 at 124.

119 Anderfelt, *supra* note 46 at 6.

120 *Ibid.* For the early German experience, see H. Pohlman, "The Inventor's Right in Early German Law" (1961) 43 Journal of the Patent Office Society 121.

121 According to one of the chief proponents of the historical necessity argument, "the patent system is not the result of inspired thinking but is a dictate of historical necessity." See H.G. Fox, *Monopolies and Patents: A Study of the History and Future of the Patent Monopoly* (Toronto: University of Toronto Studies, Legal Series, Extra Vol., 1947) at 190; but see Anderfelt, *supra* note 46 at 27.

122 John Needham, *Science and Civilization in China*, Vols. 1-2 (Cambridge: Cambridge University Press, 1954); John Needham, *The Grand Titration: Science and Society in East and West* (London: Allen and Unwin, 1969).

123 Drahos, *supra* note 74. Some of the seminal Chinese inventions include paper in AD 105 and printing in the sixth century, gunpowder, and the magnetic compass; but see Liwei Wang, "The Chinese Traditions Inimical to the Patent Law" (1993) 14 Northwestern Journal of International Law and Business 15 (arguing that the Chinese contributions were not "science" but "technique"!).

124 According to patent historian Erich Kaufer, "In Egypt and other ancient civilizations, no patent-like institutions have been discovered, and it is likely that none existed." See, Harold Wegner, *supra* note 44.

125 Harold Dorn, *The Geography of Science* (Baltimore: Johns Hopkins University Press, 1971).

126 Drahos, *supra* note 74 at 15.

127 *Report of the President's Commission on the Patent System,* reproduced in Hearings Before Subcommittee No. 3 of the Committee on the Judiciary House of Representatives, 90th Congress on H.R. 5924, H.R. 13951; and related *Bills for the General Revision of the Patent Laws, Title 35 of the United States Code, and for Other Purposes,* Serial No. 11, Part 1 (Washington, DC: US Government Printing Office, 1968) at 170. Compare with Jay Erstling, "The Protection of Intellectual Property: Of Metaphysics, Motivation, and Monopoly" (1991) 3 Sri Lanka Journal of International Law 51.

128 Merges, *supra* note 97 at 124 (and the texts cited therein.)

129 Anderfelt, *supra* note 46 at 30. For a fuller analysis of the template shift in the social nature of inventiveness, see I. de Sola Pool, "The Social Environment for Sustained Technological Growth" in W. Anderson et al., eds., *Patents and Progress: The Sources and Impact of Advancing Technology* (Chicago, IL: Richard Irwin, 1965).

130 Mary Holman, "An Economic Analysis of Government Ownership of Patented Inventions" in L.J. Harris, ed., *Nurturing New Ideas: Legal Reports and Economic Roles* (Washington, DC: Bureau of National Affairs, 1969) at 155. See, for example, Article 60 (1) of the European Patent Convention; Section 2 (4) *Patents and Design Decree 1970* of Nigeria. For a list of all patent laws in the world and their provisions on employer-ownership of inventions, see John Sinnott, *World Patent Law and Practice*, Vols. 2B-F (New York: Matthew Bender, 1976).

131 Anderfelt, *supra* note 46 at 30. For an expert and learned analysis of the law on employee inventions, Dr. Fredrik Neumeyer's works on the subject remain peerless. See Fredrik Neumeyer, *The Law of Employed Inventors in Europe,* Study No. 30, Senate Sub-Committee on Patents, Trademarks and Copyrights on the Committee of the Judiciary, 87th Congress, 2nd Session, Washington, 1963; Frederik Neumeyer, "Employees' Rights in Their Inventions: A Comparison of National Laws" (1962) 44 Journal of Patent Office and Society 674; Fredrik Neumeyer, *The Employed Inventor in the United States: R and D Policies, Law, and Practice* (Cambridge, MA: MIT Press, 1971).

132 The regime of corporate ownership of employee inventions is traceable to the *Patent Act of 1897* of the Austro-Hungarian Empire. See Ladas, *supra* note 44 at 324.

133 Anderfelt, *supra* note 46 at 101. Lending further weight, Melman's studies conclude that "the resulting condition of interdependence in inquiry renders the concept of inventor obsolete to a considerable extent." See S. Melman, "The Impact of the Patent System on Research" Study No. 11, United States Senate, Committee on the Judiciary, Studies of the Sub-Committee on Patents, Trademarks and Copyrights (Washington, DC: Government Printing Office, 1958) at 18.

134 Soltsinski, *supra* note 108 at 16. Ironically, whether under the capitalist system or the embattled communist system, the lot of the employed inventor remains the same. In Soltysinski's words, "In a socialist, as well as in a capitalist state, the patent for an employee's invention usually belongs to the employer" (*ibid.*).

135 Chimni, *supra* note 103 at 317. Bemoaning his plight under this regime, a world-renowned scientist and prolific inventor, Dr. Charles Draper, noted: "Thirty years of work on various problems have inspired a number of innovations resulting in patents carrying my name either as inventor or co-inventor. My direct income has been substantially nothing compared to the total gross income derived from sales of devices based on the ideas covered by the patents ... the total business generated for industry amounts to several billion dollars. My personal returns have been not more than a few thousand dollars." See "The Patent System from a Scientist's Point of View" (1961) Patent, Trademark, Copyright Journal of Research and Education 64 at 74. For an analysis of how the modern regime affects the "small inventor," see Ben Hattenbach, "GATT TRIPs and the Small American Inventor: An Evaluation of the Effort to Preserve Domestic Technological Innovation" (1995) 10 Intellectual Property Journal 61; John Stedman, "The Employed Inventor, the Public Interest, and Horse and Buggy Law in the Space Age" (1970) 45 New York University Law Review 1.

136 United Nations Transnational Corporation and Management Division, United Nations Department of Economics and Social Development, Intellectual Property Rights and Foreign Direct Investment, UN Doc.ST/CTC/SER.A/24, UN Sales No. E93, II.A.10 (1993).

137 Frederick Abbott, "Commentary: The International Intellectual Property Order Enters the 21st Century" (1996) 29 Vanderbilt Journal of Transnational Law 471; Christopher Mayer, "The Brazilian Pharmaceutical Industry Goes Walking from Ipanema to Prosperity: Will the New Intellectual Property Law Spur Domestic Investment?" (1998) 12 Temple International and Comparative Law Journal 377. (Noting that, in Brazil, "despite the ban on patents, foreign investments in the pharmaceutical industry rose from 113 million USD in 1971 to 644 million USD in 1979. This increased investment tends to defeat the argument that in the absence of patent protection, foreign investment from the pharmaceutical industry would be negligible"). Although the existence of a strong patent system will naturally increase profits for the investor, it seems that a large and stable market with effective demand for the services or products of the foreign investor is crucial for FDI.

138 Abbott, *supra* note 137 at 1792. For a study of the impact of the patent system in Australia, see T.D. Mandeville, D.M. Lamberton, and E.F. Bishop "Economic Effects of the Australian Patent System" in Abbott, *ibid.* at 660.

139 Arnold Plant, "The Economic Theory Concerning Patents for Inventions" (1934) 1 Economica 30; reprinted in Sir Arnold Plant, *Selected Economic Essays and Addresses* (Boston: Routledge, 1974) at 35.

140 For a similar conclusion, see D. Vaver, "Intellectual Property Today: Of Myths and Paradoxes" (1990) 69 Canadian Bar Review 98; but see Richard Carr, "Our Patent System Works: A Reply to the Melman Report" (1960) 4 Patent, Trademark, Copyright Journal of Research and Education 5.

141 Machlup, *supra* note 53 at 19. See also Oddi, "Reality or Myth," *supra* 92 at 842; but see Adrienne Catanese, "Paris Convention, Patent Protection, and Technology Transfer" (1985) 3 Boston University International Law Journal 209.

142 Edmund Kitch, "The Nature and Function of the Patent System" (1977) 20 Journal of Law and Economics 265.

143 For a recent criticism of the prospect theory, see Roger Beck, "The Prospect Theory of the Patent System and Unproductive Competition" [as cited in Oddi, "Theories" *supra* note 82 at 269]; Douglas McFetridge and Douglas Smith, "Patents, Prospects, and Economic Surplus: A Comment" (1980) 23 Journal of Law and Economics 197.

144   Robert Merges and Richard Nelson, "On the Complex Economics of Patent Scope" (1990) 90 Columbia Law Review 839; Robert Merges, "Commercial Success and Patent Standards: Economic Perspectives on Innovation" (1988) California Law Review 803.

145   Oddi, "Theories," *supra* note 82 at 275.

146   Steven Cheung has captured the spectrum of views on the economic analyses of the patent system: "One view: advanced by Bentham (1795) and shared by Say (1803), Mill (1848), and Clark (1907): holds that patent rights are absolutely necessary to encourage inventions. A second view, advanced by Taussig (1915) and shared by Pigou (1920), maintains that a system of patents is largely superfluous. Third, Plant (1934), with modern followers, argued that a patent system is actually detrimental. Finally, Arrow (1962) ... argued that although property rights in ideas are clearly useful, they are nonetheless inferior to direct government investment in inventive activities" (as cited in Oddi, "Theories" *supra* note 82 at 268).

147   Quoted in Macleod, *supra* note 50 at 1. For an account of the migration of the patent concept to the United States of America, see Bugbee, *supra* note 46; Kenneth Burchfiel, "Revising the 'Original' Patent Clause: Pseudo-History in Constitutional Construction" (1989) 2 Harvard Journal of Law and Technology 155.

148   Frederick Abbott, Thomas Cottier, and Francis Gurry, eds., *The Making of the International Intellectual Property System* (The Hague: Kluwer Law International, 1999) at 228.

149   C. Macleod, "The Paradoxes of Patenting: Invention and Its Diffusion in 18th and 19th Century Britain, France and North America" (1991) Technology and Culture 905. For example, the institution of patents was primarily used as a means of luring foreign skills and industries without fidelity to the criterion of absolute novelty. For instance, in France article 3 of the original patent law of 1791 provided that "whoever [is] the first to *bring into France* a foreign discovery shall enjoy the same advantages as if he were the inventor" (emphasis added).

150   Sesto Vecchi and Michael Scown, "Intellectual Property Rights in Vietnam" (1992) 11 Pacific Basin Law Journal 67; Dereje Worku, "Patents and the Process of Innovation in East African Countries" (1990) 21 International Review of Industrial Property and Copyright Law 38; Gaius Ezejiofor, "The Law of Patents: A Review" (1973-74) 9 African Law Studies 39.

151   Bugbee, *supra* note 46 at 141-5. See also Dan Rosen, "A Common Law for the Ages of Intellectual Property" (1984) 38 University of Miami Law Review 769. For a brief but useful account of the migration of both the common law and the patent concept to the British Commonwealth, see A.L. Goodhart, ed., "The Migration of the Common Law" (1960) 76 Law Quarterly Review 39-90. For most of the British Commonwealth countries their original patent laws were, of course, fashioned by the British colonialists. See T. Ekua Sagoe, "Industrial Property Law in Nigeria" (1992) 14 Comparative Law Yearbook of International Business 312; Mark Sklan, "African Patent Statutes and Technology Transfer" (1978) 10 Case Western Reserve Journal of International Law 55.

152   F. De Zuluetta, ed., *The Institutes of Gaius* (Oxford: Clarendon Press, 1946), Book 11, 12-14. There is a difference between the Roman and modern conception of corporeal and incorporeal rights. The Roman law distinction is believed by some classicists to have originated with the Stoics and the Epicureans. See John Austin, *Lectures on Jurisprudence*, 5th ed. (London: J. Murray, 1885), Lecture XIII. Roman law on the subject, with minor exceptions, conflated object and ownership (which, I need hardly say, are distinct concepts).

153   Alex Castles, "The Reception and Status of English Law in Australia" (1963) 2 Adelaide Law Review 1; Liwei Wang, "The Current Economic and Legal Problems Behind China's Patent Law" (1998) 12 Temple International and Comparative Law Journal 1; William Cornish, "Patents and Innovation in the Commonwealth" (1983-85) 9 Adelaide Law Review 171.

154   Abbott, *supra* note 72 at 857; Toshiko Takenaka, "The Role of the Japanese Patent System in Japanese Industry" (1994) 13 Pacific Basin Law Journal 25; John Gadsby, "The Progress of Japanese Patent Law" (1911) 27 Law Quarterly Review 60.

155   Paul Liu, "US Industry's Influence on Intellectual Property Negotiations and Special 301 Actions" (1994) 13 Pacific Basin Law Journal 87.

156 Ruey-Long Lin, "Protection of Intellectual Property in the Republic of China" (1986-87) 6 Chinese Yearbook of International Law and Affairs 120; Chung-Sen Yang and Judy Chang, "Recent Developments in Intellectual Property Law in the Republic of China" (1994) 13 Pacific Basin Law Journal 70; Andrew Walder, "Harmonization: Myth and Ceremony: A Comment" (1994) 13 Pacific Basin Law Journal 163; William Alford, "Don't Stop Thinking About ... Yesterday: Why There Was No Indigenous Counterpart to Intellectual Property Law in Imperial China" (1993) 7 Journal of Chinese Law 3; William Alford, "Making the World Safe for What? Intellectual Property Rights, Human Rights and Foreign Economic Policy in the Post-European Cold War World" (1996-97) 29 International Law and Politics 135; Richard Baum, "Science and Culture in Contemporary China: The Roots of Retarded Modernization" (1982) 22 Asian Survey 1166; Jianyang Yu, "Protection of Intellectual Property in the P.R.C.: Progress, Problems, and Proposals" (1994) 13 Pacific Basin Law Journal 140. In ancient China, Confucian teachings on the virtue of open circulation of knowledge and social solidarity stood against the privatization of knowledge inherent in the patent concept.

157 Sang-Hyun Song and Seong-Ki Kim, "The Impact of Multilateral Trade Negotiations on Intellectual Property Laws in Korea" (1994) 13 Pacific Basin Law Journal 118. It was the American Chamber of Commerce that actively lobbied for the enactment of patent law and institution in Korea.

158 Eric Schiff, *Industrialization without National Patents: The Netherlands, 1869-1912, Switzerland, 1850-1907* (New Jersey: Princeton University Press, 1971).

159 For a list of all the patent laws of the various countries of the world, see Sinnott, *supra* note 130.

160 Lionel Bently and Brad Sherman, *The Making of Modern Intellectual Property Law: The British Experience, 1760-1911* (Cambridge: Cambridge University Press, 1999) at 209.

161 Seaborne Davies, "The Early History of the Patent Specification" (1934) 50 Law Quarterly Review 86 at 95; Jerome Reichman, "Charting the Collapse of the Patent-Copyright Dichotomy: Premises for a Restructured International Intellectual Property System" (1994) 13 Cardozo Arts and Entertainment Law Journal 475.

162 This may also be attributed to the influence of the *Calico Printers Act* of 1787 on patent law. According to Bently and Sherman, "the 1787 Act recognized the individual as the source of the design." See Bently and Sherman, *supra* note 61 at 31. As Erle J. held in *Jefferys v. Boosey* (1854) 10 ER 702, "a person to be entitled to the character of an inventor must himself have conceived the idea embodied in the improvement. It must be the product of his own mind and genius and not of another's."

163 The *Designs Laws* in England initiated the bureaucratic control of the patent system. According to Bently and Sherman, the first of such laws was the *Calico Printers' Act* of 1787. See Wyndham Hulme, "The History of the Patent System under the Prerogative and at Common Law" (1896) 12 Law Quarterly Review 141; Wyndham Hulme, "The History of the Patent System under the Prerogative and at Common Law: A Sequel" (1900) 16 Law Quarterly Review 441. As an aside, Nedd Ludd's rebellion in 1779 against the emerging Industrial Revolution in the English Midlands now marks the pejorative term "Luddites," which is reserved for those opposed to new technologies, particularly genetic modification of plant life forms.

164 The system of registration was designed to curtail arguments about originality. Prior to that, patents were being routinely granted on already patented inventions. Dundas White, "The New 'Investigation' for Patents" (1903) 19 Law Quarterly Review 307. Note also that these requirements are particularly suited to peoples with a culture of writing. As argued in subsequent chapters, the system of registration, when applied globally, enables the appropriation of inventions of peoples in cultures lacking systems of registration of inventions. For a short but informative account of the influence of the British textile laws on the fledgling patent regime, see Kathy Bowrey "Art, Craft, Good Taste and Manufacturing: The Development of Intellectual Property Laws" (1997) 15 Law in Context 78. The history of British and global industrialization would be incomplete without reference to the British textile industry.

165   (1778) 1 Web Pat Cas 53; *Bainbridge* v. *Wigley* (1810) 171 ER 636; Wyndham Hulme, "On the Consideration of the Patent Grant, Past and Present" (1897) 13 Law Quarterly Review 313.

166   J.W. Baxter, *World Patent Law and Practice,* Vol. 2 (London: Sweet and Maxwell, 1976) at 7.

167   Merges, *Intellectual Property, supra* note 97 at 125; Cruikshank and Fairweather, *The Law of Patents, Designs and Trademarks* (Glasgow: International Patent Agency, 1907) at 1. Similarly, in his letter to Thomas Cromwell, Sir Antonio Guidolti proposed a scheme to bring Italian silk-weavers to England, but only on the condition that a twenty-year patent be granted on silk-weaving. See Wyndham Hulme, "On the History of Patent Law in the Seventeenth and Eighteenth Centuries" (1902) 18 Law Quarterly Review 280; James Bakewell, "The American and British Systems of Patent Law" (1891) 7 Law Quarterly Review 364. Compare with Wood Renton, "Patent Right in England and the United States" (1891) 26 Law Quarterly Review 150.

168   *The Clothesworkers of Ipswich Case*, 78 E.R. 147. Emphasis added.

169   Penrose, *supra* note 84 at 70. Emphasis added.

170   W.S. Holdsworth, "The Commons Debates 1621" (1936) 52 Law Quarterly Review 481.

171   Merges, *supra* note 97 at 125. See the famous suit in *Darcy* v. *Allin,* otherwise known as the *Case of Monopolies,* 77 E.R. 1263 or (1602) Co. Rep. 84; Richard Gardiner, "Industrial and Intellectual Property Rights: Their Nature and the Law of the European Communities" (1972) 88 Modern Law Review 507.

172   *Statute of Monopolies* (21 Jac. 1., cap 3.) Although the *Statute of Monopolies* has received favourable reviews by scholars, it did not in fact dispense with monopolies per se. The Royalty maintained residual powers to grant monopolies for the importation of arts products in England.

173   George Francis Takach, *Patents: A Canadian Compendium of Law and Practice* (Edmonton: Juriliber, 1993) at 3.

174   For a judicial account of the patent controversy of the nineteenth century see *Attorney-General* v. *Adelaide Steamship Co.,* [1913] A.C. 781.

175   Schiff, *supra* note 158 at 21.

176   Abbott, *supra* note 72 at 228; Fritz Machlup and Edith Penrose, "The Patent Controversy in the 19th Century" (1950) 10 Journal of Economic History 1-29.

177   Schiff, *supra* note 158 at 87.

178   Early patent law grappled with the distinction between "invention" and "discovery." As a leading thinker then argued, "a discoverer is one thing and an inventor is another. The discoverer is one who discloses something which exists in nature, for instance, coal fields, or a property of matter, or a natural principle: such discovery never was and never ought to be the subject of a patent ... however much effort may have gone into the discovery ... no one could be said to have invented these." See T. Webster, on *Property in Designs and Inventions in the Arts and Manufactures* (London: Chapman and Hall, 1853) at 7 (as cited in Bently and Sherman *supra* note 61 at 45). According to R. Godson, "a principle was a mere idea, and therefore could not be a fit subject for a patent." See R. Godson, "Law of Patents" (19 February 1833) 15 Hansard col. 977 (as quoted in Bently and Sherman, *supra* note 61 at 45.) However, there was no agreement as to what constituted "principle." As Rooke J. observed in the steam engine case instituted by the inventor James Watt, "the term principle is equivocal: it might be used to refer to radical elementary truths of a science: such as the natural properties of steam, its expansiveness and condensability." See *Boulton and Watt* v. *Bull* (1795) 126 E.R. 651. See also C.J. Hamson, *Patent Rights for Scientific Discoveries* (Indianapolis: Bobbs-Merril, 1930). But see Lawson Mckenzie, "Scientific Property" (1953) 118 Science at 767. The doctrinal crisis in early patent law was so pervasive "that it was said in 1835 that there was no law of patents in England." See Bently and Sherman, *supra* note 61 at 82.

179   Blanco White, *Patents for Inventions,* 5th ed. (London: Stevens and Sons, 1983) at 156-68; Donald Banner, ed., *Developments: 1988: The John Marshall Law School Center for Intellectual Property Law* (Chicago: The John Marshall Law School, 1988) at 155-390.

180   For example, while purified vitamin A was held patentable, purified Tungsten was held not patentable. Neither of them occurs naturally in the pure state. See Aldous et al., *supra* note 52 at 14.

181 For example, the inclusion of process patents in the realm of patentable subjects in the United States and Western Europe was a judicial creation. See Aldous et al. *supra* note 52 at 72. In theory, however, the courts have always protested that patent grants are construed like all other documents. See William Schuyler Jr., "Recent Developments and Future Prospects on the National Level in the United States of America" in WIPO Lectures, Montreux 1971 (Geneva: WIPO Press 1971) at 66. The reality, however, is that the patent system has a cultural life and an ideology that it seeks to expound and expand. See Robert Merges and Richard Nelson, "On the Complex Economics of Patent Scope" (1990) 90 Columbia Law Review 839.

182 For a detailed account of the making of the modern international patent system and its ramifications, see Anderfelt, *supra* note 46. For a contemporary analysis of some of the issues raised by the emerging regime of a globalizing Eurocentric patent model, see Keith Aoki, "Neocolonialism, Anti-Commons Property, and Biopiracy in the (Not-so-Brave) New World Order of International Intellectual Property Protection" (1998) 6 Indiana Journal of Global Legal Studies 11.

183 Abbott, *supra* note 72 at 235; Julie Park, "Pharmaceutical Patents in the Global Arena: Thailand's Struggle between Progress and Protectionism" (1993) 13 Boston College Third World Law Journal 121; Glenn Butterton, "Pirates, Dragons and US Intellectual Property Rights in China: Problems and Prospects of Chinese Enforcement" (1996) 38 Arizona Law Review 1081; Carolyn Corn, "Pharmaceutical Patents in Brazil: Is Compulsory Licensing the Solution?" (1991) 9 Boston University International Law Journal 71.

184 Evan Ackiron, "Patents for Critical Pharmaceuticals: The AZT Case" (1991) 17 American Journal of Law and Medicine 145. For a comparison of the instrumentalist nature of various patent laws, see, for example, Article 4, Argentina Law No. 111 on Patents on Invention, 11 October 1864; Argentina Law No. 20.794, Transfer of Technology Decree-Law 19.231; Argentina Foreign Investments Law No. 21.382, Australian Patents Act 1952-69; Sections 1 and 2, Austrian Patent Act, 1970; Bahamas Industrial Property Act of 1965, Belgian Patent Act of 24 May 1854. See, generally, Sinnott, *supra* note 130.

185 Abbott, *supra* note 72 at 235; Mark Lemley, "The Economic Irrationality of the Patent Misuse Doctrine" (1990) 78 California Law Review 599.

186 Anderfelt, *supra* note 46 at 100. For example, in 1880, when Great Britain was the undisputed technological giant of the world, it supported the abolition of the obligation for domestic working of foreign patents. By 1911, when it had lost its preeminent position, it advocated a policy of domestic working of foreign patents or, at least, of compulsory licensing of foreign patents. France and Italy have also been known to engage in similar turnarounds.

187 The term "North" refers to the countries of North America, Europe, New Zealand, Japan, and Australia. They are also called the "rich" or the "advantaged" countries of the world. For the purposes of convenience, they may further be categorized as members of the Organization for Economic Cooperation and Development (OECD), which has these member countries, namely: Australia, Austria, Belgium, Canada, Denmark, Finland, France, Germany, Greece, Iceland, Ireland, Italy, Japan, Luxembourg, the Netherlands, Norway, Portugal, Spain, Sweden, Switzerland, Turkey, the United Kingdom, and the United States. See OECD in *Figures: Statistics on the Member Countries,* 1988 edition: Supplement to the OECD Observer No. 152, at 4-5 (June/July 1988); Helen Weidner, "The United States and North-South Technology Transfer: Some Practical and Legal Obstacles" (1982-83) 1-2 Wisconsin International Law Journal 205; Akilagpa Sawyerr, "Marginalisation of Africa and Human Development" (1993) 5 RADIC 176. I am aware of the crudeness of this distinction. There are nuances to the North-South paradigm. There is a "North" in the "South," as is exemplified by the privileged and "Westernized" elites of the Third World.

188 The term "South" refers to Africa, Asia (excluding Japan), Latin America, and Oceania. These areas are also referred to as "developing countries," "less-developed countries," or "Third World countries." Considering the experiences of indigenous minorities of North America, Australia, and Europe, it cannot be denied that there is a "South" in the North. See Melville Watkins, "North-South Relations" (1975) 5 Alternatives: Perspectives on Society and Environment 33. For an excellent analysis of the nature of the economic and cultural divide between the North and South, see Nassau Adams, *Worlds Apart: The North-South Divide and*

the International System (London: Zed Books, 1993); Geir Lundestad, *East, West, North and South: Major Developments in International Politics, 1945-1996* (Oslo: Scandinavian University Press, 1988). It should be noted that "the third world is far from a homogenous group. There are no strict criteria for qualifying as a developing country "In the United Nations; they form the Group of 77 States, although the group at present consists of more than 130 states. See Ignaz Seidl-Hohenveldern, *International Economic Law* (Dordrecht: Martinus Nijhoff, 1989) at 4; Winston Langley, "The Third World: Towards A Definition" (1981) 2 Boston College Third World Law Journal 1.

189 Robert Gutowski, "The Marriage of Intellectual Property and International Trade in the TRIPs Agreement: Strange Bedfellows or a Match Made in Heaven?" (1999) 47 Buffalo Law Review 713.

190 Abbott, *supra* note 72 at 223; Michael Gadbaw and Leigh Kenny, "India" in Gadbaw and Richards, *supra* note 79 at 1. This phenomenon is not limited to India. Michael Gadbaw's studies show that most countries of Southeast Asia and Latin America generally favour laws that contribute to their respective economies, even when such a stance may result in losses to nationals of other countries. The obvious inference is that patent laws are largely designed to serve the peculiar values and national interests of respective states, hence the struggle by the North to "tighten the noose." On the competing interests of states in patents, see also Simon Broder, *A Comparative Study of the United States Patent Office and the German Patent Office* (Ann Arbour, MI: University Microfilms, 1960, unpublished doctoral dissertation) at 224; Laurence Harrington, "Recent Amendments to China's Patent Law: The Emperor's New Clothes?" (1994) 17 Boston College International and Comparative Law Review 337. See O.J. Firestone, *Economic Implications of Patents* (Ottawa: University of Ottawa Press, 1971).

191 While most states profess to share the common criteria of novelty, utility, inventive step, and industrial applicability, legal definitions as to scope, entitlements, fields of technology, and exceptions still vary considerably and have not been harmonized in international law (at least until the TRIPS Agreement). Indeed, the differences in both the law and practice of patents are largely economic. For example, the United States' isolated use of the first-to-file system is not a "doctrinal eccentricity but an economic policy; this policy practically discriminates against foreign inventors." See Alford, *supra* note 156 at 46. On the Japanese experience, see Edmund Kitch, "The Japanese Patent System and US Innovators" (1996-97) 29 International Law and Politics 177.

192 Jeffrey Berkman, "Intellectual Property Rights in the P.R.C.: Impediments to Protection and the Need for the Rule of Law" (1996) 15 Pacific Basin Law Journal 1; Liwei Wang, "China's Patent Law and the Economic Reform Today" (1991) 9 Pacific Basin Law Journal 254; Kevin Murphy, "Reform of the Patent System in Canada" (1976) Annual of Industrial Property Law 81.

193 D. Jeremy, *Transatlantic Industrial Revolution: The Diffusion of Textile Technologies between Britain and America, 1790-1830s* (Cambridge, MA: MIT Press, 1981).

194 Alford, *supra* note 156 at 147.

195 Emphasis added. See Merges, *supra* note 144 at 245; Anderfelt, *supra* note 46 at 3.

196 US Congress, Office of Technology Assessment, OTA-CIT-302 (Washington, DC: US Government Printing Office, April 1986) at 228; Dru Brenner-Beck, "Do As I Say, Not As I Did" (1992) 11 Pacific Basin Law Journal 84; Abdulqawi Yusuf, "TRIPs: Background, Principles and General Provisions" in Carlos Alberto Correa and Adulqawi Yusuf, eds., *Intellectual Property and International Trade* (The Hague: Martinus Nijhoff, 1998) at xvii.

197 Anderfelt, *supra* note 46 at 13.

198 Abbott, *supra* note 137 at xxvii; Francis Gurry, "The Evolution of Technology and Markets and the Management of Intellectual Property Rights" in Abbott, *ibid.* Hence, the number of patents filed annually all over the world has shown a rapid increase. According to Abbott, "by the end of 1993, a total of 3.9 million patents were in force throughout the world ... the total number of patent applications worldwide have risen from 1,371,806 in 1989 to 1,965,487 in 1993, an increase of 43.3 percent" (*ibid.*). See also the comments of Professor Walter Hamilton in 1941: Walter Hamilton, Senate Temporary National Economic Committee, 76th Sess., Investigation of Concentration of Economic Power: Patents and Free Enterprise 164 (Comm. Print 1941). Interestingly, there is considerable force in the percep-

tion that the United States is losing competitive advantage in heavy industries to some countries in Asia and Latin America, hence the increasing emphasis on intellectual property, particularly patents.

199 Amy Carroll, "Not Always the Best Medicine: Biotechnology and the Global Impact of US Patent Law" (1995) 44 American University Law Review 2433 at 2439; but see Gutowski, *supra* note 189 at 720.

200 Myers McDougall et al., "Theories about International Law: Prologue to a Configurative Jurisprudence" (1968) 8 Virginia Journal of International Law 188.

201 The methods of compulsion include unilateral trade sanctions, diplomatic pressure, and interlinking demands for strong patent protection with unrelated interests such as "developmental aid" programs. See Liu, *supra* note 155.

202 Thomas Mesevage, "The Carrot and the Stick: Protecting US Intellectual Property in Developing Countries" (1991) 17 Rutgers Computer and Technology Law Journal 421; Stefan Kirchanski, "Protection of US Patent Rights in Developing Countries: US Efforts to Enforce Pharmaceutical Patents in Thailand" (1994) 16 Loyola LA International and Comparative Law Journal 569; Edgar Asebey and Jill Kempenaar, "Biodiversity Prospecting: Fulfilling the Mandate of the Biodiversity Convention" (1995) 28 Vanderbilt Journal of Transnational Law 703 (arguing that, "in fact, the international patent law system is often deceptively harmful to developing countries, causing them to exchange real rights for rights which are mostly theoretical"). See also Oddi, *supra* note 92.

203 Paul Lin, "U.S. Industry's Influence on Intellectual Property Negotiations and Special 301 Actions" (1994) 13 Pacific Basin Law Journal 87.

204 Carroll, *supra* note 199. Peter Nanyenya-Takirambudde, *Technology Transfer and International Law* (New York: Praeger, 1980); David Haug, "The International Transfer of Technology: Lessons that East Europe can Learn from the Failed Third World Experience" (1992) 5 Harvard Journal of Law and Technology 209; Paul Haar, "Revision of the Paris Convention: A Realignment of Private and Public Interests in the International Patent System" (1982) 8 Brooklyn Journal of International Law 17.

205 J.H. Reichman, "Intellectual Property in International Trade: Opportunities and Risks of a GATT Connection" (1989) 29 Vanderbilt Journal of Transnational Law 747. This situation is not limited to the South. An overwhelming majority of the patents in Australia and Canada are owned by American and Japanese multinational corporations who have no real interest in the domestic working of those patents. See Christopher Arup, *Innovation, Policy and Law: Australia and the International High Technology Economy* (Cambridge: Cambridge University Press, 1993) at 52-74. An obvious consequence of this phenomenon, especially for states like Canada and Australia, with an educated workforce in the sciences and possessed of the necessary supporting infrastructure for technological innovations, is that inventors (potential and actual) who probably would have hit upon those inventions and ideas independently are precluded from putting them to work because the foreign patents have priority.

206 As Abbott notes, lost economic opportunities operate on the hypothetical plane and are almost impossible to quantify accurately. However, estimates range from $43 billion to $61 billion for the year 1986 in the United States. The industries most affected are chemicals, pharmaceuticals, and computer software. See Frederick Abbott, "Protecting First World Assets in the Third World: Intellectual Property Negotiations in the GATT Multilateral Framework: (1989) 22 Vanderbilt Journal of Transnational Law 689.

207 Penrose, *supra* at note 84 at 116-17; Alan Deardoff, "Welfare Effects of Global Patent Protection" (1992) 59 Economica 35. For counter-arguments, see Edmund Kitch, "The Patent Policy of Developing Countries" (1994) 13 Pacific Basin Law Journal 166; Richard Tapp and Richard Posek, "Benefits and Costs of Intellectual Property Protection in Developing Countries" (1990) 24 Journal of World Trade 75. While admitting a lack of "expertise in the structure and strategies of or challenges faced by the countries" of the South, these authors nonetheless argue that it is in their self-interest to participate in the international patent system.

208 Carroll, *supra* note 199 at 2474; Michael Gadbaw, "Intellectual Property and International Trade: Merger or Marriage of Convenience?" (1989) 22 Vanderbilt Journal of Transnational Law 223; Kevin McCabe, "The January 1999 Review of Article 27 of the TRIPs Agreement:

Diverging Views of Developed and Developing Countries toward the Patentability of Biotechnology" (1998) 6 Journal of Intellectual Property Law 41.

209 Abbott, "Protecting First World Assets," *supra* note 206 at 699; but see Robert Sherwood, "The TRIPS Agreement: Implications for Developing Countries" (1996-97) 37 IDEA: The Journal of Research and Technology 491. According to an UNCTAD Report, "the nationals of developing countries hold in their own countries no more than 1 percent of the world stock of patents, and in other countries two-thirds of 1 percent of foreign-owned patents." See "The Role of The Patent System in the Transfer of Technology to Developing Countries," UN Doc. TD/B/AC.11/19.

210 "Something Old, Something New" Special Survey on World Trade, The Economist, 22 September 1990 at 34-35; but see Jerome Reichman, "From Free Riders to Fair Followers: Global Competition under the TRIPS Agreement" (1996-97) International Law and Politics 11.

211 Carroll, *supra* note 199 at 2440.

212 Ruth Gana, "Has Creativity Died in the Third World? Some Implications of the Internationalization of Intellectual Property" (1995) 24 Denver Journal of International Law and Policy 109 at 113; Ruth Gana, "Prospects for Developing Countries under the TRIPs Agreement" (1996) 29 Vanderbilt Journal of Transnational Law 735 (arguing that "without the specific conditions of strong property systems, free market capitalism, and the zealous protection of corporate interests, it is unlikely that modern intellectual property in and of itself has the potential to transform developing countries into the technology producers they aspire to become").

213 Penrose, *supra* note 84 at 233. According to the UN Report, "the role of patents is limited ... partly because much of the technology required by these [developing] countries is not at that latest stage of technological advance which is covered by patents." See UN Report *supra* note 85.

214 But see Jerome Reichman, "GATT Opportunities and Risks" (1989) 22 Vanderbilt Journal of Transnational Law 777.

215 *Ibid.*, at 747. See also Frank Emmert, "Intellectual Property in the Uruguay Round: Negotiating Strategies of the Western Industrialized Countries" (1990) 11 Michigan Journal of International Law 1317; Carlos Correa, "Harmonization of Intellectual Property Rights in Latin America: Is There Still Room for Differentiation?" (1996-97) 29 International Law and Politics 109; but see Kirstin Petersen, "Recent Intellectual Property in Developing Countries" (1992) 33 Harvard International Law Journal 277.

216 Mark Weisbrot, "Globalization for Whom" (1998) 31 Cornell International Law Journal 631. The phenomenon of globalization refers to the multifaceted process of global liberalization in which international issues, mainly economic, political, and sociocultural, are as prominent as are national and local matters with regard to the integration of markets across the world. See Alex Seita, "Globalization and the Convergence of Values" (1997) 30 Cornell International Law Journal 429. Seita locates the origins of globalization in the immediate postwar creation of the United Nations; the "Bretton Woods" institutions of the International Monetary Fund (IMF) and the International Bank for Reconstruction and Development (World Bank); and the General Agreement on Trade and Tarriffs (GATT). For an analysis of the ideological character and basis of globalization, see Geisinger, *supra* note 8, Chapter 1, at 40.

217 The concept of harmonization and the concept of unification differ from one another. The former refers to the establishment of a uniform set of laws for a certain area of territories; the latter refers a search for uniform solutions to specific international problems transcending national territories. See Ladas, *supra* note 44 at 14-15.

218 No-Hyoung Park, "The Third World as an International Legal System" (1987) 7 Boston College Third World Law Journal 37. The term "Third World" was first coined in France in the early 1950s at a time of bipolarity in the world security paradigm. Thus, the United States and its Western European allies were regarded as the First World. The defunct Soviet Union, with its client states in Eastern Europe, was regarded as the Second World. The so-called "non-aligned" states of Africa (excluding apartheid South Africa), Asia (excluding Japan), and Latin America constituted the Third World. See W. Tieya, "The Third World and International Law" in R. MacDonald and Douglas Johnston, eds., *The Structure and Process of International Law* (Dordrecht: Martinus Nijhoff, 1986); Paul Brietzke, "Insurgents

in the 'New' International Law" (1994-95) 13 Wisconsin International Law Journal 1 (discussing the role of conflicts and divides, such as the North-South phenomenon, in influencing international law); Douglas Matthews, "International Inequality: Some Global and Regional Perspectives" (1988-99) 7 Wisconsin International Law Journal 261.

219 David Schiff, "Socio-Legal Theory: Social Structure and Law" (1976) 39 Modern Law Review 287; Irene Watson, "Law and Indigenous Peoples: The Impact of Colonialism on Indigenous Cultures" (1996) Cross Currents (Australia: La Trobe University Press.) 107; Richard Guest, "Intellectual Property Rights and Native American Tribes" (1995-96) 20 American Indian Law Review 111.

220 Robert Gordon, "Critical Legal Histories" (1984) 36 Stanford Law Review 57 (arguing that the concepts of law and society are not necessarily separate). As Ivan Head has also argued, "some of the applications of legal principles, designed as they often were in the industrialized countries, are not always in the best interest of developing countries." See Ivan Head, "The Contribution of International Law to Development" (1987) 25 Canadian Yearbook of International Law 29 at 30. See also Philip Jessup, "Diversity and Uniformity in the Law of Nations" (1964) 58 American Journal of International Law 343.

221 Ladas, *supra* note 44 at 14.

222 *Ibid.* (arguing persuasively that, apart from legislative texts, there are three other basic components of harmonization or unification of laws that "global legislators" must confront. The three factors are: (1) judicial and administrative decisions that construe the legislative texts; (2) traditional techniques and modes of handling legal materials; and (3) philosophical, political, and ethical ideas as to the purpose of the law by which the texts, decisions, or techniques are continuously shaped).

223 For further analysis of the phenomenon of "interculture" in international law, see Park *supra* note 218 at 39. See also the late Polish jurist Manfred Lachs, "The Development and General Trends of International Law in Our Time" (1980) 169 Recueil Des Cours 240-1; Josef Kunz, "Pluralism of Legal and Value Systems and International Law" (1955) 49 American Journal of International Law 370; Compare with Article 9 of Statute of the International Court of Justice; Barcelona Traction (*Belgium* v. *Spain*), (1970) ICJ 273-74 (Gros. J., dissenting).

224 Rosemary Coombe, "The Cultural Life of Things: Anthropological Approaches to Law and Society in Conditions of Globalization" (1995) 10 American University Journal of International Law and Policy 791 at 792; Rosemary Coombe, "Critical Cultural Legal Studies" (1998) 10 Yale Journal of Law and the Humanities 463.

225 On the various legal norms governing this concept in traditional paradigms of the South, see WIPO *supra* note 15. The cultural question relates to the "values and attitudes that bind the system together and determine the place of the legal system in the culture of the system as a whole." Park, *supra* note 218 at 38. The role of culture in international law has its own share of juridical and scholarly controversy. See Kotaro Tanaka, "The Character of World Law in the International Court of Justice" (1971) 15 Japanese Annual of International Law 1; Philip Jessup, "Non-Universal International Law" (1973) 12 Columbia Journal of Transnational Law 415.

226 Oliver Wendell Holmes Jr., "The Path of the Law" (1897) 10 Harvard Law Review 469 ("if you want to know why a rule of law has taken its particular shape, and more or less if we want to know why it exists at all, we go to tradition").

227 Dan Rosen and Chikako Usui, "The Social Structure of Japanese Intellectual Property Law" (1994) 13 Pacific Basin Law Journal 32; Samson Helfgott, "Cultural Differences Between the US and Japanese Patent Systems" (1990) 72 Journal of Patent and Trademark Office Society 231-8. For Japan and some countries of Asia, the attitude in question is a reflection of the strong Confucian influence on social behavior and philosophy. Information is largely construed as a public good in Confucian thinking and philosophy. Confucianism also posits a good society based on mutual consideration for the needs of others. In contrast, Western individualism expounds a philosophy of "every person for himself or herself." According to Rosen and Usui, "to say that Japan is a group-oriented culture and America is an individualistic one is a cliché, but there is enough truth in the stereotypes to retain them despite the exceptions that can be found."

228 Merges, *supra* 97 at 242.

229 Rosen and Usui, *supra* 227 at 33-34. See also Toshiko Takenaka, "Does a Cultural Barrier to Intellectual Property Trade Exist? The Japanese Example" (1996-97) 29 International Law and Politics 153.

230 Gana, "Creativity," *supra* note 212 at 113; Brian Barron, "Chinese Patent Legislation in Cultural and Historical Perspective" (1991) 6 Intellectual Property Journal 313. For an excellent analysis of the cultural and ideological conflicts embedded in interculturalism, see J.M. Balkin, *Cultural Software: A Theory of Ideology* (New Haven: Yale University Press, 1998).

231 Boaventura De Sousa Santos, *Toward a New Common Sense: Law, Science and Politics in the Paradigmatic Transition* (New York: Routledge, 1995) at 263. For an analytical perspective on interculturalism and globalization, see William Twining, "Globalization and Legal Theory: Some Local Implications" (1996) 49 Current Legal Problems 1.

232 Gana, "Creativity," *supra* note 212 at 112. It is therefore not surprising that some scholars have proposed a human rights solution to the problem of appropriation of TKUP. See Rosemary Coombe, "Intellectual Property, Human Rights and Sovereignty: New Dilemmas in International Law Posed by the Recognition of Indigenous Knowledge and the Conservation of Biodiversity" (1998) 6 Indiana Journal of Global Legal Studies 59. For a thorough examination of the human rights implications of the globalization trend, see Robert McCorquodale and Richard Fairbrother, "Globalization and Human Rights" (1999) 21 Human Rights Quarterly 735.

233 For a recent compendium of non-Western philosophies and ideologies on ownership and property, see WIPO *supra* note 15. For an analysis of the jurisdictional problems associated with intellectual property in an age of globalization, see Keith Aoki, "(Intellectual) Property and Sovereignty: Notes toward a Cultural Geography of Authorship" (1996) 48 Stanford Law Review 1293.

234 For a brilliant analysis of the evolution and maturation of the jurisprudence of property in the Western world and its utilitarian ideology, see Morris Cohen, "Property and Sovereignty" (1927) 13 Cornell Law Quarterly 8. The essence of privacy is always the right to exclude others. The concept of property has, however, displayed amazing capacities for fluidity and flexibility. As Keith Aoki has pertinently noted, "property" in the "West has been and continues to be an empty vessel into which we have poured an incredibly wide variety of meanings." Aoki, *supra* note 233 at 1319.

235 Gana, "Creativity," *supra* note 212 at 132, emphasis added; Sherry Hut and Timothy Mckeown, "Control of Cultural Property as Human Rights" (1999) Vol. 31, No. 2 Arizona State Law Journal 363; Adrienne Van Nieherk, "Indigenous Law and Narrative: Rethinking Methodology" (1999) 32 Comparative and International Law Journal of South Africa 208.

236 Amanda Park, "Cultural Appropriation and the Law: An Analysis of the Legal Regimes Concerning Culture" (1993-94) 8 Intellectual Property Journal 82.

237 Gana, "Creativity," *supra* note 212 at 142. Compare with Fred Cate, "Sovereignty and the Globalization of Intellectual Property" (1998) 6 Indiana Journal of Global Legal Studies 1.

238 For a thorough analysis of the interface between Western property regimes and traditional societies, see Rosemary Coombe, "The Properties of Culture and the Politics of Possessing Identity: Native Claims in the Cultural Appropriation Controversy" (1993) Vol. 6, No. 2 Canadian Journal of Law and Jurisprudence 249.

239 Rodolpho Sandoval and Chung-Pok Leung, "A Comparative Analysis of Intellectual Property Law in the United States and Mexico, and the Free Trade Agreement" (1993) 17 Maryland Journal of International Law and Trade 145.

240 The number of international instruments on patent law and institutions is legion. However, the most notable include the Paris Convention for the Protection of Industrial Property of 20 March 1883, as revised at Stockholm in July 1967, 828 U.N.T.S. 305; Patent Cooperation Treaty, 19 June 1970, 1160 U.N.T.S. 231; Convention Establishing the World Intellectual Property Organization, opened for Signature 14 July 1967. Reprinted in 828 U.N.T.S. 3. For a justification of the international patent system, see Warren Wolfeld, "International Patent Cooperation: The Next Step" (1983) 16 Cornell International Law Journal 229.

241 Paris Convention, *supra* note 240. For a detailed history and analysis of the Paris Convention, see G.H.C. Bodenhausen, *Guide to the Application of the Paris Convention for the Protec-*

tion of Intellectual Property, as Revised at Stockholm in 1967 (Geneva: BIRPI, 1968); Friedrich-Karl Beier, "One Hundred Years of International Cooperation: The Role of the Paris Convention in the Past, Present and Future" (1984) 15 International Review of Industrial Property and Copyright Law 1.

242 Belgium, Brazil, France, Guatemala, Italy, the Netherlands, Portugal, Salvador, Serbia, Spain, and Switzerland. As of April 1999, 153 states were parties to the Paris Convention.

243 Bodenhausen, *supra* note 241 at 13.

244 For an analysis of this phenomenon, see Peter Drahos, "Global Property Rights in Information: The Story of TRIPs at the GATT" (1995) 13 Prometheus 12.

245 Article 2, Paris Convention, *supra* note 240. The principle of the right of priority prescribes that, if a patent application is filed in any union state within one year of the "home country" or other first union filing (and assuming simple formalities are met), the application is effectively back-dated to the first filing. In other words, there is a one-year grace period for patentees to file their patents outside their home countries. On paper the principle of national treatment seems fair, but the capacity of citizens of different countries to exploit a reciprocally globalized set of duties and rights for intellectual property will vary dramatically. It is akin to saying that an Olympian athlete and a cripple are all potential gold medalists in a compulsory 100-metre dash.

246 Article 4, Paris Convention, *supra* note 240. On the other hand, the "national treatment" principle ensures that foreigners receive the same privileges under national patent laws as do nationals of the home country of a member state. But see Harold Wegner and Jochen Pachenberg, "Paris Convention Priority: A Unique American Viewpoint Denying 'The Same Effect' to the Foreign Filing" (1974) 5 International Review of Industrial Property and Copyright Law 361; Robert Pritchard, "The Future Is Now: The Case for Patent Harmonization" (1995) 20 North Carolina Journal of International Law and Commercial Regulation 291.

247 Wegner, *supra* note 44 at 539.

248 Patent Cooperation Treaty, *supra* note 240. Prior to the PCT an inventor seeking to patent an invention would have to file the application in a first country, and, if that country was a party to the Paris Convention, then the application was entitled to a twelve-month priority (grace period) over other similar applications for the same invention. If, within this period, the applicants failed to conclude the formalities for the grant of the patent in those states where the right of priority existed, then they lost the right of novelty and other persons might proceed to apply for a patent on that invention. It was an all-or-nothing system, but the PCT changed that regime. The PCT is a clearinghouse of sorts. As is already evident, patents remain substantively governed by national and regional patent laws and arrangements.

Thus, under the PCT, when applying for a patent the applicant files the application at a Receiving Office (RO), which is a national or regional patent office where the filing would ordinarily take place. The inventor specifies or designates the countries in which he seeks a patent. A copy of the application is remitted to the International Search Agency (ISA), which conducts a search of the prior art on the application in the affected states. While the search for priority art is on, a regime of priority for that patent application automatically begins. See Edward Mckie, "Patent Cooperation Treaty: A New Adventure in the Internationality of Patents" (1978-79) 4 North Carolina Journal of International Law and Commercial Regulation 249.

249 A search report is sent to the applicant within sixteen months of the application. The application is published eighteen months after the priority date, but if the applicant withdraws the application before publication, then publication will not take place. At the end of twenty months following the priority date, the application enters the national phase, at which time fees are paid in the designated states where the applicant seeks a patent.

250 Before filing for patent protection in a foreign country for an American invention, the US inventor must obtain a licence from the Commissioner of Patents. See 35 U.S.C. 184; *Beckman Indus., Inc.,* v. *Coleman Instruments, Inc.,* 338 F. 2d 573 (7th Cir. 1964); R. Carl Roy, "The History of the Patent Harmonization Treaty: Economic Self-Interest as an Influence" (1993) 26 John Marshall Law Review 457. Interestingly, because many more patent applications

are filed in the North than in the South, the principles of national treatment and right of priority that underpin patent harmonization legislation inexorably favour the North.

251   Convention on the Unification of Certain Points of Substantive Law on Patents for Inventions, 27 November 1963, European Treaty Series No. 47 (otherwise known as the Strasbourg Convention). The Strasbourg Convention is the blueprint for the EPC. See also David Bainbridge, *Intellectual Property* (London: Pitman Publishing, 1992) at 8.

252   Convention on the Grant of European Patents, done 5 October 1973, reprinted in (1973) 13 I.L.M. 270. All EC member states except Portugal have signed this treaty. The EPC is a procedural agreement that allows applicants to apply through the EPO to receive multiple national patents for each member of the EPC. See Cynthia Ho, "Building a Better Mousetrap: Patenting Biotechnology in the European Community" (1992) 3 Duke Journal of Comparative and International Law 173.

253   Convention for the European Patent for the Common Market, done 15 December 1975, reprinted in 19 O.J. Eur. Comm. (No. L. 17) 1 (1976). For a negotiating history of the CPC see Romuald Singer, "The European Patent Enters a New Phase" (1970) 1 International Review of Industrial Property and Copyright Law 19; Friedrich- Karl Beier, "The European Patent System" (1981) 14 Vanderbilt Journal of Transnational Law 1; Daniel Lachat, "The Luxembourg Conference on the Community Patent Convention" in 1978 Annual of Industrial Property Law (Commonwealth Law Reports, 1978) at 13.

254   Note that Article 36 of the Treaty of Rome, which established the European Union, provides a framework for the protection of patents in the European Union. See Treaty Establishing the European Community, 25 I.L.M. 503 (1957). See also European Convention Relating to the Formalities Required for Patent Applications, done Paris, 11 December 1953, Treaty Series No. 43 (1955); European Convention on the International Classification of Patents for Invention, Paris, 19 December 1954, reproduced in Treaty Series No. 12 (1963).

255   Abbott, *supra* note 72 at 779. James Fawcett and Paul Torremans, *Intellectual Property and Private International Law* (Oxford: Clarendon Press, 1998) at 27.

256   Agreement on Andean Sub-Regional Integration, 26 May 1969, reprinted in 8 I.L.M. 910 (1969). See also Decision 344 regarding Common Provisions on Industrial Property, reproduced at page 825 of Abbott, *supra* note 72. The Andean Group has Bolivia, Colombia, Ecuador, Peru, and Venezuela as members. There is also the fledgling ASEAN Framework Agreement on Intellectual Property Co-operation of 15 December 1995, concluded in Bangkok, Thailand. See text of agreement reprinted in (1996) 6 Asian Yearbook of International Law 505. This regional framework comprises member states of the Association of Southeast Asian Nations.

257   Wilfrido Fernandez, "MERCUSOR and Its Implications for Industrial Property" in Abbott, *supra* 72 at 443.

258   Treaty Establishing a Common Market between the Argentine Republic, the Federative Republic of Brazil, the Republic of Paraguay, and the Eastern Republic of Uruguay, 30 I.L.M. 1041 (1991) [hereinafter the Asuncion Treaty]. The Asuncion treaty establishing MERCUSOR is comprised of twenty-four articles and four annexes.

259   North American Free Trade Agreement, 32 I.L.M. 612 (1992).

260   The agreement entered into force on 15 February 1978. Its original name was Industrial Property Organization for English-Speaking Africa (ESARIPO). In December 1985 it changed its name to ARIPO by decision of its council. ARIPO has its headquarters in Harare, Zimbabwe.

261   Similarly, a protocol on patents within the framework of ARIPO, adopted at a special meeting held in Harare in December 1982, entered into force on 25 April 1984. This protocol provides for the processing and granting of patent application on behalf of contracting states designated. See, Abbott, *supra* note 72 at 477.

262   This is probably the closest to an automatic extraterritorial patent.

263   See Agreement to Revise the Bangui Agreement on the Creation of an African Intellectual Property Organization of 2 March 1977, Bangui, 24 February 1999. See also Tshimanga Kongolo, "The New OAPI Agreement as Revised in February 1999" (2000) 3 Journal of World Intellectual Property 5; Tshimanga Kongolo, "Towards a More Balanced Coexistence

of Traditional Knowledge and Pharmaceuticals Protection in Africa" (2001) Vol. 35, No. 2 Journal of World Trade 349.

264 Abbott, *supra* note 72 at 482. These include Partnership Agreement between the African, Caribbean and Pacific States and the European Community and Member States, as concluded in Brussels on 3 February 2000. EU Document CE/TFN/GEN/23-OR ACP/00/0371/ 00 and the Scandinavian Patent Community.

265 While the World Bank is increasingly getting involved in the question of building infrastructure for the takeoff and maintenance of patent regimes in the South, UNCTAD is concerned with the role of trade and investment in enhancing economic development in the developing countries – a role tied up with patents. In the same vein, UNEP and the UNDP are both concerned with achieving sustainable development in the international community and in promoting the interests of developing countries. Of course, this has implications for the patent concept, particularly regarding the contemporary question of patents on plant resources. High on UNEP's agenda is the protection of biological diversity, which inevitably deals with the relationship between the patent system and biodiversity. On the other hand, the WHO seeks to promote improvements in health care around the world and, consequently, the extent of patent protection on pharmaceutical products are of interest to it since these patents affect the price and availability of drugs.

266 WIPO is a metamorphosed version of the International Bureau of the Paris Convention (BIRPI). See Faryan Afifi, "Unifying International Patent Protection: The World Intellectual Property Organization Must Coordinate Regional Patent Systems" (1993) 15 Loyola LA International and Comparative Law Journal 453.

267 In addition to the Paris Convention, these include the Strasbourg Agreement Concerning the International Patent Classification, the Budapest Treaty on the International Recognition of the Deposit of Microorganisms for the Purposes of Patent Procedure, and the International Union for the Protection of New Varieties of Plants (UPOV) 2 December 1961, 815 U.N.T.S. 89.

268 In November 1961 Brazil introduced a draft resolution, entitled "The Role of Patents in the Transfer of Technology to Underdeveloped Countries" (UN Doc. A/C.2/L.565, 8 November 1961), which heavily criticized the international patent system.

269 General Assembly, Official Records, 16th Sess., 2nd Committee, 778th Meeting, 7 December 1961. The Brazilian resolution was substantially amended, but its primary result was the passing of a resolution mandating the secretary-general to hold an international conference on the patent question.

270 Doc. E/3861/Rev. 1, March 1964.

271 Doc. E/Conf. 46/141. Vol. 1, Part 3, Annex A. IV.26. See also, Anderfelt, *supra* note 46 at 185.

272 Doc. E/Conf. 46/141. Vol. 1, Part 3, Annex F, Report of III Committee.

273 See, Doc. A/C.2/L.824, 26 November 1965. The co-sponsors were Austria, the Dominican Republic, Mexico, and Peru.

274 For a thorough analysis of this trend, see generally Anderfelt, *supra* note 46 at 175-240. Anderfelt's magisterial work on the subject concludes with the observation that "the present international patent system is inadequate for the needs and requirements of developing countries." Anderfelt, *supra* at 278.

275 The perception by the North that WIPO was leaning towards the interests of the South may be attributed to the following reasons. First, the WIPO framework, unlike the WTO framework, lacks the coercive abilities to enforce IPRs. Second, WIPO has a history of assisting the South in preparing laws and developing personnel in patent matters. Third, WIPO, unlike the WTO, has always operated on the principle of one-state-one-vote, and the balance of power inside WIPO thus favours the more numerous states of the South. See Abbott, *supra* note 72 at 366.

276 Agreement Establishing the Multilateral Trade Organization [World Trade Organization], 15 December 1993, reprinted in (1993) 33 I.L.M. 13; see also William Walker, "Uruguay Round TRIPs: A Bibliographic Essay" (1989) 22 Vanderbilt Journal of Transnational Law 911; Braga, *supra* note 26; Thomas Dillon, "The World Trade Organization: A New Legal Order for World Trade?" (1995) 16 Michigan Journal of International Law 349; Michael

Doane, "TRIPS and International Intellectual Property Protection in an Age of Advancing Technology" (1994) American University Journal of International Law and Policy 465.

277   General Agreement on Tariffs and Trade, 30 October 1947, 55 U.N.T.S. 187.

278   Agreement on Trade-Related Aspects of Intellectual Property Rights; reprinted, 33 I.L.M. 1197 (1994).

279   Friedrich Beier and Gerhard Schricker, eds., *GATT or WIPO? New Ways in the International Protection of Intellectual Property* (Munich: Max Planck Institute, 1989); Jagdish Bhagwati and Mathias Hirsch, eds., *The Uruguay Round and Beyond: Essays in Honor of Arthur Dunkel* (Ann Arbor, MI: University of Michigan Press, 1999); John Croome, *Reshaping the World Trading System: A History of the Uruguay Round* (The Hague: Kluwer Law International, 1999); Asif Qureshi, *The World Trade Organization: Implementing International Trade Norms* (Manchester: Manchester University Press, 1996); David Hartridge and Arvind Subramanian, "Intellectual Property Rights: The Issues in GATT" (1989) 22 Vanderbilt Journal of Transnational Law 893; Hans Kunz-Hallstein, "The United States Proposal for a GATT Agreement on Intellectual Property and the Paris Convention for the Protection of Industrial Property" (1989) 22 Vanderbilt Journal of Transnational Law 265.

280   Article 27, TRIPS. In addition, member states are not allowed to grant compulsory licences for non-working of the patents. Note that for purposes of defining "non-working," an importation or marketing of the patented product is construed as sufficient local working of the invention.

281   Gana, *Creativity, supra* note 212 at 113; Alan Gutterman, "The North-South Debate Regarding the Protection of Intellectual Property Rights" (1993) 28 Wake Forest Law Review 89; Marshall Leafer, "Protecting United States Intellectual Property Abroad: Toward A New Multilateralism" (1991) 76 Iowa Law Review 273.

282   Convention on Biological Diversity, 29 December 1993, opened for signature 5 June 1992 (entered into force 29 December 1993) reprinted in (1993) I.L.M. 813.

283   Cheryl Hardy, "Patent Protection and Raw Materials: The Convention on Biological Diversity and Its Implications for US Policy on the Development and Commercialization of Biotechnology" (1994) 15 University Pa. Journal of International Business Law 299.

284   For an illuminating account of the sources of general international law, see the dated but classic work of Clive Parry, *The Sources and Evidences of International Law* (Manchester: Manchester University Press, 1965).

285   Louis B. Sohn, "The Shaping of International Law" (1978) 8 Georgia Journal of International and Comparative Law 1.

286   Article 38, ICJ Statute; Statute of the International Court of Justice. Concluded at San Francisco, 26 June 1945. Entered into force 24 October 1945. 1976 Y.B.U.N. 1052. Article 38 is probably an approximation of the sources of international law and largely provides a legal construct in which the ICJ has to operate. Although the committee of jurists that drafted the Statute of the Court in 1920 included in a draft provision that the items listed in the first paragraph of Article 38 should be applied in *ordre successif*, it is not clear whether the said article actually intends a hierarchical order. For an excellent analysis of this interesting point of law, see Michael Akehurst, "The Hierarchy of the Sources of International Law" (1974-75) 47 British Yearbook of International Law 273; M. Dixon, *Textbook on International Law* (London: Blackstone Press, 1993) at 19. There are also instances where public unilateral declarations by the high authorities of a state may constitute a source of international law. See "Nuclear Tests Case" (1971) ICJ Reports 15.

287   "North Sea Continental Shelf Cases" (1969) ICJ Reports at 41-4; "Asylum Case" (1950) ICJ Reports 266; "Rights of Passage Case" (1960) ICJ Reports 6. For an excellent philosophical inquiry into the nature of customary law and its resilience and flexibility, see Lon Fuller, "Human Interaction and the Law" (1969) 14 American Journal of Jurisprudence 1; Lon Fuller, "Law as an Instrument of Social Control and Law as a Facilitation of Human Interaction" (1975) 1 Brigham Young University Law Review 89; Anthony D'Amato, *The Concept of Custom as a Source of International Law* (Ithaca: Cornell University Press, 1971).

288   This is a belief by the state that its compliance is required by international law. See "*Nicaragua v. United States* (Merits)" (1986) ICJ Reports 98; Lotus Case (1927) PCIJ Series A, No.

10. The concept of *opinio juris* has been criticized by Lauterpacht on the grounds that it is circular: to wit, states are only bound by custom if they believe themselves to be so bound. See H. Lauterpacht, "Sovereignty Over Submarine Areas" (1950) 27 British Yearbook of International Law 376.

289 For an excellent analysis of the normative and legal character of resolutions and their declaratory function as customary international law, see Jorge Castaneda, *Legal Effects of United Nations Resolutions* (New York: Columbia University Press, 1969); Rosalyn Higgins, *The Development of International Law through the Political Organs of the United Nations* (London: Oxford University Press,1963); F. Blaine Sloan, "The Binding Force of a Recommendation of the General Assembly of the United Nations" (1948) 25 British Yearbook of International Law 1; F. Blaine Sloan, "General Assembly Resolutions Revisited (Forty Years Later)" (1987) 57 British Yearbook of International Law 39; Jochen Frowein, "The Internal and External Effects of Resolutions by International Organizations" (1989) 49 Heidelberg Journal of International Law 778; D.N. Saraf, "Resolutions of International Organizations: Binding Norms?" (1990) 14 Cochin University Law Review 1.

290 Nirmala Naganathan, "What Constitutes Custom in International Law?" (1971) 1 Colombo Law Review 68; Benedetto Conforti, *International Law and the Role of Domestic Legal Systems* (Dordrecht: Martinus Nijhoff Publishers, 1993) at 61.

291 "Anglo-Iranian Oil Company Case," [1952] ICJ Reports 93. On reservations to bilateral treaties, see "Delimitation of the Continental Shelf Between the United Kingdom and France" (1979) 18 United Nations Report of International Arbitral Awards 3. On reservations to multilateral treaties see the advisory opinion of the International Court of Justice in the case of "Reservations on the Convention on Genocide" (1951) ICJ Reports 15. In *"Belilos* v. *Switzerland"* 132 Eur. Ct. H.R. (Ser. A) (1988), the European Court decided that a reservation entered into by the Swiss government on the European Convention on Human Rights was not permissible because it was of a general character. The ratio here is that, where a reservation or interpretative declaration to a treaty undermines the essence of that treaty, such a reservation would probably be construed as invalid.

292 "South West Africa Cases" (1962) ICJ Reports 310.

293 See the jurisdictional phase of the decision of the ICJ in "Aegean Sea Continental Shelf Cases" (1978) ICJ Reports 3. See also Eduardo Jimenez de Arechaga, "The Work and the Jurisprudence of the International Court of Justice, 1947-1986" (1987) 58 British Yearbook of International Law 1.

294 Vienna Convention on the Law of Treaties, reprinted in 1155 U.N.T.S. 339.

295 R.Y. Jennings, "The Progressive Development of International Law and Its Codification" (1947) 24 British Yearbook of International Law 301. See also Chin Lin and Olufemi Elias, "The Role of Treaties in the Contemporary International Legal Order" (1997) 66 Nordic Journal of International Law 1.

296 Abbott, *supra* note 72 at 488.

297 On judicial precedent as sources of international patent law, according to the "Appellate Body Report on Japan: Taxes on Alcoholic Beverages," 1 November 1996, WT/DS8/AB/R, WT/DS10/AB/R, WT/DS13/AB/R, at 14: "Adopted panel reports create legitimate expectations among all relevant parties to a dispute." In effect, the panel agreed that a certain kind of implied or expected *res judicata* may apply in the interpretation of multilateral agreements on patent law. Another doctrine of importance is the rule of consistent application. See Thomas Cottier and Krista Schefer "The Relationship between WTO Law, National and Regional Law" in Abbott, *supra* 72 at 558. This doctrine prescribes that, where a national rule allows for different interpretations, national or regional law has to be construed in accordance with international obligations. See the European Court of Justice in Werner and Leifer Case, C-70/94 [1995] ECR 1-3189.

298 Abbott, *supra* note 72 at 496.

299 For an interesting analysis of the category of "general principles of law recognized by civilized nations," see M. Cherif Bassiouni, "A Functional Approach to General Principles of International Law" (1990) 11 Michigan Journal of International Law 768. (Note citations therein.)

300 Conforti, *supra* note 290 at 4.

301   "Competence of the General Assembly for Admission of a State to the United Nations" (1950) ICJ Reports 4.

302   Quincy Wright, "The Legal Nature of Treaties" (1916) 10 American Journal of International Law 706.

303   Carlos Manuel Vazquez, "The Four Doctrines of Self-Executing Treaties" (1995) 89 American Journal of International Law 695; Ignatius Seidl-Hohenveldern, "Transformation or Adoption of International Law into Municipal Law" (1963) 12 International and Comparative Law Quarterly 88.

304   Lord Atkin seems to have summarized the position thus in the locus classicus of *Attorney-General for Canada* v. *Attorney-General for Ontario*, [1937] A.C. 326. In his words, "within the British Empire there is a well-established rule that the making of a treaty is an executive act, while the performance of its obligations, if they entail alteration of the existing domestic law, requires legislative action." See also "The Parlement Belge" 4 P.D. 129 (1879); *Rayner* v. *Department of Trade*, [1990] 2 A.C. 419; *Arab Monetary Fund* v. *Hashim*, [1991] 1 All E.R. 316. For further disquisition on the subject, see F.A. Mann, *Studies in International Law* (London: Clarendon Press, 1973) 328-33.

305   *Foster* v. *Neilson*, 27 US 253 (1829). See also, *United States* v. *Percham*, 32 US (7 Pet.) 51 (1833); Richard Wilder, "The Effect of the Uruguay Round Implementing Legislation on US Patent Law" (1995-6) 36 IDEA: The Journal of Research and Technology 33.

306   *General Sani Abacha and Ors.*, v. *Chief Gani Fawehinmi* [2000] Part 660 Vol. 6 Nigerian Weekly Law Report 228.

307   Sec 102 (a) (1), An Act to Approve and Implement the Trade Agreements Concluded in the Uruguay Round of Multilateral Trade Negotiations, the Uruguay Round Agreements Act 1994 19 U.S.C. 3512. As the US Court of Appeals held in *Suramerica de Aleaciones Laminadas, C.A.* v. *United States*, 966 F 2d. 660 (Federal Circuit. 1992) "the GATT does not trump domestic legislation; if the statutory provisions at issue here are inconsistent with the GATT, it is a matter for Congress and not this Court to decide and remedy." According to Vazquez, research by Professor Hudec on American case law reveals that, in fourteen cases where the issue came up for consideration, the courts consistently affirmed that local law trumped the GATT. Eight of such cases avoided the issue and six affirmed the supremacy of federal law over the GATT. See Vazquez, *supra* note 303 at 111.

308   Henry Schermers, "Some Recent Cases Delaying the Direct Effect of International Treaties in Dutch Law" (1989) 10 Michigan Journal of International Law 266.

309   Myers McDougal, "The Impact of International Law upon National Law: A Policy Oriented Perspective" (1959) 4 South Dakota Law Review 25; John Jackson, "Status of Treaties in Domestic Legal Systems: A Policy Analysis" (1992) 86 American Journal of International Law 2.

310   Vazquez, *supra* note 303 at 105. There has been considerable scholarly activity in this vexed issue of international law. However, neither scholarly exertions nor international trade diplomacy has brought any resolution to the matter. See Frederick Abbott, "GATT and the European Community: A Formula for Peaceful Coexistence" (1990) 12 Michigan Journal of International Law 1.

311   Pierre Pescatore, "The Doctrine of Direct Effect: An Infant Disease of Community Law" (1983) 8 European Law Review 158. See also, Vazquez *supra* note 303 at 105.

312   Vazquez, *supra* note 303 at 114. For further analysis of this aspect of European Community law, see David Demiray, "Intellectual Property and the External Power of the European Community: The New Extension" (1994) 16 Michigan Journal of International Law 187.

313   See, for example, *International Fruit Company* v. *Produktschap voor Groenten en Fruit* (Case 21-24/72); Preliminary Ruling of 12 December 1972; [1972] ECR 1219; GA Mayras Case, [1972] ECR 1219. For an empirical analysis of the situation in Germany, see Ernst Pakuscher, "The Patent Court of the Federal Republic of Germany and the European Patent Conventions" (1980) 4 Comparative Law Yearbook 167. But see the European Court of Justice in *Fa. Alfons Lutticke* v. *Haupzollant Sarrelouis*, Case 57/65, (1971) 10 Com. Mark. Law Report 674. (Holding that: "the obligation [of the Paris Convention] is not qualified by any condition

nor made subject, in its carrying out or its effects, to the intervention of any act of either of the Community institutions or of the member States. The prohibition is thus complete, legally complete, and consequently capable of producing direct effects in the legal relationship between the member States and their subjects." *Ibid.*

314 Council Decision Concerning the Conclusion on Behalf of the European Community, as Regards Matters within Its Competence, of the Agreements Reached in the Uruguay Round of Multilateral Negotiations (1986-94) (94/800/EC), OJ No L 336/.23rd December 1994 at 2.

315 Vazquez, *supra* note 303 at 117. For a general analysis of the relationship between individual and international law, see J.M. Udochi, "The Individual as a Subject of International Rights and Duties" (1963) 2 Columbia Journal of Transnational Law 54.

316 John Jackson, "The Great 1994 Sovereignty Debate: United States Acceptance and Implementation of the Uruguay Round Results" (1997) 36 Columbia Journal of Transnational Law 157.

317 Vazquez, *supra* note 303 at 121; Schermers *supra* note 308. Compare with Jordan Paust, "Self Executing Treaties" (1988) 82 American Journal of International Law 760.

318 For a comprehensive and excellent analysis of the various methods of and schools of thought on the domestic applicability of treaties, see the collection of essays in Stefan Riesenfeld and Frederick Abbott, eds., *Parliamentary Participation in the Making and Operation of Treaties: A Comparative Study* (Dordrecht: Martinus Nijhoff, 1994) at xi.

319 Advisory Opinion No. 17, Interpretation of the Convention between Greece and Bulgaria Respecting Reciprocal Emigration, (1930) P.C.I.J (ser. B) No. 17, at 32 (July 31).

320 *Society for the Propagation of the Gospel in Foreign Parts* v. *New Haven*, 8 Wheat, 464 (1823) at 490; *Volkwagen A.G.* v. *Schlunk*, 486 US 694, 700 (1988).

Chapter 3: Implications of Biopiracy for Biological and Cultural Diversity

1 Article 2, Convention on Biological Diversity, entered into force on 29 December 1993, reprinted in (1992) 31 I.L.M 813; Jonathan Charney, "Biodiversity: Opportunities and Obligations" (1995) 28 Vanderbilt Journal of Transnational Law 613. Harvard biologist Edward O. Wilson has defined biodiversity as "the variety of organisms considered at all levels, from genetic variants belonging to the same species through arrays of species of genera, families and still higher taxonomic levels; includes the variety of ecosystems, which comprise both the communities or organisms within particular habitats and the physical conditions under which they live." See Edward Wilson, *The Diversity of Life* (Cambridge, MA: Harvard University Press, 1992) at 393. There are no accurate data on the immensity of biodiversity. It has been estimated that there are probably 1.4 million known species or organisms on this planet. Of this number, 750,000 are insects, 41,000 are vertebrates, and 250,000 are plants; the rest correspond to a complex of invertebrates, fungi, and microorganisms. See Ranee Panjabi, "International Law and the Preservation of Species: An Analysis of the Convention on Biological Diversity Signed at the Rio Earth Summit in 1992" (1993) 11 Dickinson Journal of International Law 187; compare with Cyril de Klemm, "The Convention on Biological Diversity: State Obligations and Citizens Duties" (1989) 19 Environmental Policy and Law 50.

2 Genetic diversity refers to the variety of genes. Inside the nucleus of every animal or plant cell are chromosomes, thread-shaped bodies consisting largely of four strings of DNA (Deoxyribonucleic acid), each of which comprises about a billion nucleotide pairs. If stretched out fully, the DNA would be roughly one metre long. The full information contained in DNA if translated into ordinary-size letters of printed text would fill all fifteen editions of the *Encyclopaedia Britannica*. Genes confer particular characteristics on the organisms that inherit them. Each gene is, in effect, a chemical instruction controlling a particular characteristic. These characteristics may include resistance to disease, rapid growth, an environmental adaptation to a particular factor, capacity to grow straight, and so on. Genes vary, and the variant of the same gene is called an allele. Thus, differences in the genetic makeup of a species are caused by the different alleles of each gene. A particular combination of genes is known as a genotype, and a given set of chromosomes is called a genome. The

term "gene pool" refers to the total number of genes within a group of interbreeding plants or animals (i.e., the pool of genes within a population). The population encompasses its wild and cultivated relatives.

3 Species diversity refers to the variety of species within an ecosystem.

4 An ecosystem can be defined as a given physical environment and all the organisms therein, together with the web of interactions those organisms have with each other as well as with the environment.

5 Kalyan Chakrabarti, *Conservation versus Development* (Calcutta: Darbari Prokashan, 1994) at 153.

6 Human cultural diversity is almost synonymous with biodiversity. The extinction of human culture is mutually implicated in the loss of biological diversity and vice versa. See Neil Gunningham and Mike Young, "Toward Optimal Environmental Policy: The Case of Biodiversity Conservation" (1997) 24 Ecology Law Quarterly 243.

7 Compare with Charlotte De Fontaubert et al., "Biodiversity in the Seas: Implementing the Convention on Biological Diversity in Marine and Coastal Habitats" (1998) 10 Georgetown International Environmental Law Review 753.

8 Wilson, *supra* note 1 at 281; Holly Doremus, "Patching the Ark: Improving Legal Protection of Biological Diversity" (1991) 18 Ecology Law Quarterly 265.

9 Katharine Baker, "Consorting with Forests: Rethinking Our Relationship to Natural Resources and How We Should Value Their Loss" (1995) 22 Ecology Law Quarterly 677.

10 James Salzman, "Valuing Ecosystem Services" (1997) 24 Ecology Law Quarterly 887.

11 Erin Newman, "Earth's Vanishing Medicine Cabinet: Rain Forest Destruction and Its Impact on the Pharmaceutical Industry" (1994) 20 American Journal of Law and Medicine 479; Amy Guerin Thompson, "An Untapped Resource in Addressing Emerging Infectious Diseases: Traditional Healers" (1998) 6 Indiana Journal of Global Legal Studies 257.

12 Kenton Miller et al., "Deforestation and Species Loss: Responding to the Crisis" in J.T. Matthews, ed., *Preserving the Global Environment: The Challenge of Shared Leadership* (New York: W.W. Norton, 1991) at 97. For an account of the despoliation of the Amazon, see Georges Landau, "The Treaty for Amazonian Cooperation: A Bold New Instrument for Development" (1980) 10 Georgia Journal of International and Comparative Law 463.

13 The rosy periwinkle is a plant found only in Madagascar. It has well-publicized curative properties with respect to childhood leukemia and Hodgkin's disease.

14 The Yew tree (Taxus brevifolia), found in the old-growth forests of the Pacific Northwest, contains taxon, which is an anti-cancer drug. See Panjabi on the CBD, *supra* note 1 at 193; Doremus, *supra* note 8 at 277.

15 This rare Cameroonian vine contains chemicals like Michellamine B, which is capable of blocking the reproduction of the AIDS virus.

16 The roots of tropical shrubs of the Rauwolfa genus produce reserpine, which is used as a sedative and for the treatment of high blood pressure.

17 The stupendous potential of tropical plants to yield complex chemicals is a function of their evolutionary circumstances as they have had to develop complex chemical arsenals in order to repel predators.

18 Digitalis, a drug derived from the purple foxglove, is used to treat congestive heart failure and other cardiac ailments. Similarly, curare, a poison used by Yanomami Indians for hunting and fishing, is used as a muscle relaxant.

19 According to the Chiang Mai Declaration, "we recognise the vital importance of medicinal plants in healthcare ... the value of the medicinal plants used today and the great potential of the plant kingdom to provide new drugs" ("The Chiang Mai Declarations: Saving Lives by Saving Plants" <http://www.wwf-uk.org/filelibrary/pdf/guidesonmedplants. pdf>.

20 Kathryn Rackleff, "Preservation of Biological Diversity: Toward a Global Convention" (1992) 3 Colorado Journal of International Environmental Law and Policy 405.

21 Robert Prescott-Allen and Christine Prescott-Allen, *Genes from the Wild: Using Wild Genetic Resources for Food and Raw Materials* (London: International Institute for Environment and Development, 1983) at 17.

22 *Global Outlook 2000* (New York: United Nations, 1990) at 95.

23 GDP measures the total output of goods and services for final use produced by residents and non-residents, regardless of the allocation to domestic and foreign claims. GDP is calculated without making deductions for depreciation of "manmade" assets or depletion and degradation of natural resources. For further critique of the GDP model as a parameter for measuring human development, see Sections 3 (2) (a) and (b), *infra*.

24 World Bank, *World Development Report: Development and the Environment* (1992) at 222.

25 Paul Roberts, "International Funding for the Conservation of Biological Diversity: Convention on Biological Diversity" (1992) 10 Boston University International Law Journal 303.

26 Kerry Ten Kate and Sarah Laird, *The Commercial Use of Biodiversity: Access to Genetic Resources and Benefit Sharing* (London: Earthscan Publications Ltd., 1999) at 1.

27 Royal Gardner, "Diverse Opinions on Biodiversity" (1999) 6 Tulsa Journal of Comparative and International Law 303.

28 Contemporary ecology has embraced the theory that there is an inherent and abiding conflict in nature manifested in the struggle by organisms for survival. Thus, nature is characterized by change, not constancy. See Daniel Botkin, "Adjusting Law to Nature's Discordant Harmonies" (1996) 7 Duke Environmental Law and Policy Forum 25; Bryan Norton, "Change, Constancy, and Creativity: The New Ecology and Some Old Problems" (1996) Duke Environmental Law and Policy Forum 49.

29 Roberts, *supra* note 25 at 307.

30 P. Leelakrishnan et al., "Law Fiddles While Forest Habitat Burns" (1988) 12 Cochin University Law Review 1 at 3.

31 Krishan Iyer, "Wounded Nature versus Human Future" (1995) 19 Cochin University Law Review 1 at 2; James Harding, "Ecology as Ideology" (1973-74) 3 Alternatives 18.

32 See generally, Fran Trippett, "Towards a Broad-Based Precautionary Principle in Law and Policy: A Functional Role for Indigenous Knowledge Systems (TEK) within Decision-Making Structures" (Dalhousie University, unpublished LLM thesis, 2000) at 2-55.

33 In Hinduism some trees have divine status (e.g., the Hindu god, Vanadevata). See Iyer, *supra* note 31 at 45-48; Rana Singh ed., *Environmental Ethics: Discourse and Cultural Traditions* (Varanasa: National Geographical Society of India, 1993).

34 Iyer, *supra* note 31 at 4.

35 *Ibid.*; Lloyd Burton, "Indigenous Peoples and Environmental Policy in the Common Law Nation-States of the Pacific Rim: Sovereignty, Survival, and Sustainability" (1998) Colorado Journal of International Law and Policy 156.

36 Patricia Fry, "A Social Biosphere: Environmental Impact Assessment, the Innu, and Their Environment" (1998) 56 University of Toronto Faculty of Law Review 177.

37 Padmasiri de Silva, *Environmental Philosophy and Ethics in Buddhism* (London: Macmillan Press, 1998). On Confucianism, see Wei-Bin Zhang, *Confucianism and Modernization-Industrialization and Democratization of the Confucian Regions* (London: Macmillan Press, 1999).

38 Elias Carreno Peralta, "A Call for Intellectual Property Rights to Recognize Indigenous Peoples' Knowledge of Genetic and Cultural Resources" in Anatole Krattiger et al., eds., *Widening Perspectives on Biodiversity* (Gland, Switzerland: IUCN, 1994) at 288; Janet McDonnell, *The Dispossession of the American Indian, 1887-1934* (Indianapolis, IN: Indiana University Press, 1991) 1.

39 Roger Moody, ed., *The Indigenous Voice: Visions and Realities* (London: Zed Press, 1988) at 40-6. See also Barry Boyer, "Building Legal and Institutional Frameworks for Sustainability" (1993) 1 Buffalo Environmental Law Journal 63; Allison Mitcham, "The Wild Creatures, the Native People, and Us: Canadian Literary-Ecological Relationships" (1977-78) 7 Alternatives 20; Dorothy Spencer, *Disease, Religion and Society in the Fiji Islands* (Seattle: University of Washington Press, 1941). Interestingly, the Maori of New Zealand use the same word for "land" as for "placenta" – "whenuia" – to symbolize human relationship with "Mother Earth."

40 A. Tunks, "Tangata Whenua Ethics and Climate Change" (1997) 1 New Zealand Journal of Environmental Ethics 67-123.

41 The word "aborigine" comes from the Latin *ab origine*, which means "from the beginning." See Chris Cunneen and Terry Libesman, *Indigenous People and the Law in Australia* (Sydney:

Butterworths, 1995) at 3. See generally, Michael Goldman, ed. *Privatizing Nature: Political Struggles for the Global Commons* (London: Pluto Press, 1998).

42 According to the spokesperson of the San people of Botswana: "Once upon a time, humans, animals, and the wind and the sun and stars were all able to talk together. God changed this, but we are all still part of a wider community. We have the right to live, as do the plants, animals, wind, sun and stars; but we have no right to jeopardise their existence." See Tracy Dobson, "Loss of Biodiversity: An International Environmental Policy Perspective" (1992) 17 North Carolina Journal of International Law and Commercial Regulation 277. For an examination of Igbo conceptions of plants, see Bede Okigbo, *Plants and Food in Igbo Culture* [Ahiajoku Lecture] (Owerri: Government Printer, 1980).

43 Johan Colding and Carl Folke, "The Taboo System: Lessons about Informal Institutions for Nature Management" (2000) 12 Georgetown International Environmental Law Review 413. "Taboo" is an Anglicization of the Polynesian word "tapu." It refers to a socio-religious prohibition imposed as a protective measure.

44 Taboos need not be permanent. For example, they may exist against the collection of certain species at certain periods of the life cycle. Many plants are "taboo" in different parts of Africa (e.g., the peepal tree [Ficus religiosa] and the Khejri tree [Prosopis cineraria]. Colding and Folke, *supra* note 43 at 452.

45 Omar Bakhashab, "Islamic Law and the Environment: Some Basic Principles" (1988) 3 Arab Law Quarterly 287.

46 The Holy Quran, 6:165. [An'am-the cattle] [English Translation].

47 Bakhashab, *supra* note 45 at 289.

48 Richard Bond, "Salvationists, Utilitarians, and Environmental Justice" (1977) 7 Alternatives 31.

49 Genesis 1:26-27.

50 Jack Morrell and Arnold Thackray, *Gentlemen of Science: Early Correspondence of the British Association for the Advancement of Science* (London: Royal Historical Society, 1986); Evelyn Keller, *Reflections on Gender and Science* (New Haven: Yale University Press, 1985); Carolyn Merchant, *The Death of Nature: Women, Ecology and the Scientific Revolution* (New York: Harper and Row, 1980).

51 For example, "Bachelor of Science" and "Master of Science" are the degrees awarded in science disciplines, irrespective of gender differences. On the demonization of women's scientific contributions, see Brian Easlea, *Witch-Hunting, Magic, and the New Philosophy: An Introduction to Debates of the Scientific Revolution, 1450-1750* (Brighton, Sussex: Haverfield Press, 1980). In addition to its gender bias, science, especially in its early stages, also had inbuilt cultural biases, sometimes manifesting themselves as undisguised racism. For example, the famous British scientist Robert Boyle vowed to annihilate New England Indians for their "ridiculous notions about the workings of nature." In his view the American Indians were a "discouraging impediment to the empire of man over the inferior creatures of God." See Easlea, *ibid.*

52 For example, the preamble to the CBD recognizes the "vital role that women play in the conservation and sustainable use of biological diversity."

53 Thomas Kuhn, *The Structure of Scientific Revolutions* (Chicago: University of Chicago Press, 1970).

54 Rio Declaration on Environment and Development, reproduced in 31 I.L.M. 874 (1992).

55 Stockholm Declaration of the United Nations Conference on the Human Environment, reproduced in 11 I.L.M. 1416 (1972).

56 Judith McGeary, "A Scientific Approach to Protecting Biodiversity" (1998-99) 14 Journal of Natural Resources and Environmental Law 85.

57 See Preamble, African Convention on the Conservation of Nature and Natural Resources, reprinted in 1001 U.N.T.S. 3.

58 Lynn White, Jr., "The Historical Roots of Our Ecological Crisis" 155 Science (10 March 1967) 1203.

59 Preamble, World Charter for Nature, reprinted in 21 I.L.M. 455 (1983).

60 Preamble, CBD, *supra* note 1.

61 Malcolm Shaw, *International Law* (Cambridge: Grotius Publications Ltd., 1986) at 12. Be that as it may, history records the existence of notions of "international law" as in treaties between the city-states of Lagash and Umma in 2000 BC. Ramses II of Egypt and King of the Hittites also signed treaties.

62 For instance, the weakest or the smallest European states or principalities were immune from colonization by members of the same Christian, Caucasian family. On the other hand, some huge empires and kingdoms of Africa and the Americas were laid to waste and colonized – a situation that would never have come into being had their inhabitants been Caucasian Christians. In the case of nomadic peoples of Australia and North America, the territories where they had lived for thousands of years before colonization were construed as terra nullius and, in some cases, the peoples were either exterminated or charitably "civilized" into extinction. Yet nomadic Europeans like the Saamis in the Nordic countries were, comparatively speaking, left alone. See Colin Samson et al., eds., *Canada's Tibet: The Killing of the Innu* (London: Survival, 1999). Compare with Dudmundur Alfredsson, "The Rights of Indigenous Peoples with a Focus on the National Performance of the Nordic Countries" (1999) 59 Heidelberg Journal of International Law 529.

63 Mark Lindley, *The Acquisition and Government of Backward Territory in International Law: Being a Treatise on the Law and Practice Relating to Colonial Expansion* (New York: Negro Universities Press, 1969).

64 Shaw, *supra* note 61 at 12.

65 See, however, Mark Jarvis, ed., *The Influence of Religion on the Development of International Law* (Dordrecht: Martinus Nijhoff Publishers, 1991)

66 Vandana Shiva, *Staying Alive: Women, Ecology and Development* (London: Zed Books, 1988) at 31.

67 Makau wa Mutua, "Savages, Victims and Saviours: The Metaphor of Human Rights" (2001) 42 Harvard International Law Journal 201; Sandra Harding, *Whose Science? Whose Knowledge? Thinking from Women's Lives* (Ithaca, NY: Cornell University Press, 1991); Nazrul Islam et al., *Environmental Law in Developing Countries: Selected Issues* (2001, IUCN) 81.

68 David Slater, "Contesting Occidental Visions of the Global: The Geopolitics of Theory and North-South Relations" (1994) [December] Beyond Law at 97; Chakrabarti, *supra* note 5; Anita Chistina Butera, "Assimilation, Pluralism and Multiculturalism: The Policy of Racial/ Ethnic Identity in America" (2001) 2 Buffalo Human Rights Law Review 1. As famous Nigerian author Chinua Achebe warns, "to those who believe that Europe and North America have already invented universal civilization, and all the rest of us have to hurry up and enrol, what I am proposing will appear unnecessary if not downright foolish. But for others who may believe with me that universal civilization is no where yet in sight, the task will be how to enter the preliminary conversations." C. Achebe, *Home and Exile* (New York: Oxford University Press, 2000) at 104.

69 For example, Article 12 of ILO Convention 107 of 1957 provides that "measures should be taken to *facilitate the adaptation of workers belonging to the population concerned (traditional and indigenous peoples) to the concepts and methods of industrial relations in a modern society*" (emphasis added). See the Convention Concerning the Protection and Integration of Indigenous and Other Tribal and Semi-Tribal Populations in Independent Countries (otherwise known as ILO 107), 26 June 1957, reprinted in 328 U.N.T.S. 247.

70 Shiva, *supra* note 66 at 32; R.G.A. Dolby, "The Transmission of Science" (1977) 15 History of Science 1; B. Barnes and D. Edge, eds., *Science in Context: Readings in The Sociology of Science* (Cambridge, Mass; MIT Press, 1982; L. Salter, *Mandated Science: Science and Scientists in the Making of Standards* (Boston, Mass; Kluwer, 1993), Sandra Harding, ed., *Racial Economy of Science: Toward a Democratic Future* (Bloomington: Indiana University Press, 1993).

71 International Labour Organization (No. 169) Concerning Indigenous and Tribal Peoples in Independent Countries; concluded at Geneva, 27 June 1989, 28 I.L.M. 1382 (1989). Emphasis added.

72 Michael Balick and Paul Alan Cox, *Plants, People and Culture: The Science of Ethno-botany* (New York: Freeman and Company, 1996). In many European myths the forest was considered to be the abode of witches and gremlins – a place of evil.

73 George Foster, *Traditional Societies and Technological Changes* (New Delhi: Allied Publishers, 1973).

74 But see Charles, Prince of Wales, "We Must Go with the Grain of Nature" 2000 Reith Lecture, reprinted in the *Times* (London) 18 May 2000; John Passmore, *Man's Responsibility for Nature: Ecological Problems and Western Traditions* (New York: C. Scribner's and Sons, 1974) at 4.

75 Shiva, *supra* note 66 at xvii.

76 Rosa Luxemberg, *The Accumulation of Capital* (London: Routledge, 1951).

77 Phrases like "polluter pays," "incentive-based approach," and so on have taken their place in international environmental law. These econo-juridical concepts practically equate damage to the environment with the supposed financial costs of repairing that damage. This of course implies that there are monetary equivalents to environmental pollution and degradation and that all life forms have a monetary value. See, for example, Council Recommendation on the Implementation of the Polluter-Pays Principles, 14 November 1974, C(74) 223; reprinted in Philippe Sands, Richard Tarasofsky, and Mary Weiss, eds., *Documents in International Environmental Law* (Manchester: Manchester University Press, 1994). See also Catherine O'Neill and Cass Sunstein, "Economics and the Environment: Trading Debt and Technology for Nature" (1992) 17 Columbia Journal of Environmental Law 93. This philosophy largely underpins such international environmental instruments as the Montreal Protocol on Substances that Deplete the Ozone Layer, 26 I.L.M. 1550 (1987); Council Recommendation on the Uses of Economic Instruments in Environmental Policy, 31 January 1991, C(90) 177. Reproduced in Sands, Tarasofsky, and Weiss, *ibid.*, at 1185.

78 It is becoming increasingly usual to see phrases such as "market-based," "demand-side management," "technological optimism," "non-adversarial dialogue," and similar coded language in contemporary literature on international environmental law.

79 Article 4, African Convention on Conservation of Nature, *supra* note 57.

80 *Ibid.*

81 Christopher Stone, "Should Trees Have Standing? Toward Legal Rights for Natural Objects" (1972) 45 Southern California Law Review 450; see also Olga Moya, "Adopting an Environmental Justice Ethic" (1996) 5 Dickinson Journal of Environmental Law and Policy 215.

82 Neil Evernden, *The Natural Alien: Humankind and the Environment* (Toronto: University of Toronto Press, 1985) at 23.

83 McGeary, *supra* note 56 at 94. See generally Frederick Ferre and Peter Hartell, eds., *Ethics and Environmental Policy: Theory Meets Practice* (Athens and London: University of Georgia Press, 1994); Robert Repetto, ed., *The Global Possible: Resources, Development, and the New Century* (New Haven: Yale University Press, 1985).

84 Chakrabarti, *supra* note 5 at 30; see also B. Devall and G. Sessions, *Deep Ecology: Living as if Nature Mattered* (Utah: Gibbs Smith, 1985); C. Merchant, *Radical Ecology: The Search for A Liveable World* (London: Routledge, 1992); M. Bookchin, *The Ecology of Freedom: The Emergence and Dissolution of Hierarchy* (New York: Cheshire Books, 1991).

85 Chakrabarti, *ibid.*

86 Bruce Pardy, *Environmental Law: A Guide to Concepts* (Toronto: Butterworths, 1996) Book review by Ikechi Mgbeoji (2000) 9 Dalhousie Journal of Legal Studies 344.

87 Klaus Bosselman, "Plants and Politics: The International Legal Regime Concerning Biotechnology and Biodiversity" (1995) 7 Colorado Journal of International Environmental Law and Policy 111 at 116.

88 Rackleff, *supra* at 411.

89 Newman, *supra* note 11 at 486.

90 Rackleff. Some of the food crops can, in their respective capacity, fulfill diverse roles. For example, the winged bean, grown in New Guinea, has been called a "one species-supermarket: the entire plant – roots, seeds, leaves, stems and flowers – is edible, and a coffee-like beverage can be made from its juice. It has a nutritional value equal to that of soybeans."

91 Jack Kloppenburg Jr., *First the Seed: The Political Economy of Plant Biotechnology, 1492-2000* (Cambridge: Cambridge University Press, 1988) at 47-48.

92 William Lesser, *Sustainable Use of Genetic Resources under the Convention on Biological Diversity: Exploring Access and Benefit Sharing Issues* (Oxford: CAB International, 1997) at 14. A

short list of major plant resources shows their sources of origin: (1) America = sunflower, tepary bean; (2)Meso-America = maize, tomato, sieva bean, scarlet runner bean, cotton, avocado, papaya, cacao, cassava, sweet potato, common bean; (3) Lowland South America = yam, pineapple, cassava, sweet potato, cotton; (4) Highland South America = potato, peanut, lima bean, cotton bean, cotton, llama, alpaca; (5) Northern Europe = oats, sugarbeet, rye, cabbage; (6) Africa = African rice, yam, sorghum, pearl millet, watermelon, cowpea, coffee, cotton, sesame; (7) Near East = wheat, barley, onion, pea, lentil, chickpea, fig, date, flax, pear, pomegranate, grapes, olive, apple; (8) Central Asia = common millet, buckwheat, alfalfa, hemp, foxtail millet, grapes, broad bean, pea, eggplant, cucumber, cotton, sesame; (9) China = soybean, cabbage, onion, peach, foxtail millet; (10) Southeast Asia = Oriental rice, banana, citrus, yam, mango, sugar cane, taro, tea; (11) South Pacific = coconut, breadfruit, noble sugar cane.

93 Nikolai Vavilov, *The Origin, Variation, Immunity and Breeding of Cultivated Plants,* trans. K. Chester (New York: Ronald Press, 1951).

94 Anthony Stenson and Tim Gray, *The Politics of Genetic Resource Control* (London: Macmillan Press, 1999).

95 Nikolai Vavilov, "Studies on the Origin of Cultivated Plants" (1925) Vol. 16, No. 2 Bulletin of Applied and Plant Breeding 1-248.

96 Panjabi, *supra* note 1 at 196.

97 Colding and Folke, *supra* note 43 at 415.

98 Erich Isaac, *Geography of Domestication* (Englewood Cliffs, NJ: Prentice-Hall, 1970); Henry Bailey Stevens, *The Recovery of Culture* (New York: Harper and Brothers, 1949).

99 Lara Ewens, "Seed Wars: Biotechnology, Intellectual Property, and the Quest for High Yield Seeds" (2000) 23 Boston College International and Comparative Law Review 285.

100 Annie Patricia Kameri-Mbote and Philippe Cullet, "Agro-Biodiversity and International Law: A Conceptual Framework" (1999) 11 Journal of Environmental Law 257 at 260.

101 Elias Carreno Peralta, "A Call for Intellectual Property Rights to Recognize Indigenous Peoples' Knowledge of Genetic and Cultural Resources" in Krattiger et al., *supra* note 38 at 288.

102 For example, "On one uphill traverse of a rice field in Liberia, farmers grew fourteen different varieties of rice, each matched to the degree of slope, amount of insulation, and type of soil in the particular paddies." see Craig Jacoby and Charles Weiss, "Recognizing Property Rights in Traditional Biocultural Contribution" (1997) 16 Stanford Environmental Law Journal 74 at 84. Similarly, recent studies show that the Epugao of Luzon in the Philippines can identify 200 varieties of sweet potatoes and that Andean farmers cultivate thousands of clones of potatoes, more than 1,000 of which have names. See Walter Reid and Kenton Miller, *Keeping Options Alive: The Scientific Basis for Conserving Biodiversity* (Washington, DC: World Resources Institute, 1989) at 57.

103 Kloppenburg, *supra* note 91 at 185.

104 Norman Simmonds, *Principles of Crop Improvement* (New York: Longman, 1979) at 11.

105 Cited in Kloppenburg, *supra* note 91 at 185. Other authorities, such as Jack Harlan, credit the American Indian with a "magnificent performance" in the improvement of maize, potato, manioc, sweet potato, peanut, and the common bean.

106 According to Kate and Laird, "the majority of the world's diversity is closely tied to traditional management and livelihood practices and many 'natural' areas bear the mark of the interconnection between cultural and biological diversity." Kate and Laird, *supra* note 26 at 3.

107 Harriet Ketley, "Cultural Diversity versus Biodiversity" (1994) 16 Adelaide Law Review 99. Thus naturalness may in some cases be a culturally constructed concept. Policy makers should therefore exercise caution before demarcating any particular habitat as a "nature" park devoid of human interaction with "nature." See Dan Perlman and Glenn Adelson, *Biological Diversity: Exploring Values and Priorities in Conservation* (Massachusetts: Blackwell, 1997).

108 David Wood, "Conservation and Agriculture: The Need for a New International Network of Biodiversity and Development Institutes to Resolve Conflict" in Krattiger et al., *supra* note 38 at 425.

109 Ghillean Prance, "The Amazon: Paradise Lost" in Les Kaufman and Kenneth Mallory, eds., *The Last Extinction, 2nd Edition* (Cambridge, MA: MIT Press, 1993) at 88.

110 David Takacs, *The Idea of Biodiversity: Philosophies of Paradise* (Baltimore: Johns Hopkins University Press, 1996).

111 Jeffrey McNeely et al., eds., *Protecting Nature: Regional Reviews of Protected Areas* (Gland: IUCN, 1994) at 5.

112 W.L. Thomas Jr., ed., *Man's Role in Changing the Face of the Earth* (Chicago: University of Chicago, 1956). For similar evidence in the Pacific and Caribbean, see John Young, *Sustaining the Earth: The Story of the Environmental Movement – Its Past Efforts and Future Challenges* (Cambridge, MA: Harvard University Press, 1990).

113 See, for example, the 1933 Convention Relative to the Preservation of Fauna and Flora in their Natural State, 172 L.N.T.S. 241. This convention spawned the "natural parks" syndrome in Africa.

114 Tiyanjana Maluwa, "Environment and Development in Africa: An Overview of Basic Problems of Environmental Law and Policy" (1989) 1 African Journal of International and Comparative Law 650.

115 The IUCN divides these protected areas into six categories: (1) areas construed as "strict nature reserve/wilderness area[s]" managed for "scientific" purposes; (2) "national parks" managed for ecosystem and recreation; (3) natural monuments and natural landmarks; (4) habitat and species management areas; (5) protected landscape/seascape areas; and (6) protected areas mainly managed for the sustainable use of natural resources. See Van Der Zoon, ed., *Biological Diversity* (The Hague: The Netherlands, 1995) at 20.

116 John Mugabe, "Technology and Biodiversity in Kenya: Technological Capabilities and Institutional Systems for Conservation" in Krattiger et al., *supra* note 38 at 81.

117 Mugabe, *ibid*. In this objectification of marginalized peoples and cultures, the experience of Kenyan Masai is particularly well known. Of course, in this clichéd approach to environmental governance, the "management" of such reserves would be under the control of government-appointed "experts" who, apart from college qualifications, would probably lack intimate knowledge of the environment in question.

118 For example, the Convention on Nature Protection and Wildlife Preservation in the Western Hemisphere defines strict wilderness reserves as "a region under public control characterized by primitive conditions of flora, fauna and habitation." This convention mandates all governments of the American republics to explore the possibility of establishing this type of "wilderness" reserve as soon as possible after the effective date of the convention. See Article 1 of the Convention on Nature Protection and Wildlife Preservation in the Western Hemisphere, *ibid*.

119 Sara Dillon, "Trade and the Environment: A Challenge to the GATT/WTO Principle of 'Ever-Freer Trade'" (1996) St. John's Journal of Legal Commentary 351. [Noting that there are still strong views that "pre-technological societies are the most serious threats to the environment, in their subsistence-level economic activities."]

120 But see Van der Zoon, *supra* note 115 at 117.

121 E.P. Stebbing, *The Forests of India* (reprint) (New Delhi: J. Reprints Agency, 1982) at 62.

122 Gabriel Muyuy Jacanamejoy, "Community Participation in the Conservation of Biodiversity" in Krattiger et al., *supra* note 38 at 231.

123 But see Convention on Nature Protection and Wildlife Preservation in the Western Hemisphere, *supra* note 118; Convention on the Conservation of European Wildlife and Natural Habitats, 19 September 1979, Berne, European Treaty Series No. 104; (1982) U.K.T.S. 56; ASEAN Agreement on the Conservation of Nature and Natural Resources, 9 July 1985, Kuala Lumpur; reproduced in Sands et al., *supra* note 77 at 958.

124 Michael Wells, ed., *Investing in Biological Diversity: A Review of Indonesia's Integrated Conservation and Development Projects* (Washington, DC: The World Bank, 1999).

125 See, for example, Article 8 (j) of the CBD, *supra* note 1.

126 For example, Article 14 of ILO no. 169 provides that "state parties recognize the rights of ownership and possession of the peoples concerned over the lands which they traditionally occupy." See generally Convention (ILO no. 169) Concerning Indigenous and Tribal Peoples in Independent Countries, *supra* note 71. See also, Article 8 (j) of the CBD, *supra* note 1.

127 The natural rate of extinction is an estimate of the rate of extinction that would occur in the absence of human influence. The natural rate of extinction is approximately one to three species per 100 years. Scientists estimate that the current rate of extinction is 1,000 to 10,000 times faster than the natural rate of extinction. For a discussion of five major mass extinctions of the Phanerezoic period, see David Jablonski, "Causes and Consequences of Mass Extinctions: A Comparative Approach" in David Elliott, ed., *Dynamics of Extinction* (New York: John Wiley and Sons, 1986) at 183.

128 Reid and Miller, *supra* note 102 at 104.

129 See Dobson, *supra* note 42 at 280-1.

130 For further examples of the cascade effect, see Roberts, *supra* note 25 at 309-10.

131 June Starr and Kenneth Hardy, "Not by Seeds Alone: The Biodiversity Treaty and the Role for Native Agriculture" (1993) 12 Stanford Environmental Law Journal 85.

132 Iyer, *supra* note 31; James Karr, "Protecting Ecological Integrity: An Urgent Social Goal" (1993) Yale Journal of International Law 297; Monique Manguet and Rene Letolle, "Why Is the Environment Deteriorating? Why Is the Deterioration Speeding Up at the End of the 20th Century? (1993) 51 European Yearbook 31; David Spence, "Paradox Lost: Logic, Morality, and the Foundations of Environmental Law in the 21st Century" (1995) 20 Columbia Journal of Environmental Law 145.

133 John Ryan, *Life Support: Conserving Biological Diversity* (Worldwatch Paper 108) Washington, DC: Wordwatch Institute, 1992) at 14.

134 For example, the potato came to Europe from South America in the eighteenth century. Enjoying initial success, it helped to triple the population of Ireland. Hoever, local predators overwhelmed its narrow genetic base and unleashed the fungus *Phutophtora infestans* in 1846. The result was disastrous. Half the crop was lost, two million Irish died, two million more emigrated, and the rest were thrown into abject poverty. See David Tillford, "Saving the Blue Prints: The International Regime for Plant Resources" (1998) 30 Case Western Reserve Journal of International Law 373. Similar but less drastic examples include the "Victoria blight" of 1946 and the American corn blight of 1970.

135 As Holmes Rolston III warns, "to continue the developmental pace of the last century for another millennium will produce sure disaster." See Holmes Rolston III, "Rights and Responsibilities on the Home Planet" (1993) 18 Yale Journal of International Law 251; Mulford Sibley, "The Relevance of Classical Political Theory for Economy, Technology and Ecology" (1973-74) 3 Alternatives 14; Les Kaufman, "Why the Ark Is Sinking" in Kaufman and Mallory, *supra* note 109. See also Ryan Winter, "Reconciling the GATT and WTO with Multilateral Environmental Agreements: Can We Have Our Cake and Eat It Too" (2000) 11 Colorado Journal of International Environmental Law and Policy 223.

136 Peter Victor, "The Environmental Impact of Economic Activity: A Multi-Disciplinary View" (1971-73) 1 Alternatives 20; Richard Falk, "Toward a World Order Respectful of the Global Ecosystem" (1991-92) 19 Boston College Environmental Affairs Law Review 711; Charles Perrings, *Economy and Environment: A Theoretical Essay on the Interdependence of Economic and Environmental Systems* (Cambridge: Cambridge University Press, 1987).

137 Michael Blumm, "Wetlands Protection and Coastal Planning: Avoiding the Perils of Positive Consistency" (1978) 5 Columbia Journal of Environmental Law 69; "People and Wetlands: The Vital Link" 7th Meeting of the Conference of the Contracting Parties to the Convention on Wetlands (Ramsar, Iran, 1971), San Jose, Costa Rica, 10-18 May 1999; Resolution VII.8, Guidelines for Establishing and Strengthening Local Communities' and Indigenous Peoples' Participation in the Management of Wetlands, <http://users.ox.ac.uk/-wgtrr/ram-res.html>. The irony here is that the so-called developing nations are determined to pursue the same destructive path to "development" as the "developed world." See "Brazil Will Not Let Amazon Become 'Untouchable Sanctuary,'" Reuters News Agency, 23 January 2000, *Globe and Mail* (Toronto), 23 January 2000; Robert Carter and David Lasenby, "Values and Ecology: Prolegomena to an Environmental Ethics" (1977) 6 Alternatives 39.

138 For example, a 1980 study in Brazil showed that road construction accounted for 26 percent of deforestation in the Amazon. See Grainger, "The State of the World's Forests" (1980) 10 Ecologist 6.

139 Bosselman, *supra* note 87; Walter Reid, ed., *Biodiversity Prospecting: Using Genetic Resources for Sustainable Development* (1993) at 3.
140 Brian Nickerson, "The Environmental Laws of Zimbabwe: A Unique Approach to the Management of the Environment" (1994) 14 Boston College Third World Law Journal 189.
141 "Climate Change and Biodiversity" IUCN: The World Conservation Union. (September 1998) Report of the Ninth Global Biodiversity, 1997, Kyoto, Japan 1997 (Gland, Switzerland: IUCN, 1997).
142 John Kenneth Galbraith, *The Affluent Society* (Boston: The Riverside Press, 1958).
143 But see Theo Van Boven, "Human Rights and Development: The UN Experience" in David Forsythe, ed., *Human Rights and Development: International Views* (New York: St Martin's Press, 1989) at 121.
144 Kabita Chakma, "Development, Environment and Indigenous Women in the Chittagong Hill Tracks of Bangladesh" in Krattiger et al., *supra* note 38 at 233. In this context, Marc Williams has argued that "development is still contained within the capitalist political economy" measured in the rate of consumption, acquisition of wealth and "disposable" income. In this paradigm the perceived limitation to the continuous consumption of limited global resources is not the finiteness of the resources themselves but, rather, the "present state of technology and social organization." See Marc Williams and Lucy Ford, "The WTO, Social Movements and Global Environmental Management" in Chris Rootes, ed., *Local, National and Global* (London: Frank Cass, 1999) at 209.
145 Ibrahim Shihata, "The World Bank and the Environment: A Legal Perspective" (1992) 15 Maryland Journal of International Law and Trade 1; Sharon Venne, *Our Elders Understand Our Rights: Evolving International Law Regarding Indigenous Rights* (Penticton, BC: Theytus Books, 1998). According to Vandana Shiva, "the 'development' of Brazil by transnational banks and corporations is the primary cause of the destruction of the richness of Amazonian rainforests: the highest expression of life. Natives of Africa and Amazonia survived centuries with their ecologically evolved, indigenous knowledge systems. What local peoples had conserved through history, Western experts and knowledge destroyed in a few decades, a few years even." See Shiva, *supra* note 66 at 26.
146 For example, the World Bank supplied 30 percent of the US$1.55 billion spent in the "accelerated development" of the Rondonia region of Brazil. The money went towards building expensive highways through the biodiversity-rich Rondonia region in order to attract agricultural monoculturists. Some critics argue that the entire project was an unparalleled ecological disaster for the region and for biodiversity. Another example is the Narmada Valley Project. The Narmada is 1,312 kilometres long and is India's largest west-flowing river, serving over twenty million people for centuries. In the opinion of the World Bank, the Narmada was "least used," and the remedy for this perceived underuse was to build 2 dams across the river, 28 small dams, and 3,000 other water projects, including irrigation projects for industrial farming, pisciculture, and so on. The World Bank was to dole out US$450 million for the big dams. The environmental impact of this "modernization" project was highlighted and a successful lobby put it in abeyance.

Another World Bank project, the Carajas iron-mining project in Brazil, burned so much wood in the smelting process that the result was massive air pollution and the deforestation of the indigenous tribes environment. For further analysis of several World Bank projects, see Bruce Rich, "The Multilateral Development Banks, Environmental Policy, and the United States" (1985) 12 Ecology Law Quarterly 685-8; Stephan Schwartzman, *Bankrolling Disasters: International Development Banks and the Global Environment* (San Francisco: Sierra Club, 1986). Ironically, the process of "development" is also fuelled by the policies and indeed, the raison d'etre of the World Bank and other influential and powerful "developmental" institutions and multinational corporations. See Scott Holwick, "Transnational Corporate Behaviour and Its Disparate and Unjust Effects on the Indigenous Cultures and the Environment of Developing Nations: *Jota v. Texaco*, A Case Study" (2000) 11 Colorado Journal of International Environmental Law and Policy 183.

Recently however, the bank has apparently started to address environmental concerns arising from its "development" projects through the institution of audits and Environmental Impact Assessments (EIAs). See Cindy Buhl, *A Citizen's Guide to the Multilateral De-*

velopment Banks and Indigenous Peoples (Washington, DC: The Bank Information Centre, 1994) at 4. See also Hugo Stokkle and Arne Tostensen, eds., *Human Rights in Development: Global Perspectives and Local Issues* (The Hague: Kluwer Law International, 1999); Ronald Theodore Libby, *The Ideology and Power of the World Bank* (Ann Arbor, MI: unpublished doctoral thesis, 1985).

147 Perhaps, starting from the United Nations Conference on the Human Environment in Stockholm in 1972, and re-emphasized in the report of the World Commission on Environment and Development chaired by Grø Harlem Brundtland of Norway, the direct link between human economic pursuits under the dominant paradigm and the conservation of biological diversity assumed the status of a global creed. See Stockholm Declaration on the Human Environment, *supra* note 55; *Our Common Future: The World Commission on Environment and Development:* (Oxford, Oxford University Press, 1987); Jeffrey Kovar, "A Short Guide to the Rio Declaration" (1993) 4 Colorado Journal of International Environmental Law and Policy 119; Maurice Strong, "Beyond Rio: Prospects and Portents" (1993) 4 Colorado Journal of International Environmental Law and Policy 21; Timothy Wirth, "The Road from Rio: Defining a New World Order" (1993) 4 Colorado Journal of International Environmental Law and Policy 37.

148 Shiva, *supra* note 66 at 9; T. De la Court, *Beyond the Brundtland: Green Development in the 1990s* (London: Zed Books, 1990) at 15.

149 Dillon, *supra* note 119 at 363. In this context, recent scholarship has advocated replacing the GNP with indicators that take into account social, environmental, and other factors besides those that have only a monetary value. Kele Onyejekwe, "GATT Agriculture, and Developing Countries" (1993) 17 Hamline Law Review 77.

150 Chakma, *supra* note 144 at 233.

151 For example, when asked whether he hoped to approximate Britain's standard of living after India achieved independence, Mahatma Gandhi replied: "it took Britain half the resources of the planet to achieve this prosperity; how many planets will a country like India require?" De la Court, *supra* note 148 at 16.

152 Trippett, *supra* note 32 at 118-20.

153 Brundtland Report, *supra* note 147 at 115.

154 Declaration of the Hague, 11 March, 1989, UN Doc. A/44/340-E/1989, 28 I.L.M. 1308 (1989) (Noting that "today, the very conditions of life on our planet are threatened by the severe attacks to which the earth's atmosphere is subjected."); The Langkawi Commonwealth Heads of Government Declaration on Environment, adopted at Langkawi (Malaysia), 21 October 1989, reproduced in (1990) 5 American University Journal of International Law and Policy 589; UN Study on the International Dimensions of the Right to Development, UN Doc.E/CN.4/1334, para.27; United Nations Declaration on Social Progress and Development UNGAR 2542 (XXIV) of 11 December 1969 and UNGAR 32/130 of 1977.

155 Trends in International Environmental Law, edited by the Editors of the *Harvard Law Review* (American Bar Association, 1992).

156 Katarina Tomasevski, "The Influence of the World Bank and IMF on Economic and Social Rights" (1997) 66 Nordic Journal of International Law 385.

157 See World Bank, *International Implications for Donor Countries and Agencies of Meeting Basic Human Rights,* 1977 (unpublished). See also World Bank Operational Directive 4.00 Annex A: Environmental Assessment, 1989. Reproduced in Sands et al. *supra* note 77 at 1323.

158 John McIntyre and Daniel Papp, eds., *The Political Economy of International Technology Transfer* (New York: Quorum Books, 1986) at 47.

159 "The Realisation of the Right to Development" Consultation, Geneva, 8-12 January 1990, HR/PUB/91/2.1991. See also, Report of UNSG Doc E/CN.4/1991/11.

160 Recently, the vice-president of the United States reiterated that "Americans won't sacrifice 'way of life' to conservation." See "US Won't Cut Energy Use; Produce More – V.P. Cheney Says," *National Post,* 1 May 2001, at A1. However, a research team at the University of Sao Paulo has shown that the globalization of the American lifestyle, with its large homes, numerous appliances, and two or three cars per family, "would be impossible to sustain." Panjabi, *supra* note 1 at 214. The average adult American discards 1,429 pounds of garbage per annum.

161 Kevin Jones, "United States Dependence on Imports of Four Strategic and Critical Minerals: Implications and Policy Alternatives" (1988) B.C.L. Rev. 217. For example, in order to produce the four million tons of zinc imported into America between 1983 and 1987, 48 billion tons of rock had to be mined, processed, and disposed of in Malaysia with terrible impacts on the diversity of plants there. It is the policies and development paradigm of the North that determine the fate of plant diversity. In this context, the North's concern for the loss of plant resources in the South is considered by some scholars to be insincere. Neil Middleton, Phil O'Keefe, and Sam Moyo, *The Tears of the Crocodile: From Rio to the Reality in the Developing World* (London: Pluto Press,. 1993). Similarly, the United States consumes 25 percent of the world's energy and emits 22 percent of its carbon dioxide.

162 United States, Japan, Canada, Germany, France, Italy, and Britain. For juridical support, see Chapter 5 of Agenda 21. Adopted by the UN Conference on Environment and Development (UNCED) at Rio de Janeiro, 13 June 1992. UN Doc.A/CONF. 151/26 (Vols. 1, 2, and 3) (1992).

163 Norman Myers, *The Primary Source: Tropical Forests and Our Future* (New York: Norton, 1984). Contemporary data show that "Haiti, Sri Lanka, El Salvador, Ghana, Nigeria, and Bangladesh have lost all or nearly all of their primary rainforest. Latin America has lost 37 percent of its tropical forests and Africa has lost 52 percent." See O'Neill and Sunstein, *supra* note 77 at 98-99.

164 Takacs, *supra* note 110. Anil Agarwal, "Human-Nature Interactions in a Third World Country" in George James, ed., *Ethical Perspectives on Environmental Issues in India* (New Delhi: A.P.H. Publishing Corporation, 1999) at 31. Agarwal observes that more than a quarter of all Central American forests have been destroyed since 1960 for cattle-ranching. While domestic consumption of beef in Central America has fallen dramatically, between 85 and 95 percent of the beef produced is shipped to the United States for hamburgers and pet food. A French appetite for peanut oil caused the great Sahelian drought of 1968-74 that claimed more than 100,000 thousand lives in Africa.

165 Morris Cohen, "Property and Sovereignty" (1927-28) Cornell Law Quarterly 30.

166 See, O'Neill and Sunstein, *supra* note 77 at 94.

167 *Ibid.*; Walther Lichem, "From North-South to One World: The Challenge for Europe" (1988) 36 European Yearbook 44; Stephen Vosti and Thomas Reardon, eds., *Sustainability, Growth and Poverty Alleviation: A Policy and Agro-Ecological Perspective* (Baltimore: Johns Hopkins University Press, 1997) at xviii.

168 John Williamson, "The Outlook for Debt Relief or Repudiation in Latin America" (1986) 2 Oxford Review of Economic Policy 1; George Ayittey, "How the Multilateral Institutions Compounded Africa's Economic Crisis" (1999) 30 Law and Policy in International Business 585.

169 For an analysis of the North-South divide on these issues, see Ranee Panjabi, "The South and the Earth Summit: The Development/Environment Dichotomy" (1992) 11 Dickinson Journal of International Law 77 at 88; Ranee Panjabi, *The Earth Summit at Rio: Politics, Economics, and the Environment* (Boston: Northeastern University Press, 1997).

170 The literature on this is immense. According to O'Neill and Sunstein, "In 1987 to 1988, primary, non-fuel commodities comprised 39.3 percent of exports from severely indebted middle-income countries and 52.5 percent of exports from severely indebted low-income countries." O'Neil and Sunstein, *supra* note 77, *ibid*. There are varied estimates on the extent of forest cover left. See Matthew Royer, "Halting Neo-Tropical Deforestation: Do the Forest Principles Have What It Takes?" (1996) 6 Duke Environmental Law and Policy Forum 105.

171 Nathalie Chalifour, "Global Trade Rules and the World's Forests: Taking Stock of the World Trade Organization's Implications for Forests" (2000) 12 Georgetown International Environmental Law Review 575.

172 Between 1980 and 1989, commodity prices declined 33 percent and there has not been any noticeable improvement in prices. *Ibid.* See also Todd Johnston, "The Role of International Equity in a Sustainable Future: The Continuing Problem of Third World Debt and Development" (1998) 6 Buffalo Environmental Law Journal 35.

173 Various suggestions, including debt forgiveness, debt-for-nature swap, debt-for-equity-swap, and so on, have been proposed as possible solutions to this crisis; Catherine Fuller and Douglas Williamson, "Debt-for-Nature Swaps: A New Means of Funding Conservation in Developing Nations" (1988) 11 International Environmental Reporter 301.

174 Dillon, *supra* note 119 at 366. For further analysis of the prevailing trading order, see Batram Brown, "Developing Countries in the International Trade Order" (1994) 14 Northern Illinois University Law Review 347; Bradley Boyd, "The Development of a Global Market-Based Debt Strategy to Regulate Lending to Developing Countries" (1988) 30 Georgia Journal of International and Comparative Law 461. Further exacerbating the crisis, the World Bank has long been engaged in creating "export-based" economies in the biodiversity-rich countries of the South.

175 Obiora Chinedu Okafor, "The Status and Effect of the Right to Development in Contemporary International Law: Towards a South-North 'Entente'" (1995) 7 RADIC 865. See also Jan Tinbergen, *Rio Reshaping the International Order: A Report to the Club of Rome* (New York: E.P. Dutton, 1976).

176 *International Tropical Timber Agreement 1994*, <http://sedac.ciesin.org/entri/texts/ITTA.1994>; Canadian Council on International Law, *Global Forests and International Environmental Law* (Boston: Kluwer Law, 1996) at 361; Brian Johnson, *Responding to Deforestation: An Eruption of Crisis, an Array of Solutions* (Washington, DC: World Wildlife Fund and the Conservation International, 1991).

177 David VanderZwaag and Douglas Mackinlay, "Towards a Global Forests Convention: Getting Out of the Woods and Barking Up the Right Tree" in Canadian Council on International Law, *Global Forests and International Environmental Law* (Boston: Kluwer Law, 1996) at 1. See also Non-Legally Binding Authoritative Statement of Principles for a Global Consensus on the Management, Conservation, and Sustainable Development of all Types of Forests. Adopted by the UN Conference on Environment and Development at Rio de Janeiro, 13 June 1992. UN Doc. A/CONF. 151/26 (Vol. 3) (1992), reprinted in 31 I.L.M. 881 (1992).

178 ECOSOC Resolution, E/2000/L/32, Resumed Substantive Session of 2000, New York; Report of the Inter-governmental Forum on Forests, E/CN.17/2000/14 of March 2000. For a broad treatment of the problems associated with the global forest regimes, see Canadian Council on International Law, *Global Forests and International Environmental Law* (Boston: Kluwer Law, 1996), *ibid.*

179 Henry Steck, "Power and the Liberation of Nature: The Politics of Ecology" (1971-72) 1 Alternatives 4; Anup Shah, *Ecology and the Crisis of Overpopulation: Future Prospects for Global Sustainability* (Cheltenham, UK: Edward Elgar, 1998).

180 O'Neill and Sunstein, *supra* note 77 at 102; Chapter 5 of Agenda 21, *supra* note 162.

181 Steck, *supra* note 179 at 8.

182 Joel Cohen, "Conservation and Human Population Growth: What Are the Linkages?" in S.T.A. Picket et al., eds., *The Ecological Basis of Conservation: Heterogeneity, Ecosystems, and Biodiversity* (New York: Chapman and Hall, 1997).

183 Shekah Singh, "Sovereignty, Equity and the Global Environment" in James, ed., *supra* note 164 at 131. Shekah also notes that the attention placed on the population of the South is misplaced as one person from the North consumes as much as do forty people from the South. See also Andrew Ringel, "The Population Policy Debate and the World Bank: Finity v. Supply-Side Demographics" (1993) 6 Georgetown International Environmental Law Review 213.

184 Cohen, *supra* note 182 at 33; Statement by Dr. Mahathir Mohammad, Prime Minister of Malaysia, UNCED, Rio, 13 June 1992.

185 Newman, *supra* note 11 at 489. Jeffrey McNeely, *Economics and Biological Diversity: Developing and Using Economic Incentives to Conserve Biological Resources* (Gland, Swit.: IUCN 1988) at 49.

186 In the United States this partly caused the American dust bowl of the 1930s, ably depicted in John Steinbeck's classic, *The Grapes of Wrath*.

187 Vandana Shiva, *The Violence of the Green Revolution* (London: Zed Books, 1991).

188 *Ibid.*, at 203.

189 UNEP/CBD/COP/3/22, 22 September 1996 at 12.
190 The Green Revolution refers to the unprecedented increase in wheat yield as a result of Norman Borlaug's discovery that a wheat variety known as Norin 10, when crossed into some Mexican wheat lines, produced a dwarf wheat strain that responds dramatically to heavy fertilizer application. The impact of this discovery was so radical that it affected the food supply of one-quarter of the global human population and earned Borlaug a Nobel Prize for peace. Tillford, *supra* note 134 at 393.
191 Conservation International News, Vol. 5, No. 1, Spring 2000 at 3.
192 Cary Fowler and Pat Mooney, *Shattering: Food, Politics and the Loss of Genetic Diversity* (Tucson: University of Arizona Press, 1990) at 63.
193 *Biodiversity Prospecting: Using Genetic Resources for Sustainable Development* (Washington, DC: WRI, 1993); Judith Jones, "Regulating Access to Biological and Genetic Resources in Australia: A Case Study of Bioprospecting in Queensland" (1998) 5 Australasian Journal of Natural Resources Law and Policy 89.
194 Edgar Asebey and Jill Kempenaar, "Biodiversity Prospecting: Fulfilling the Mandate of the Biodiversity Convention" (1995) 28 Vanderbilt Journal of Transnational Law 703; but see John Adair, "The Bioprospecting Question: Should the United States Charge Biotechnology Companies for the Commercial Use of Public Wild Genetic Resources" (1997) 24 Ecology Law Quarterly 131.
195 Christopher Hunter "Sustainable Bioprospecting: Using Private Contracts and International Legal Principles and Policies To Conserve Raw Materials" (1997) 25 Environmental Affairs 129.
196 Doremus, *supra* note 8 at 266.
197 Newman, *supra* note 11 at 482. The pharmaceutical giant Eli Lily now has farms in Texas, where it cultivates rosy periwinkles. See Naomi Roht-Arriaza, "Of Seeds and Shamans: The Appropriation of the Scientific and Technical Knowledge of Indigenous and Local Communities" (1996) 17 Michigan Journal of International Law 940.
198 Newman, *ibid.*
199 This refers to the number of samples that a laboratory can screen for bioactivity in any given period.
200 *Population, Environment, and Development in Tanzania* (NY: United Nations, 1993). In one particularly egregious case, the entire adult population of maytenus buchananni – the source of the anti-cancer compound maytansine – was harvested when a mission sponsored by the United States National Cancer Institute collected 27,215 kilograms in Kenya for testing in its drug development program.
201 This is dealt with in Chapter 4.
202 Stenson and Gray, *supra* note 94 at 38.
203 IUCN: The World Conservation Union, October 1999. Report of the Eleventh Global Biodiversity Forum: Exploring Synergy Between the UN Framework Convention on Climate Change and the Convention on Biological Diversity (Gland, Swit.: IUCN) at 7. This estimate is indeed very conservative; the latest estimates are much higher.
204 IUCN, 1997 Kyoto, *supra* note 141 at 19. See also Convention on the Protection of the Ozone Layer, 22 March 1985, Vienna. 26 I.L.M. 1529 (1987). Since 1860, the ten warmest years in recorded history occurred within fifteen years.
205 Another significant contributor to the greenhouse effect is the petroleum sector, particularly gas flaring in Nigeria, which produces over one-quarter of such sources of greenhouse gas. See IPIECA Report, *Biodiversity and the Petroleum Industry: A Guide to the Biodiversity Negotiations* (London: IPIECA, 2000).
206 IUCN, 1997 Kyoto, *supra* note 141 at 20. Increases in global temperature also cause rises in the ocean volumes, thus submerging coastlines and wetlands along with their plant diversity (e.g., mangroves).
207 United Nations Framework Convention on Climate Change, concluded at Rio de Janeiro, 29 May 1992. Entered into force 21 March 1994. 31 I.L.M. 849 (1992). It has been signed by 154 governments, but the convention remains inoperative. On 11 December 1997 delegates from 160 nations agreed to the Kyoto Protocol to the Climate Convention, which calls for the first ever legally binding commitments to reduce carbon dioxide and other

greenhouse gases. See Ranee Panjabi, "Can International Law Improve the Climate? An Analysis of the United Nations Framework Convention on Climate Change Signed at the Rio Summit in 1992" (1993) North Carolina Journal of International Law and Commercial Regulation 491.

208 Simone Bilderbeek et al., eds., *Biodiversity and International Law: The Effectiveness of International Environmental Law* (Amsterdam: IOS Press, 1992).

209 Timothy Swanson, *Global Action for Biodiversity* (London: Earthscan Publications Ltd., 1997).

210 See, Article 8 (j) of the CBD, *supra* note 1.

211 Timothy Swanson, *supra* note 209 at 5.

212 Curtis Horton, "Protecting Biodiversity and Cultural Diversity under Intellectual Property Law: Toward a New International System" (1995) 10 Journal of Environmental Law and Litigation 1 at 4. In Brazil alone, one Indian ethnic group disappears each year. This process has continued since 1900. At least "90 percent of the 6,000 languages now being spoken are expected to die out within roughly 100 years." *Ibid.* Another obvious consequence of this is an incalculable loss of knowledge spanning millennia. See also The Declaration of Belem, <http://users.ox.ac.uk/~wgtrr/belem.htm>; *USA Today,* Thursday, 18 July 2002.

213 For further analysis of the linkage between culture and plant diversity, see D.M. Warre, L.J. Slikkerveer, and D. Brokensha, eds., *The Cultural Dimensions of Development: Indigenous Knowledge Systems* (London: Intermediate Technology Publications); Daniel Sitarz, ed., *Agenda 21: The Earth Summit Strategy to Save Our Planet* (Boulder, CO: Earth Press, 1993).

214 National Research Council, *Lost Crops of Africa: Grains,* Vol. 1 (Washington, DC: National Academy Press, 1996).

215 Victor Orsinger, "Natural Resources of Africa: Conservation by Legislation" (1971-73) 5 African Law Studies 29.

216 Daniel Esty, *Greening the GATT: Trade, Environment, and the Future* (Washington, DC: Institute for International Economics, 1994).

217 P. Van Heijnsbergen, *International Legal Protection of Wild Fauna and Flora* (Amsterdam: Ohmsha Press, 1997). It is a matter of philosophical conjecture whether the current trend has given rise to the phenomenon of "biojurisprudence," as some scholars would posit. See Roman Tokarczysk, "Biojurisprudence: A Current in Jurisprudence" (1996) 7 Finnish Yearbook of International Law 341.

218 International Convention for the Protection of Plants, 1929,126 L.N.T.S. 305.

219 International Plant Protection Convention, 5 December 1951, reprinted in 150 U.N.T.S. 67.

220 See, Ian Brownlie, *Principles of Public International Law* (Oxford: Clarendon Press, 1982) at 19.

221 Convention on International Trade in Endangered Species of Wild Fauna and Flora, 3 March 1973, 993 U.N.T.S. 243.

222 Gurdip Singh, *Global Environmental Change and International Law* (New Delhi: Aditya Books Ltd., 1991).

223 W.E. Hall, *A Treatise on International Law* (London: Butterworths, 1977) at 41.

224 Article 31 (2), Vienna Convention on the Law of Treaties, 1155 U.N.T.S. 331.

225 Heijnsbergen, *supra* note 217 at 76.

226 R.R. Baxter, "Multilateral Treaties as Evidence of Customary International Law" (1965-66) 41 B.Y.I.L. 278.

227 Ian Brownlie, *Principles of International Law* (Oxford: Clarendon Press, 1979) 7.

228 Georg Schwazenberger, *Manual of International Law,* Vol. 1 (London: Stevens and Sons, 1960) at 286; P. Birnie and A.E. Boyle, *International Law and the Environment* (Oxford: Clarendon Press, 1992) at 487; M.J. Glennon, "Has International Law Failed the Elephant?" (1990) 84 American Journal of International Law 30.

229 This section deals only with those aspects of the CBD that bear upon the issues raised in this book, particularly access to plant resources, equitable sharing of the benefits of plant resources, and the general implications of the international patent system on plants and plant-based traditional knowledge. For a general overview and analysis of the CBD, see

Cyrille de Klemm and Clare Shine, *Biological Diversity Conservation and the Law: Legal Mechanisms for Conserving Species and Ecosystems* (Gland, Swit.: IUCN, 1993).

230 CBD, *supra* note 1. For a negotiating account of the CBD, see Melinda Chandler, "The Biodiversity Convention: Selected Issues of Interest to the International Lawyer" (1993) 4 Colorado Journal of International Environmental Law and Policy 141; compare with International Treaty on Plant Genetic Resources for Food and Agriculture. Adopted on 3 November 2001, http://fao.org/ag/cgrfa/IU.

231 Lyle Glowka et al., eds., *A Guide to the Convention on Biological Diversity* (Gland, Swit.: IUCN, 1994) at 1.

232 *Ibid.*, at ix; Francoise Burhenne-Guilmin and Susan Casey-Lefkowitz, "The Convention on Biological Diversity: A Hard Won Global Achievement" (1992) 3 Yearbook of International Environmental Law 44.

233 Article 1, CBD, *supra* note 1.

234 There is thus a controversial debate as to whether the CBD is a "framework" convention or not. Commentators like Catherine Tinker have argued that "the CBD is not a framework convention, which requires further elaboration to be operational. It is a specific treaty with duties and obligations structured within a fully-operational system." See Catherine Tinker, "Responsibility for Biological Diversity Conservation under International Law" (1995) 28 Vanderbilt Journal of Transnational Law 777 at 7. Other scholars, however, disagree and argue that the convention is couched in generalities, leaving states with ample room to flesh out the specifics as they may deem fit, taking into account the broad objectives of the convention. However, because the CBD does not contain many hard law rules but, rather, broad policy objectives that are to be fleshed out by future protocols or by national initiatives, it may be described as a framework convention. See Swanson, *supra* note 209; Lee Kimball, "The Biodiversity Convention: How to Make It Work" (1995) 28 Vanderbilt Journal of Transnational Law 763; Charney, *supra* note 1 at 619.

235 Catherine Tinker, "Introduction to Biological Diversity: Law, Institutions, and Science" (1994) 1 Buffalo Journal of International Law 1.

236 Other conventions on biological diversity protection adopt a sectoral, or species-based, approach. For example, the Ramsar Convention deals specifically with wetlands. See Convention on Wetlands of International Importance, Especially as Waterfowl Habitat, Ramsar, 996 U.N.T.S. 245; (1972) 11 I.L.M. 963; M.J. Bowman, "The Ramsar Convention Comes of Age" (1995) 42 Netherlands International Law Review 1. Similarly, Convention on International Trade in Endangered Species, *supra* note 221, addresses trade in endangered species. Other conventions may even be limited by geography. See, for example, the Convention on Nature Protection and Wildlife Preservation in the Western Hemisphere, *supra* note 118. See Kathleen Rogers and James Moore, "Revitalizing the Convention on Nature Protection and Wild Life Preservation in the Western Hemisphere" (1995) 36 Harvard International Law Journal 465. Others include the African Convention on the Conservation of Nature and Natural Resources 1968, *supra* note 57; Convention on the Conservation of European Wildlife and Natural Habitats, *supra* note 123.

237 Rackleff, *supra* note 20; Robert Ward, "Man or Beast: The Convention on Biological Diversity and the Emerging Law on Sustainable Development" (1995) 28 Vanderbilt Journal of Transnational Law 823.

238 Report of the Fourth Global Biodiversity Forum, 1996, Montreal, Canada (Gland, Swit.: 1996) at 35.

239 Oliver Houck, "On the Law of Biodiversity and Ecosystem Management" (1997) Minnesota Law Review 869; Martin Belsky, "Using Legal Principles to Promote the 'Health' of an Ecosystem" (1996) 3 Tulsa Journal of Comparative and International Law 183.

240 Deborah Brosnan, "Ecosystem Management: An Ecological Perspective for Environmental Lawyers" (1994) 4 Journal of Environmental Law 135.

241 For example, World Charter for Nature, *supra* note 59.

242 Stockholm Declaration, Report of the UN Conference on the Human Environment, UN Doc. A/CONF.48/14 reprinted in 11 I.L.M. 11 (1972), *supra* note 55.

243 Article 3, CBD, *supra* note 1.

244 Article 15, CBD, *ibid.*

245 Panjabi, *supra* note 1 at 222.
246 Walter Reid, Vice President, World Resources Institute, Testimony before Canadian House of Commons Standing Committee on the Environment, 23 November 1992, House of Commons, Issue No. 47, 3rd Session, 34th Parliament, 1991-92.
247 For a fuller discussion of this concept in international law, see Kemal Baslar, *The Concept of the Common Heritage of Mankind in International Law* (The Hague: Martinus Nijhoff, 1998).
248 Paragraph 3 of the CBD preamble provides that the conservation of biological diversity is a common concern of mankind. See Preamble, CBD, *supra* note 1. See Article 8, African Convention on the Conservation of Nature and Natural Resources 1968, *supra* note 57. See also Abdulqawi Yussuf, "International Law and Sustainable Development: The Convention on Biological Diversity" (1995) African Yearbook of International Law 109.
249 The term "exclusive sovereign right" should be used advisedly as the limits and contours of the concept are matters of intense debate in international law, particularly with respect to the limits imposed by the principles of state responsibility, sustainable use of shared resources, and so on. The modalities for working out the finer details of domestic state sovereignty are not devoid of complex constitutional and property law issues. For a consideration of this issue as it affects an African country, see Amokaye Oludayo, "The Convention of Biological Diversity, Access to and Exploitation of Genetic Resources and the Land Tenure System in Nigeria" (1999) 11 African Journal of International and Comparative Law 86. The short point is that ownership of plant resources is inextricably tied to control and ownership of land in the domestic legal order. Further, there are unexplored complications relating to the question of the quantum of international "interest" in domestic abuse or use of plant resources. See Jackton Ojwang, "Kenya's Place in International Environmental Law Initiative" (1993) 5 African Journal of International and Comparative Law 781.
250 Tinker, *supra* note 234. Traditionally, the concept of state responsibility was developed to address state-alien relationships.
251 Marian Miller, "Sovereignty Reconfigured: Environmental Regimes and Third World States" in Karen Litfin, ed., *The Greening of Sovereignty in World Politics* (Cambridge, MA: MIT Press, 1998). The new FAO treaty on plant use also reaffirms the concept of CCM over plants.
252 This is the injunction to use one's property in a manner that does not injure another's property. See *Barcelona Traction, Light and Power Company, Ltd.*, (*Belg.* v. *Spain*), 1970 ICJ 4.
253 The Diversion of Water from the Meuse Case (*Netherlands* v. *Belgium*), 1937 P.C.I.J. (Ser. A/B) No. 70, at 4. (Reproduced in Lakshman Guruswamy et al. eds., *Supplement of Basic Documents to International Environmental Law and World Order* [Saint Paul, MN: West Publishing Co., 1994] at 1232.)
254 Ian Brownlie, *Principles of Public International Law*, 4th ed. (Oxford: Oxford University Press, 1990) at 437-40; *Jean-Baptiste Caire* v. *United Mexican States* (1929) 5 R.I.A.A. 516; *Report of the International Law Commission on the Work of Its Forty-Fifth Session*, UN GAOR, 48th Sess., Supp. No. 10, at 79, UN Doc. A/48/10 (1993).
255 Some scholars dispute the validity of this distinction. See Phillip Allott, "State Responsibility and the Unmaking of International Law" (1988) 29 Harvard International Law Journal 1; Allan Boyle, "State Responsibility and International Liability for Injurious Consequences of Acts Prohibited by International Law: A Necessary Distinction?" (1990) 39 International and Comparative Law Quarterly 1; Daniel Magraw, "Trans-Boundary Harm: The International Law Commission's Study of 'International Liability'" (1986) 80 American Journal of International Law 305. But see J. Combacau and D. Alland, "Primary and 'Secondary' Rules in the Law of State Responsibility: Categorizing International Obligations" (1985) 16 Netherlands Yearbook of International Law 81; Gunther Handl, "Liability as an Obligation Established by a Primary Rule of International Law" (1985) 16 Netherlands Yearbook of International Law 49.
256 Shabtai Rosenne, ed., *The International Law Commission's Draft Articles on State Responsibility* (1991); Marina Spinedi and Bruno Simma, eds., *United Nations Codification of State Responsibility* (New York: Oceana Publications, 1987). For further analysis of the so-called secondary rules, see *Report of the International Law Commission on the Work of Its Forty-Second Session*, UN GAOR, 42nd Sess., Supp. No. 10. UN Doc. A/42/10 (1990). From this distinction, it

seems clear that, unlike state responsibility, liability will not attach unless there is actual fault and harm. See Andrea Gattini "Smoking/No Smoking: Some Remarks on the Current Place of Fault in the ILC Draft Articles on State Responsibility" (1999) 10 European Journal of International Law 397. For further analysis, see James Crawford, "Revising the Draft Articles on State Responsibility" (1999) 10 European Journal of International Law 435.

257   A revised version of the ILC Draft Articles was recently adopted at the 53rd Session of the ILC. For the text and articles and commentaries, see Official Records of the General Assembly, Fifty-Sixth Session, Supplement No. 10 (A/56/10) Chap. 5. These are reproduced with a critical apparatus in James Crawford, *The ILC's Articles on State Responsibility: Introduction, Text and Commentaries* (Cambridge: CUP, Forthcoming). See also ILC Draft Articles on State Responsibility, A/CN.4/L.602/Rev., 26 July 2001, http://law.cam.ac.uk/rcil/ILCSR. See also Pierre-Marie Dupuy, "The International Law of State Responsibility: Revolution or Evolution?" (1989) 11 Michigan Journal of International Law 105; Pierre-Marie Dupuy, "Reviewing the Difficulties of Codification: On Ago's Classification and Obligations of Means and Obligations of Result in Relation to State Responsibility" (1999) 10 European Journal of International Law 371.

258   Allott strongly critiques this point. See Allot, "On State Responsibility" *supra* note 255; Vaughan Lowe, "Precluding Wrongfulness or Responsibility: A Plea for Excuses" (1999) 10 European Journal of International Law 405.

259   Tinker, *supra* note 234 at 785; Alain Pellet, "Can a State Commit a Crime? Definitely Yes!" (1999) 10 European Journal of International Law 425.

260   The term "soft law" has no clear-cut definition in international law. It may, however, be used to categorize those ambiguous obligations found in binding international agreements. It may also be used for those non-binding standards of conduct found in declarations, resolutions, and other such non-binding instruments. However, see Christine Chinkin, "The Challenge of Soft Law: Development and Change in International Law" (1989) 38 International and Comparative Law Quarterly 580. On the relationship between soft law and treaty law obligations, see Alan Boyle, "Some Reflections on the Relationship of Treaties and Soft Law" (1999) 48 International and Comparative Law Quarterly 901.

261   Rebecca Hoelting, "After Rio: The Sustainable Development Concept Following the United Nations Conference on Environment and Development" (1994) 24 Georgia Journal of International and Comparative Law 117; David Hunter et al., "Environment and Trade Concepts and Principles of International Law: An Introduction" (1995) 3 Global Environmental Law Annual 99.

262   The Brundtland Commission defined sustainable development as "development that meets the needs of the present without compromising the ability of future generations to meet · their own needs." See *Our Common Future, supra* note 147 at 43.

263   Binding international legal instruments that affirm the place of the concept of sustainable development in the jural field include the Convention on Biological Diversity, *supra* note 1, Climate Change Convention, *supra* note 207; Agreement Establishing the World Trade Organization (1994) 33 I.L.M 1. Regional instruments that also recognize the concept of sustainable development include the North American Agreement on Environmental Co-operation, reprinted in (1993) 32 I.L.M. 1480; Treaty on European Union, Maastricht, 7 February 1992, Official Journal, 1992, C 191 (29 July 1992). International judicial imprimatur may also be found in the advisory opinion of the ICJ in the case of *Legality of the Threat of Use of Nuclear Weapons* [Advisory opinion] (1997) 35 I.L.M. 809; the separate opinion of Justice Weeramantry in the Case Concerning the Gabcikovo-Nagymaros Project (*Hungary* v. *Slovakia*) (1998) 37 I.L.M. 162 at para. 29. For a broader account of the role of the International Court of Justice in the environmentalist trend, see E. Valencia-Ospina, "The International Court of Justice and International Environmental Law" (1992) 2 Asian Yearbook of International Law 1.

264   For a succinct account and analysis of the literature on the concept of sustainable development, see Philippe Sands, "International Law in the Field of Sustainable Development" (1994) 65 British Yearbook of International Law 303; Patricia Birnie, "Environmental Pro-

tection and Development" (1995) 20 Melbourne University Law Review 66; Edith Brown Weiss, "In Fairness to Future Generations" (1992) 8 American University Journal of International Law and Policy 19.

265 Interestingly, the concept itself has been the subject of terminological anarchy. See Sands, *supra* note 264, at 305. There are also doubts whether the concept means an "objective," a "process," a "principle," or all of these things put together.

266 This presupposes that the prevailing ideology of development is a universal truth. But see Priya Kurian, "International Environmental Policy: Redefining for Theory and Practice" in Robert Bartlett et al., eds., *International Organizations and Environmental Policy* (Westport, CT: Greenwood Press, 1995) at 1 (arguing that "the progressive, secular, and materialist philosophy on which modern life rests ... is deeply flawed and ultimately destructive to ourselves"). But see Richard Stewart, "Economics, Environment, and the Limits of Legal Control" (1985) 9 Harvard Environmental Law Review 1 (arguing that environmentalism is not necessarily opposed to economic activity); Peter Lallas et al., "Environmental Protection and International Trade: Toward Mutually Supportive Rules and Policies" (1992) 16 Harvard Environmental Law Review 271.

267 Dillon, *supra* note 119 at 371. But see Jayne Daly, "Toward Sustainable Development: In Our Common Interest" (1995) Pace Law Review 153; J.B. Ruhl, "Sustainable Development: A Five Dimensional Algorithm for Environmental Law" (1999) 18 Stanford Environmental Law Journal 31. At the core of the problem is the fact that the concept of sustainable development is an improbable compromise seeking to satisfy divergent and opposed values and interests. For example, on the matter of intragenerational equity, the question may be asked whether the concept will require a redistribution of resources so that those who maintain a position of privilege will offset the need for further ravaging of the earth by industrial activities in order to satisfy the needs of the needy.

268 Kurian, *supra* note 266 at 1.

269 Michael McCloskey, "The Emperor Has No Clothes" (1999) 9 Duke Environmental Law and Policy 153 at 157.

270 Bedrich Moldan and Suzanne Billharz, eds., *Sustainability Indicators: Report on Indicators of Sustainable Development* (Toronto: John Wiley and Sons., 1997).

271 *Ibid.,* at 71.

272 Donald Brown, "After the Earth Summit: The Need to Integrate Environmental Ethics into Environmental Science and Law" (1992) 2 Dickinson Journal of Environmental Law and Policy 1; Dan Tarlock, "Environmental Law: Ethics Or Science?" (1996) 7 Duke Environmental Law and Policy Forum 193; Walter Kuhlman, "Making the Law More Eco-Centric: Responding to Leopold and Conservation Biology" (1996) 7 Duke Environmental Law and Policy Forum 133.

273 Moldan and Billharz, *supra* note 270 at 18.

274 Dillon, *supra* note 119 at 372. Compare with Mukul Sanwal, "Sustainable Development, the Rio Declaration and Multilateral Cooperation" (1993) 4 Colorado Journal of International Environmental Law and Policy 45.

275 Ronnie Lischutz, "Wasn't the Future Wonderful? Resources, Environment, and the Emerging Myth of Global Sustainable Development" (1991) 2 Colorado Journal of International Environmental Law and Policy 35.

276 Abdulqawi Yusuf, "International Law and Sustainable Development: The Convention on Biological Diversity" (1995) African Yearbook of International Law 109 at 125. See also Article 10 (c) of the CBD, *supra* note 1. For an interesting analysis of a practical inquiry into the mechanics of sustainable development in a sub-Saharan African community, see Philippa England, "Tree Planting, Sustainable Development and the Roles of Law in Bongo, North-East Ghana" (1995) 39 Journal of African Law 138.

277 While some international instruments ostensibly emphasize conservation of plant resources, others emphasize sovereignty. Compare the African Convention on the Conservation of Nature and Natural Resources, *supra* note 57, with the CBD, *supra* note 1.

278 Sun Lin, ed., *UNEP's New Way Forward: Environmental Law and Sustainable Development* (Nairobi, Kenya: UNDP, 1995).

279   *Ibid.,* at 12.
280   *Juan Antonio Oposa et al.* v. *The Honourable Fulgencio Factoran Jr., Secretary of the Department of the Environment and Natural Resources, et al.,* Supreme Court of the Philippines, G.R. No. 101083; reprinted in, 33 I.L.M. 173 (1994).
281   The children were suing to stop large-scale leasing of forests for timber-logging, particularly original rainforest tracts. Since the decision granted standing, an executive order has cancelled sixty-five of the original leases, including those in the old-growth rainforests. See Ted Allen, "The Philippine Children's Case: Recognizing Standing for Future Generations" (1993) 6 Georgetown Environmental law Review 713.
282   The question of precautionary principle in international law is of particular importance to the issue of biopiracy as it relates to the question of release of patented genetically modified plants into the environment.
283   Edward Christie, "The Eternal Triangle: The Biodiversity Convention, Endangered Species Legislation and the Precautionary Principle" (1993) 10 Environmental and Planning Law Journal 470; Christian Dominice, "The International Responsibility of States for Breach of Multilateral Obligations" (1999) 10 European Journal of International Law 353.
284   Edward McWhinney, "The International Law-Making Process and the New International Economic Order" (1976) 14 Canadian Yearbook of International Law 57.
285   *Ibid.*
286   Article 26, Vienna Convention, *supra* note 224.
287   This is otherwise known as the principle of common but differentiated responsibilities. This elasticity was first employed in the 1982 UNCLOS. See United Nations Convention on the Law of the Sea (With Annex V). Concluded at Montego Bay, 10 December 1982; entered into force 16 November 1994. UN Doc. A/CONF.62/122; reprinted in 21 I.L.M 1261 (1982). However, this device has been criticized by a host of writers as importing inconsistencies into the implementation of environmental treaties.
288   The mechanisms of avoidance include non-compliance with their treaty obligations or defiance. See Jung-Gum Kim and John Howell, *Conflict of International Obligations and State Interests* (The Hague: Martinus Nijhoff, 1972). Of course, the entering of reservations to treaties and conventions is well known and may be held unacceptable under certain treaties. See Articles 19, 20, 21, 22, and 24, Vienna Convention, *supra* note 224. In the cases where reservations are disallowed, states may employ the use of interpretive statements or clauses. More often than not, this is largely designed to appease local forces. For example, in signing the CBD, the United States attached an interpretive statement approved by its powerful biotechnology and pharmaceutical industries.
289   Article 18, Vienna Convention on the Law of Treaties, *supra* note 224.
290   Glowka, *supra* note 231 at 4.
291   John Mugabe et al., eds., *Access to Genetic Resources: Strategies for Sharing Benefits* (Bonn: IUCN-ELC, 1997).
292   Panjabi, "On the CBD," *supra* note 1 at 220. The financial interests are staggering. The World Resources Institute estimates that the retail value of drugs derived from plants approximates to $43 billion per year and that the pharmaceutical industry grows at more than ten percent per year. See Al Gore, "Essentials for Economic Progress: Protect Biodiversity and Intellectual Property Rights" (1992) October, Journal of NIH Research 5.
293   The ILO Convention No. 169 had blazed the trail as per paragraph 7 of the preamble of the convention.
294   Preamble, CBD, *supra* note 1.
295   Article 8 (j), CBD, *ibid.*
296   Daniel Jenks, "The Convention on Biological Diversity: An Efficient Framework for the Preservation of Life on Earth?" (1995) 15 Northwestern Journal of International Law and Business 636 at 642; Jeffrey Kushan, "Biodiversity: Opportunities and Obligations" (1995) 28 Vanderbilt Journal of Transnational Law 755.
297   Gregory Maggio, "Recognizing the Vital Role of Local Communities in International Instruments for Conserving Biodiversity" (1997-98) 16 Journal of Environmental Law 179.
298   In-situ conservation refers to the conservation of ecosystems and natural habitats and the maintenance and recovery of viable populations of species in their natural surroundings

(and, in the case of domesticated or cultivated species, in the surroundings in which they have developed their distinctive properties). See Article 2, CBD, *supra* note 1.

299 Ex-situ conservation refers to the conservation of components of biological diversity outside their natural habitats.

300 *Report of the Fourth Global Biodiversity Forum, supra* note 238 at 35.

301 Chandler, *supra* note 230 at 161. Chandler observes that, during the negotiation of the CBD, the gene-rich-technology-poor countries of the South sought to use the convention and their plant genetic wealth as a means of gaining access to and transfer of technology, the handling of biotechnology, and the sharing of its benefits. This may be evident from Articles 9 and 15 of the CBD, *supra* note 1. The North resisted this move.

302 Article 16 (5) of the CBD, *ibid.*

303 Article 16 (2), CBD, *ibid.*

304 See, for example, *The Impact of Intellectual Property Rights Systems on the Conservation and Sustainable Use of Biological Diversity and on the Equitable Sharing of Benefits from Its Use* UNEP/CBD/COP/3/23, 1996; *The Relationship between Intellectual Property Rights and the Relevant Provisions of the Agreement on Trade-Related Aspects of Intellectual Property Rights (TRIPS Agreement) and the Convention on Biological Diversity.* UNEP/CBD/ISOC/5, 11 May 1999.

305 Report of the Intergovernmental Committee for a Convention on Biological Diversity, UN Environmental Programme, 7th Negotiating Sess., 5th Session of the International Negotiating Committee, UN Doc. UNEP/Bio.Div/N7-INC.5/4 (1992). The Government of Ethiopia thus suggested that, at a later date, the following paragraph be added to Article 16:

> Where a technology, an organism or genetic material which is patented or legally protected in any other way as an intellectual property has incorporated an organism or organisms, a genetic material or materials, a technology or technologies or any other traditional practice or practices originating in another country or countries, the patent or other intellectual property right shall not be valid in the country or countries of origin of any of its component parts; and the benefits accruing from the application of the patent or other intellectual property right in other countries shall be equitably shared between the holder or holders of the protected right and the country or countries of origin. (*ibid.*)

306 Declaration of Malta, done at the 25th Pacem in Maribus conference held in Malta in November 1997; reproduced in R. Rajagopalan, ed., *Common Heritage and the 21st Century* (Msida, Malta: International Ocean Institute, 1998).

307 *Ibid.*, at 7-11.

308 *Ibid.*, at 15.

309 Gary Meyers, "Surveying the Lay of the Land, Air, and Water: Features of Current International Environmental and Natural Resources Law, and Future Prospects for the Protection of Species Habitat to Preserve Global Biological Diversity" (1992) 3 Colorado Journal of International Environmental Law and Policy 479.

310 George Mitchell, *World on Fire: Saving an Endangered Earth* (New York: Charles Scribner's Sons, 1991) at 116.

311 Lynn White Jr., *supra* note 58; but see Thomas Derr, "Religious Responsibility for the Ecological Crisis: An Argument Run Amok" (1975) Vol. 18, No. 1 Worldview 43.

312 Cynthia Giagnocavo and Howard Goldstein, "Law Reform or World Re-Form: The Problem of Environmental Rights" (1990) 35 McGill Law Journal 345. In the same vein, former Vice President Al Gore has argued that "it will not be enough to change our laws, policies and programs. The solutions we seek must stem from a new faith in the future, a faith that justifies sacrifices in the present, and from a new courage to choose higher values in the conduct of human affairs. We must also display new reverence for our place in the natural world." Quoted in Dobson, *supra* note 42 at 308. See also M.P.A. Kindall, "Talking Past Each Other at the Summit" (1993) 4 Colorado Journal of International Environmental Law and Policy 69 at 71.

313 David Schlosberg, *Environmental Justice and the New Pluralism: The Challenge of Difference for Environmentalism* (Oxford: Oxford University Press, 1999) at 3.

314 Alexander Gillespie, *International Environmental Law, Policy and Ethics* (Oxford: Clarendon Press, 1997) at 1.
315 George James, ed., *Ethical Perspectives on Environmental Issues in India* (New Delhi: A.P.H. Publishing Corporation, 1999).

Chapter 4: The Appropriative Aspects of Biopiracy

1 Bellagio Declaration, <http://users.ox.ac.uk/~wgtrr/bellagio.htm>.
2 Jack Kloppenburg, *First the Seed: The Political Economy of Plant Biotechnology, 1492-2000* (Cambridge: Cambridge University Press, 1988) at 189.
3 Valentina Tejera, "Tripping over Property Rights: Is It Possible to Reconcile the Convention on Biological Diversity with Article 27 of the TRIPs Agreement?" (1999) 33 New England Law Review 967 at 971.
4 Vandana Shiva, "Biopiracy: Need to Change Western IPR Systems" The Hindu, Wednesday, 28 July 1999, at 3; Luiza Villaema, "Indians Want Patent: Chiefs Prepare International Law Suit against Scientist Who Registered Indigenous Knowledge" ISTOE Magazine, No. 1581, Sao Paulo, 19 January 2000.
5 Lakshmi Sarma, "Biopiracy: Twentieth Century Imperialism in the Form of International Agreements" (1999) 13 Temple International and Comparative Law Journal 107; David Orr, "India Accuses US of Stealing Ancient Cures" The Times (London), 31 July 1999, at 2; Mario Osava, "Brazil Biodiversity: Crackdown on Eco-Pirates" IPS, 14 August 1999.
6 William Lesser, *Sustainable Use of Genetic Resources under the Convention on Biological Diversity: Exploring Access and Benefit-Sharing Issues* (Oxford: CAB International, 1997) at 113.
7 Others, like Bartolome las Casas, opined that the natives were creatures of God and were endowed with the same rational capacities as were the invading Europeans. Similarly, Pope Paul III, in the Papal Bull of 1537, clearly noted that the Indians were human beings with the same rational abilities as the Europeans. See Felix Cohen "The Spanish Origin of Indian Rights in the Law of the United States" (1942) 31 Georgetown Law Journal 12.
8 Guillermo Floris Margadant, "Official Mexican Attitudes towards the Indians: An Historical Essay" (1980) 54 Tulane Law Review 967; Edwin Williamson, *The Penguin History of Latin America* (London: Penguin, 1992).
9 *Ibid.*, at 964.
10 The term "ethnobotany" was coined by the American botanist John W. Harshberger in 1895 to describe studies of plants used by "primitive" and aboriginal peoples.
11 Martha Johnson, ed., *Lore-Capturing Traditional Environmental Knowledge* (Ottawa: Dene Cultural Institute and International Development Research Centre, 1992).
12 According to eminent anthropologist Virgil Vogel, until 1884, when Kohler distilled cocaine from coca leaves, Native American accounts of the immense powers of the coca leaves were dismissed. Physicians using coca to relieve pain were ridiculed as being incapable of judging a remedy's qualities. Pharmacists making preparations of the drug were looked upon as being involved in fraud, while the natives who employed it in their daily life, as well as the travellers who were impressed by what they had observed of its effects, were regarded as ignorant or as superstitious. See Virgil Vogel, "American Indian Influence on the American Pharmacopoeia," in Johnson, *supra* note 11 at 103.
13 Rajiv Sinha, *Ethnobotany: The Renaissance of Traditional Herbal Medicine* (Jaipur, India: Ina Shree Publishers, 1996).
14 *Ibid.* For example, years before James Linds's "experiment" with scurvy, the American Indians had known its cause and cure.
15 George Meyer, Kenneth Blum, and John Cull, eds., *Folk Medicine and Herbal Healing* (Springfield, IL: Charles Thomas Publisher, 1981).
16 Reflecting the general, off-hand dismissal of TKUP in those times, as late as 1936 the scientist Norman Hines dismissed the claims of the Shoshone women as useless. He maintained that "no drug has yet been discovered which, when taken by mouth, will induce temporary sterility." See, Johnson *supra* note 11 at 111. However, other doctors were not so blinded. Dr. Benjamin Barton of the medical faculty of the University of Pennsylvania declared in 1798 that the Indians had knowledge of some of the most inestimable medicines. He wrote a treatise on sixty indigenous plant remedies, and fourteen of his students wrote disserta-

tions on native remedies and practice. It is equally interesting to note that the discoverer of insulin, Dr. Frederick Banting, was tremendously inspired by the Indians of British Columbia, who used extracts from devil's club for the treatment of diabetes. See *ibid.*

17 Johnson, *ibid.* See also Erwin Ackerknecht, *Medicine and Ethnology* (Baltimore, MD: Johns Hopkins University Press, 1991).

18 N. Ezeabasili, *African Science: Myth or Reality* (New York: Vantage Press, 1977).

19 Carl-Martin Edsman, ed., *Studies in Shamanism* (Uppsala: Almqvist and Boktryckeri, 1962); Andreas Lommel, *Shamanism: The Beginnings of Art* (Toronto: McGraw-Hill Book Company, 1967); J.B. Loudon, ed., *Social Anthropology and Medicine* (London: Academic Press, 1976).

20 In January 1975 the Board of the American Association for the Advancement of Science passed the following Resolution: "Be it resolved that the Council of the Association (a) formally recognizes the contributions made by Native Americans in their own traditions of inquiry to the various fields of science, engineering, and medicine, and (b) encourages and supports the growth of natural and social programs in which traditional Native American approaches and contributions to science, engineering, and medicine are the subject of serious study and research." See David Landy, ed., *Culture, Disease, and Healing: Studies in Medical Anthropology* (New York: Macmillan Publishing, 1977).

21 *Ibid.,* at 252.

22 Michael Balick and Paul Alan Cox, *Plants, People and Culture: The Science of Ethnobotany* (New York: Freeman and Company, 1996).

23 *Ibid.*

24 *Ibid.,* at 36.

25 *Ibid.,* at 2.

26 Peter Morley, "Culture and the Cognitive World of Traditional Medical Beliefs: Some Preliminary Considerations" in Peter Morley and Roy Wallis, eds., *Culture and Curing: Anthropological Perspectives on Traditional Medical Beliefs and Practices* (Pennsylvania: University of Pennsylvania Press, 1978) at 1.

27 *Ibid.,* at 4.

28 *Ibid.,* at 6. But see O.N. Muchena and E. Vanek, "From Ecology through Economics to Ethno-Science: Changing Perceptions on Natural Resource Management" in D.M. Warren et al., eds., *The Cultural Dimensions of Knowledge: Indigenous Knowledge Systems* (London: Intermediate Technology Publications, 1995).

29 Lesser, *supra* note 6 at 114.

30 Cheryl Hardy, "Patent Protection and Raw Materials: The Convention on Biological Diversity and Its Implications for US Policy on the Development and Commercialization of Biotechnology" (1994) 15 University of Pennsylvania Journal of International Business Law 299; John Adair, "The Bioprospecting Question: Should the United States Charge Biotechnology Companies for the Commercial Use of Public Wild Genetic Resources?" (1997) 24 Ecology Law Quarterly 131.

31 Kloppenburg, *supra* note 2 at 154.

32 In his material dialectics argument, Marx acutely observed that "the discovery of gold and silver in America, the extirpation, enslavement and entombment in mines of the indigenous population of that continent, the beginnings of the conquest and plunder of India, and the conversion of Africa into a preserve for the commercial hunting of black skins, are all things which characterize the dawn of the era of capitalist production. These idyllic proceedings are the chief moments of primitive accumulation ... These treasures captured outside Europe by undisguised looting, enslavement, and murder flowed back to the mother-country and were turned into capital there." See Karl Marx, *Capital,* Vol. 1 (New York: Vintage Books, 1977) at 915-18.

33 Kloppenburg, *supra* note 2 at 154.

34 Lesser, *supra* note 6 at 14.

35 According to Abraham Lama, "Quinine is the well-known anti-malarial compound which comes from the bark of the Peruvian cinchona tree. Andean indigenous groups supposedly learned of the bark's powers while observing feverish jaguars eating the cinchona bark. In 1636 an Incan healer cured the Spanish viceroy's wife of her recurrent malaria fever using

bark from the bush. The Countess of Chichon's wife was so excited that she distributed what she called the 'Countess'[s] powder' to the people of Lima who suffered the fever. Jesuit priests sent the remedy to Europe with the name 'Jesuit's Powder' and soon after Cardinal Lugo dispersed the miraculous medication and purveyed it under the name 'Cardinal's powder.' Rome in that era was the malarial capital of the world. Surrounded by marshes, its 'mal aire' (bad air) led to the disease's name 'malaria.' The unhealthy conditions of the Vatican meant that the seat of Christianity was nearly abandoned several times, after the deaths of various Popes and dozens of cardinals. By 1650, the mysterious remedy had become known at the Vatican and awakened interest in other European capitals. In 1679, Britain's Robert Talbot had quina plants sent from Peru and began to market the powder derivative of the bark. In 1820, French chemists Pelletier and Caventou isolated quinine or 'chinchonia,' – named in honor of viceroy Chinchon's wife – from the bark of the quina tree. They honoured the Countess, but nobody ever remembered the Inca doctors who discovered its curative properties, who genetically developed the plant, and used it for many years." See Abraham Lama, "Law to Protect Native Intellectual Property" IPS News Bulletin, 12 January 2000, <http://www.ips.org>. See also *Merrel Dow* v. *HN Norton*, [1996] Report of Patent Cases 76.

36   Kloppenburg, *supra* note 2 at 155.
37   As a result of the abundance of food potatoes, the European population nearly doubled in one century (1750-1850). The arguments in this section are substantially derived from an earlier work. See Ikechi Mgbeoji, "Beyond Rhetoric: State Sovereignty, Common Concern, and the Inapplicability of the Common Heritage Concept to Plant Genetic Resources" (2003) Vol. 16, No. 4 Leiden Journal of International Law 821-37. I am grateful to the Leiden Journal for permission to reproduce the article.
38   Kloppenburg, *supra* note 2 at 155. The Irish and indeed all the British practically relied on potatoes for subsistence.
39   The imperial botanical gardens stretched from Australia, Africa, the Caribbean, and India to virtually all corners of the globe. See Lucile Brockway, *Science and Colonial Expansion: The Role of the British Botanic Gardens* (New York: Academic, 1979).
40   Pat Mooney, *Seeds of the Earth: A Private or Public Resource?* (Ottawa: Inter Pares, 1979).
41   Karl Hammer, "A Paradigm Shift in the Discipline of Plant Genetic Resources" (2003) 50 Genetic Resources and Crop Evolution 3-10.
42   John Mugabe et al., eds., *Access to Genetic Resources: Strategies for Sharing Benefits* (Bonn: IUCN-ELC, 1997) at 7.
43   Christopher Joyner, "Legal Implications of the Concept of the Common Heritage of Mankind" (1986) 35 International and Comparative Law Quarterly 190. As earlier indicated, this section is a reproduction of my earlier published work in the Leiden Journal of International Law.
44   Rudiger Wolfrum, "The Principle of the Common Heritage of Mankind" (1983) 43 Heidelberg Journal of International Law 312. While some scholars attribute the origins of the common heritage concept to Ambassador Pardo of Malta in 1967, others point to Aldo Cocca's statement some months earlier at the deliberations for peaceful uses of outer space in 1967. It seems, however, that Ambassador Pardo was the first to articulate the concept of CHM as a potential principle of international law. In any event, the notion of CHM does not predate 1967. The implication of this is that it could not have governed transactions on plants prior to its debut.
45   For an examination of the confusion surrounding the concept of CHM, see Stephen Gorove, "The Concept of Common Heritage of Mankind: A Political, Moral or Legal Innovation" (1972) San Diego Law Review 390.
46   Aldo Cocca, "Mankind as the New Legal Subject: A New Juridical Dimension Recognized by the United Nations" (1971) Proceedings of the 13th Colloquium on the Law of Outer Space 211. The notion of "mankind" as a full-fledged legal entity has not yet come into juridical existence.
47   There are, however, some differences between CHM and the notion of *res communis humanitatis*. For further analysis of this issue, see Malcolm Shaw, *International Law*, 2nd ed. (Cambridge: Grotius Publications, 1986) 276 [hereinafter, Shaw]; Cheng Bin, "The Legal

Regime of Airspace and Outer Space: The Boundary Problem, Functionalism versus Spatialism – The Major Premises" (1980) 5 Annals of Air and Space Law 323. Attempts to extend the concept of CHM to the vexed issue of transfer of technology have, however, failed. See the Draft International Code of Conduct on the Transfer of Technology of 6 May 1980 (1978) 17 I.L.M. 462.

48 Ian Brownlie, *Principles of Public International Law,* 3rd ed. (Oxford: Clarendon Press, 1972) at 258-86.

49 J. Van Dyke and C. Yuen, "Common Heritage v. Freedom of the Seas: Which Governs the Seabed?" (1982) 19 San Diego Law Review 493.

50 L. Tennen, "Outer Space: A Preserve for All Humankind" (1979) 1 Houston Journal of International Law 145.

51 C. Christol, "The Common Heritage of Mankind Provision in the 1979 Agreement Governing the Activities of States on the Moon and Other Celestial Bodies" (1980) 14 International Lawyer 429.

52 C. Joyner, "Antarctica and the Law of the Sea: Rethinking the Current Legal Dilemmas" (1981) San Diego Law Review 415.

53 J. Crawford, *The Creation of States at International Law* (Oxford: Clarendon Press, 1979); *Rainbow Warrior Incident* (1985) International Law Reports 74.

54 Jennings and Watts, eds., *Oppenheim's International Law,* Vol. 1, 9th ed. (London: Longman, 1992) at 563-80.

55 B. Larschan and B. Brennan, "The Common Heritage of Mankind Principle in International Law" (1982-83) 21 Columbia Journal of Transnational Law 305.

56 Kloppenburg, *supra* note 2 at 167.

57 Josef Henry Vogel, "An Economic Analysis of the Convention on Biological Diversity: The Rationale for a Cartel" (on file with the author). Persons interested in this article may reach Professor Vogel at henvogel@earthling.net.

58 See Article 38 of the Statute of the International Court of Justice. Concluded at San Francisco, 26 June 1945. Entered into force 24 October 1945 (1976 Y.B.U.N. 1052).

59 Most of these characteristics are derived from Ambassador Pardo's historic statement. See "Declaration and Treaty Concerning the Reservation Exclusively for Peaceful Purposes of the Seabed and the Ocean Floor, Underlying the Seas beyond the Limits of Present National Jurisdiction, and the Use of Their Resources in the Interests of Mankind," UN Doc. A/AC.105 /C.2/SR (17 August 1967); A. Pardo, *The Common Heritage: Selected Papers on Oceans and World Order 1967-1974* (1975).

60 For a comprehensive, albeit debatable, analysis of the concept of CHM in international law, see Kemal Baslar, *The Concept of the Common Heritage of Mankind in International Law* (The Hague: Martinus Nijhoff, 1998).

61 D. Shraga, "The Common Heritage of Mankind: The Concept and Its Application" (1986) 15 Annales D'Etudes Internationales 45.

62 For example, Part II of the 1982 Law of the Sea Convention provides that "the Area [i.e., the sea-bed and ocean floor beyond the limits of national jurisdiction] and its resources are the common heritage of mankind." See Article 137, United Nations Convention on the Law of the Sea (With Annex V). Concluded at Montego Bay, 10 December 1982. Entered into force, 16 November 1994, UN Doc. A/CONF.62/122; reprinted in 21 I.L.M 1261 (1982); Declaration of Principles Governing the Sea-Bed and the Ocean Floor, and the Subsoil Thereof, Beyond the Limits of National Jurisdiction; adopted by the United Nations General Assembly, 17 December 1970. UN Doc. A/RES/2749 (XXV), 10 I.L.M. 220 (1971).

63 Agreement Concerning the Activities of States on the Moon and other Celestial Bodies, 5 December 1979, (1979) 18 I.L.M. 1434.

64 Treaty on Principles Governing the Activities of States in the Exploration and Use of Outer Space, Including the Moon and Other Celestial Bodies, 610 U.N.T.S. 205.

65 Antarctic Treaty, concluded at Washington, 1 December 1959; entered into force 23 June 1961, reprinted in 40 U.N.T.S. 71.

66 See, for example, Joyner, *supra* note 43 at 198.

67 Anthony D'Amato, *The Concept of Custom in International Law* (Ithaca: Cornell University Press, 1971).

68 *Supra* note 65.
69 *South West Africa Cases* (1960) I.C.J. Rep. 6 at 323.
70 The 1933 Montevideo Convention on Rights and Duties of States, 26 December 1933. 165 L.N.T.S 19.
71 Jennings and Watts, *supra* note 54 at 563-80. See, for example, International Convention for the Protection of Plants, 16 April 1929, reprinted in 126 L.N.T.S. 305 (1931-32).
72 Michael Akehurst, "Custom as a Source of International Law" (1974-75) 47 British Yearbook of International Law 12 at 36.
73 D'Amato *supra* note 67 at 74-75. See also Myers McDougall, Lasswell, and Reisman, "The World Constitutive Process of Authoritative Decision" (1967) 19 Journal of Legal Education 403.
74 See, for example, Justice Gray in the Paquette Habana, 175 US 677 (1900).
75 Gregori Tunkin, "Remarks on the Juridical Nature of Customary Norms of International Law" (1961) 49 California Law Review 419; Lazare Kopelmanas, "Custom as a Means of the Creation of International Law" (1937) 28 British Yearbook of International Law 127.
76 Kloppenburg, *supra* note 2 at 154; Cyrille de Klemm, *Biological Diversity Conservation and the Law: Legal Mechanisms for Conserving Species and Ecosystems* (Gland, Swit.: IUNC, 1993) at 56.
77 Kloppenburg, *supra* note 2 at 154.
78 Klaus Bosselman, "Plants and Politics: The International Legal Regime Concerning Biotechnology and Biodiversity" (1995) 7 Colorado Journal of International Environmental Law and Policy 111.
79 See, for example, *The Indian Wildlife (Protection) Act* of 1972, the *Papua New Guinea Fauna (Protection and Control) Act* of 1976, the *Ugandan Forest Act* and the *Kenya Forest Act* of 1942.
80 International Convention for the Protection of Plants, *supra* note 71. Article 4 of the Plant Convention provided thus: "The Contracting States undertake to enact all necessary measures both to prevent and combat plant diseases and pests and to supervise the importation of plants and parts of plants, in particular those consigned from countries not as yet possessing any official organisation for the protection of plants. When Contracting States require that plants or parts of plants to be imported shall be accompanied by a health certificate issued by a competent official agent duly authorized by the exporting state, the Contracting States must conform to the provisions of the present Convention." Article 6 proceeds further by providing that "each state retains the right to inspect and place in quarantine plants or parts of plants, or temporarily and exceptionally to prohibit their importation, even when the consignments are accompanied by a health certificate.
81 Mugabe, *supra* note 42.
82 PSNR emerged and developed after the end of the Second World War to affirm and assert the sovereignty of developing countries over their own natural resources. It seems to have matured from a "fundamental principle of the New International Economic Order (NIEO) to the same status of jus cogens similar to the right of self-determination in the present international order." Baslar, *supra* note 60 at 137.
83 See Permanent Sovereignty over Natural Resources, UN General Assembly Resolution 1803 (XVII); Principle 21 of the 1972 Stockholm Declaration, Article 30 of the Charter of Economic Rights and Duties of States of 1974 (UNGAOR 3281, XXIV); Principle 2 of the 1992 Rio Declaration. For an excellent and exhaustive treatment of the subject, see N. Schrijver, *Permanent Sovereignty over Natural Resources: Balancing Rights and Duties* (Cambridge: Cambridge University Press, 1997). Some scholars have thus opined that the CHM regime can only start where it is agreed that the PSNR ends.
84 N. Schrijver, "Permanent Sovereignty over Natural Resources versus the Common Heritage of Mankind: Complementary or Contradictory Principles of International Economic Law?" in Paul Waart et al., *International Law and Development* (Dordrecht: Martinus Nijhoff, 1988); Stephen Schwebel, *Justice in International Law* (Cambridge: Cambridge University Press, 1994) at 401.

85 R.M. McClearly, "The International Communities Claim to Right in Brazilian Amazonia" (1991) 39 Political Studies 691; J. Goldenberg and E.R. Durham, "Amazonia and National Sovereignty" (1992) 2 International Environmental Affairs 22.

86 Hence, Cocca's argument that "the oxygen produced by the Amazonic forest ... should be declared common heritage of mankind." See A.A. Cocca, "The Common Heritage of Mankind: Doctrine and Principle of Space Law: An Overview (1986) 29th Coll. On the Law of Outer Space 17, at 22.

87 K. Zimmerman, "The Deforestation of the Brazilian Amazon: Law, Politics, and International Cooperation" (1990) 21 Inter-American Law Review 513.

88 McClearly, *supra* note 85 at 692.

89 Internally, the "development" of the Amazon by Brazilian has literally destroyed the Indians. As the governor of Roraima declared in 1975, "an area as rich as this cannot afford the luxury of conserving half a dozen Indian tribes who are holding back the development of Brazil."

90 Baslar, *supra* note 60 at 154.

91 Lesser, *supra* note 6 at 99.

92 It is significant that Vice President Wallace was also the founder of one of the largest seed firms in the world – Pioneer Hi-Bred. He was also an ardent advocate of patents on plants.

93 Kloppenburg, *supra* note 2 at 158.

94 Similar projects were started in Guatemala, El Salvador, Venezuela, Brazil, Uruguay, Argentina, Costa Rica, Cuba, Colombia, Peru, and Chile under the auspices of the United States Department of Agriculture. See, Kloppenburg, *supra* note 2 at 158.

95 *Ibid.*

96 Lesser, *supra* note 6 at 14.

97 The indigenous and diverse stock of rice species and cultivars in the Philippines is very significant.

98 This country and the region is the centre of origin of such economically important cereals as wheat.

99 Liberia contains a majority of African indigenous rice species.

100 Potatoes originated in the Andes.

101 Other IARCs include the International Institute of Tropical Agriculture (IITA) in Nigeria, the International Crops Research Institute for the Semi-Arid Tropics (ICRISAT) in India, and the International Maize and Wheat Improvement Center (CIMMYT) in Mexico. See Kloppenburg, *supra* note 2 at 161.

102 *Ibid.*

103 *Ibid.*

104 "Institutes Conserving Crop Germ Plasm: The IBPGR Global Network of Genebanks." International Board for Plant Genetic Resources (Rome, 1984). Fifty of the ninety IBPGR designated gene banks are located in the North.

105 See, UNEP/CBD/IC/2/13, 1994 par. 16.

106 Lesser, *supra* note 6 at 99.

107 Naomi Roht-Arriaza, "Of Seeds and Shamans: The Appropriation of the Scientific and Technical Knowledge of Indigenous and Local Communities" (1996) 17 Michigan Journal of International Law 940 at 945.

108 Kloppenburg, *supra* note 2 at 166.

109 Anthony Stenson and Tim Gray, *The Politics of Genetic Resource Control* (London: Macmillan Press, 1999) at 13.

110 Kloppenburg *supra* note 2 at 160. It deserves mentioning that the doomed Green Revolution of the 1960s was not devoid of North-South intrigues, particularly during the Cold War. In effect, the Green Revolution was not completely altruistic as it constituted a major pawn in both the economic and ideological conflicts between the East and the West in the struggle for the soul of the South. The appropriated plant germplasm from the South was thus modified and subjected to heavy doses of agricultural inputs such as fertilizers and herbicides for the production of "bumper" harvests, which were then used to dramatize the wonders of capitalist science and technology. The objective was to delegitimize the

competing ideology of communism, which was then appealing to many countries of the South, particularly in Asia.

111 Kloppenburg, *supra* note 2 at 161 (emphasis added).
112 Naomi Arriaza, "Of Seeds and Shamans: The Appropriation of the Scientific and Technical Knowledge of Indigenous and Local Communities" (1996) Michigan Journal of International Law 919 at 944.
113 The erosion of plant genetic diversity was partly fuelled by the successes of the HYVs. In effect, since the elite commercial varieties on which agribusiness is based show a high degree of genetic uniformity, they are susceptible to disease and pest attack in a way that more heterogeneous land races are not. Needless to add, the effects of narrow plant genetic bases can be quite disastrous. The Irish potato blight of the nineteenth century is a notorious case. Similarly, the American corn blight of 1970 wiped out 15 percent of the American corn harvest for that year.
114 Robert Prescott-Allen, *Genes from the Wild: Using Genetic Resources for Food and Raw Materials* (London: Earthscan, 1988) at 17.
115 Kloppenburg, *supra* note 2 at 163.
116 The IBPGR and Its Policy on in Situ Conservation, FAO Doc/AGPG: IBFGRI/83/143 at 1. Note that for many plants there is no alternative to in situ conservation. These include cocoa, breadfruit, and rubber. Their seeds are not storable.
117 The IBPGR is housed in the FAO headquarters in Rome. The IBPGR secretariat is in the FAO headquarters, and its financial resources are provided by the CGIAR. The IBPGR's goal is to encourage and coordinate efforts to conserve, document, evaluate, and use plant germplasm. This organization is the primary international coordinator of worldwide genetic resource activities. Since 1993 the IBPGR has been reorganized and renamed the International Plant Genetic Resources Institute (IPGRI).
118 "Report of the Quinquennial Review of the International Board for Plant Genetic Resources," Consultative Group on International Agricultural Research (Rome, 1980).
119 "International Board for Plant Genetic Resources Annual Report 1983." (Rome, 1983).
120 Kloppenburg, *supra* note 2 at 104.
121 As quoted in Kloppenburg, *supra* note 2 at 172 (emphasis added).
122 A. Putter, ed., *Safeguarding the Genetic Basis of Africa's Traditional Crops* (The Netherlands: CTA, 1994).
123 The implication here is that all plant germplasm from the South consists of wild raw materials. See Abdulqawi Yussuf, "International Law and Sustainable Development: The Convention on Biological Diversity" (1995) 1 African Yearbook of International Law 109.
124 "Plant Genetic Resources: Report of the Director-General." Document C 83/25, August (Rome, FAO). The debate on the 1983 undertaking was acrimonious. The fact that a numerical vote was forced on the question was by itself no mean feat. At the end of the day, the numerical advantage of the states of the South and their common outrage at the appropriation of plant germplasm withstood the formidable attempts of the North to thwart a change of the status quo.
125 International Undertaking on Plant Genetic Resources, Resolution 8/83 of the Twenty-Second Session of the FAO Conference, 23 November 1983; Agreed Interpretation of the International Undertaking Resolution 4/89 and Farmers Right, Resolution 5/89, of Twenty-Fifth session, 1989. One hundred and two countries have adhered to the undertaking, which is a non-binding instrument.
126 Pat Mooney, *The Seeds of the Earth: A Private or Public Resource?* (Ottawa: Inter Pares, 1979). Mooney's brilliant work detailed and analyzed the mechanisms of appropriation and privatization of the South's germplasm by the multinational seed merchants of the North.
127 1983 Undertaking, *supra* note 125.
128 David Tilford, "Saving the Blue Prints: The International Legal Regime for Plant Resources" (1998) 30 Case Western Reserve Journal of International Law 373.
129 *Ibid.*
130 Denmark, Finland, France, Norway, Sweden, the United Kingdom, and New Zealand officially opposed the 1983 undertaking.

131 Lesser, *supra* note 6 at 105.
132 Tilford, *supra* note 128 at 424.
133 UNEP/CBD/IC/2/13, 1994, Par. 22, *supra* note 105.
134 Tilford, *supra* note 129 at 42.2. See Revision of the International Undertaking on Plant Genetic Resources, Analysis of Some Technical, Economic and Legal Aspects for Consideration in Stage II: Access to Plant Genetic Resources and Farmers' Rights, Commission on Plant Genetic Resources, 6th Sess., FAO Doc. CPGR-6/95/8 Supp. (19-30 June 1995). The new FAO treaty settles the question of legal ownership of plant germplasm in the gene banks.
135 Tilford, *supra* note 128.
136 Report of the Conference of FAO, Genetic Resources and Biological Diversity, Annex 3, at 1, UN Doc. C/91/ REP (1991).
137 FAO, Global Plan of Action for the Conservation and Sustainable Utilization of Plant Genetic Resources for Food and Agriculture and the Leipzig Declaration. Adopted by the International Technical Conference on Plant Genetic Resources (Rome: FAO, 1996).
138 *Ibid.*, at para. 57. It states that many food crops have "been consciously selected and improved by farmers since the origins of agriculture" and have "continued to be developed and improved by farmers without interruption since ancient times."
139 FAO Draft International Code of Conduct for the Collection and Transfer of Plant Germplasm, Doc. CPGR/91/10 of March 1991.
140 For a detailed account of the evolution of the FAO International Treaty on Plant Genetic Resources for Food and Agriculture, see Commission on Plant Genetic Resources for Food and Agriculture, *supra* note 81. See also, Composite Draft Text of the International Undertaking on Plant Genetic Resources, Commission on Genetic Resources for Food and Agriculture, Third Intersessional Meeting of the Contact Group, 2000, Doc. CGRFA/CG-3/00/2.
141 1983 Undertaking, *supra* note 125 (emphasis added).
142 This subsequently led to the FAO International Treaty on Plants.
143 The Keystone Center is a non-profit organization that serves to mediate on environmental issues.
144 Note that the concept of farmers rights was originally introduced by Mexico, supported by Libya, and vehemently opposed by the Netherlands. See Michael Halewood, "Indigenous and Local Knowledge in International Law: A Preface to Sui Generis Intellectual Property Protection" (1999) 44 McGill Law Journal 953 at 970.
145 Emphasis added.
146 The official reason stated by the USA was the absence of a clause on national security. See FAO Verbatim Record of the Thirty-First Session of the Conference, C 2001/PV/4, 2002.
147 CBD, *Report of the Open-Ended Ad Hoc Working Group on Access and Benefit Sharing*, UNEP/CBD/COP/6/6/, 2001), Recommendation 3, at 35.

Chapter 5: Patent Regimes and Biopiracy

1 Valentina Tejera, "Tripping over Property Rights: Is It Possible to Reconcile the Convention on Biological Diversity with Article 27 of the TRIPs Agreement?" (1999) 33 New England Law Review 967.
2 Vandana Shiva, "Biopiracy: Need to Change Western IPR Systems" The Hindu, Wednesday, 28 July 1999, at 3; Luiza Villaema, "Indians Want Patent: Chiefs Prepare International Law Suit against Scientist Who Registered Indigenous Knowledge" ISTOE Magazine, No. 1581, Sao Paulo, 19 January 2000.
3 Lakshmi Sarma, "Biopiracy: Twentieth Century Imperialism in the Form of International Agreements" (1999) 13 Temple International and Comparative Law Journal 107; David Orr, "India Accuses US of Stealing Ancient Cures" The Times (London), 31 July 1999, at 2; Mario Osava, "Brazil Biodiversity: Crackdown on Eco-Pirates" IPS, 14 August 1999.
4 William Lesser and Anatole Krattiger, "Marketing Genetic Technologies in South-North and North-South Exchanges: The Proposed New Facilitating Organization" in Anatole Krattiger et al., eds., *Widening Perspectives on Biodiversity* (Gland, Swit.: IUCN, 1994) at 291.

5 Newscientist, http://www.newscientist.com/ns/980214/editorial.html.
6 A machine is any instrument used to transmit force or to modify its application. For an excellent treatment of this crucial issue, see Benoit Joly and Marie-Angele de Looze, "An Analysis of Innovation Strategies and Industrial Differentiation through Patent Applications: The Case of Plant Biotechnology" (1996) 25 Research Policy 1028. As Rachel Carson reminds hubristic scientists, notwithstanding the strides of modern science, no scientist can make a blade of grass. Also see Tamsen de Valour, "The Obviousness of Cloning" (1995) 9 Intellectual Property Journal 349. However, in 1873 Louis Pasteur was awarded US Patent No. 141,072 for a yeast organism free from organic germs of contagion as an article of manufacture. This was a rather anomalous case as the courts consistently rejected patents on life forms. See Herbert Jervis, "Impact of Recent Legal Developments on the Scope and Enforceability of Biotechnology Patent Claims" (1994) 4 Dickinson Journal of Environmental Law and Policy 79.
7 Graham Dutfield, *Intellectual Property and the Life Science Industries: A Twentieth-Century History* (Burlington, VT: Ashgate Publishing Company, 2003).
8 Jack Kloppenburg Jr., *First the Seed: The Political Economy of Plant Biotechnology, 1492-2000* (Cambridge: Cambridge University Press, 1988) at 261. Emphasis added.
9 Geertrui Van Overwalle, "Patent Protection for Plants: A Comparison of American and European Approaches" (1998-99) 39 IDEA: The Journal of Research and Technology 143.
10 According to Frank Press, president of the American National Academy of Sciences, "a nation with a weak base in plant biology hostages its future. It risks a serious disadvantage in world markets" (Kloppenburg, *supra* note 8).
11 *Plant Patent Act*, 35 U.S.C. 161 (1988 and Supp. 1996). The essence of this legislation was to dilute the strict requirements previously provided for patentability.
12 *Plant Variety Protection Act* 7 U.S.C. 2402 (1988 and 1996).
13 John Golden, "Biotechnology, Technology Policy, and Patentability: Natural Products and Invention in the American System" (2001) 50 Emory Law Journal 101 at 126.
14 447 US 303 (1980).
15 227 U.S.P.Q. 443 (1985).
16 1077 Official Gazette Patent Office 24 (1987), announcement dated 7 April 1987.
17 Michael Greenfield, "Recombinant DNA Technology: A Science Struggling with the Patent Law" (1993) 25 Intellectual Property Law Review 135; David Scalise and Daniel Nugent, "International Intellectual Property Protections for Living Matter: Biotechnology, Multinational Conventions and the Exception for Agriculture" (1995) 27 Case Western Reserve Journal of International Law 83.
18 Edmund Sease, "From Microbes, to Corn Seeds, to Oysters, to Mice: Patentability of New Life Forms" (1989) 38 Drake Law Review 551; Thomas Keane, "The Patentability of Biotechnological Inventions" (1992) Irish Law Times 139.
19 (1989) 25 CPR (3d) 257.
20 *Patent Act*, RSC 1970, c. P-4.
21 Pioneer-Hi-Bred, *supra* note 9 at 260.
22 Union for the Protection of New Plant Varieties, commonly known by its French acronym UPOV (Union pour la Protection des Obstentions Vegetales).
23 International Convention for the Protection of New Varieties of Plants, 2 December 1961, as revised at Geneva on 10 November 1972, on 23 October 1978, and on 19 March 1991; 815 U.N.T.S. 89 (hereinafter, UPOV Convention). This convention is administered by WIPO. At present forty-four countries are parties to UPOV, and its member states are predominantly from the North.
24 Convention on the Grant of the European Patents (EPC), 5 October 1973, 13 I.L.M. 276, or 1160 U.N.T.S. 231.
25 EPC, *ibid.*
26 Prior to that, the decisions in *Ciba-Geigy, Technical Board of Appeal* 3.3.1., 26 July 1983 (T 49/83) and *Lubrizol [hybrid plants] Technical Board of Appeal* 3.3.2., 10 November 1988 (T 320/87) had prohibited the patenting of plant varieties. In other words, the EPC prohibited the patenting of plants or their propagating material in the genetically fixed form of the plant variety but did not prohibit the patenting of plants per se. However, in the Plant

Genetic Systems case, patenting of plant cells are permissible whereas the patenting of plants per se is no longer legal in European Union patent law. See Van Overwalle, *supra* note 9 at 170. Indeed, the law on patentability of plants in Europe is recondite. It seems that, since the TRIPs era, European Union patent law has reverted to the ratio in Ciba-Geigy and Lubrizol.

27 Van Overwalle, *supra* note 9 at 172.

28 Carrie Smith, "Patenting Life: The Potential and Pitfalls of Using the WTO to Globalize Intellectual Property Rights" (2000) 26 North Carolina Journal of Law and Commercial Regulation 143. Compare with Keith Maskus, "Intellectual Property Challenges for Developing Countries: An Economic Perspective" (2001) 1 University of Illinois Law Review 457.

29 But see Margaret Boulware et al., "An Overview of Intellectual Property Rights Abroad" (1994) 16 Houston Journal of International Law 441.

30 For an exhaustive account of the negotiating history of the TRIPs agreement, see Terence Stewart, ed., *The GATT Uruguay Round: A Negotiating History (1986-1992)*, Vol. 1 (Boston: Kluwer Law and Taxation Publishers, 1993).

31 Tejera, *supra* note 1 at 970.

32 See also Michelle Gravelle and John Whalley, "Africa and the Uruguay Round" (1996) Transnational Law and Contemporary Problems 123. The implication is that the TRIPs agreement may not command much compliance at the domestic levels of these states. See Peter Gerhart, "Reflections: Beyond Compliance Theory: TRIPs as a Substantive Issue" (2000) 32 Case Western Reserve Journal of International Law 357; Jerome Reichman, "The TRIPs Agreement Comes of Age: Conflict or Cooperation with the Developing Countries" (2000) 32 Case Western Reserve Journal of International Law 441.

33 Marci Hamilton, "The TRIPs Agreement: Imperialistic, Outdated, and Overprotective" (1996) 29 Vanderbilt Journal of Transnational Law 747; Jerome Reichman, "Intellectual Property in International Trade: Opportunities and Risks of a GATT Connection" (1989) 29 Vanderbilt Journal of Transnational Law 747.

34 According to the OAU/STRC Task Force Declaration on Community Rights and Access to Biological Resources, "privatization of life forms through any Intellectual Property Rights (IPR) regime violates the African sense of respect for life." See Declaration and Draft Model Law by the OAU/STRC Task Force on Community Rights and Access to Biological Resources, March 1999, <http://users.ox.ac.uk/~wgtrr/OAU-decl.htm>.

35 Mark Ritchie et al., "Intellectual Property Rights and Biodiversity: The Industrialization of Natural Resources and Traditional Knowledge" (1996) 11 St. John's Journal of Legal Commentary 431 at 432.

36 Kenneth Krosin, "Are Plants Patentable under the Utility Patent Act?" (1985) 67 Journal of Patents, Trademark Office Society 220; Nancy Linck, "Patentable Subject under Section 101: Are Plants Included?" (1985) 67 Journal of Patents, Trademark Office Society 489.

37 Lara Ewens, "Biotechnology and Intellectual Property" (2000) 23 Boston College International and Comparative Law Review at 289.

38 Peter Goss, "Guiding the Hand That Feeds: Towards Socially Optimal Appropriability in Agricultural Biotechnological Innovation" (1996) 84 California Law Review 1395.

39 Agreement on Trade-Related Aspects of Intellectual Property Rights, 33 I.L.M. 1197 (1994) (TRIPs).

40 A process may be defined as a method involving the application of materials to produce a result. See *General Tire and Rubber Co.* v. *Phillips Petroleum Co.*, [1967] S.C.R. 664 at 671.

41 Article 27 (1), TRIPs, *supra* note 39. Emphasis added.

42 *Ibid.* With respect to the exceptions, particularly public policy and *ordre public*, it is not yet clear how the courts would interpret and apply them. Exceptions on the grounds of *ordre public* are judicially administered exceptions to the usual commitment of individual nations to recognize and give effect to foreign law in circumstances deemed appropriate by the domestic forum. See Kent Murphy, "The Traditional View of Public Policy and Ordre Public in Private International Law" (1981) 11 Georgia Journal of International and Comparative Law 591.

No country can afford to open its tribunals to the legislatures of the world without reserving for its judges the power to reject foreign law that is harmful to its domestic forum. These doctrines are mandated by the exigent forces of local morality and social order. However, while the doctrine of public policy has its origins in fifteenth-century English common law, *ordre public* is associated with civil law and has a statutory force of its own independent of custom. See W.S.M. Knight, "Public Policy in English Law" (1922) 38 Law Quarterly Review 207.

Under common law jurisdictions, courts reserve a power under the public policy doctrine to deny legal effect to claims or causes of action deemed injurious to the interests of the public or offensive to public morality or decency. See Gerhart Husserl, "Public Policy and Order Public" (1938) 25 Virginia Law Review 37; Nicholas Katzenbach, "Conflicts on an Unruly Horse: Reciprocal Claims and Tolerances in Interstate and International Law" (1956) 65 Yale Law Journal 1087; Percy Winfield, "Public Policy in the English Common Law" (1929) 42 Harvard Law Review 76; Monrad Paulsen and Michael Sovern, "Public Policy" in the Conflict of Laws" (1956) 56 Columbia Law Review 969. Needless to say, the scope of these concepts may be as wide and expansive as the judge determining the actual case allows – hence its notorious description as an "unruly horse." See Judge Burrough's remarks in *Richardson* v. *Mellish* 130 English Reports 294 at 303: "I protest arguing too strongly upon public policy. It is a very unruly horse and once you get astride it, you never know where it will carry you." There have thus been several judicial exhortations that the duty of the court is to "expound but not to expand" this area of the law. See *Fender* v. *St. John-Mildmay*, [1938] A.C. 1, per Lord Atkin. However, there is a marked restraint in the judicial employment of the public policy doctrine in limiting the juridical efficacy of international obligations at the domestic level.

43 Convention on Biological Diversity, COP Decision VI/10 on Article 8 (j), Outline of Composite Report on the Status and Trends Regarding the Knowledge, Innovations, and Practices of Indigenous and Local Communities Relevant to the Conservation and Sustainable Use of Biodiversity, and the Plan and Timetable for its Preparation, Annex 1 para. 18, UN Doc. UNEP/SCD/COP/6/20, <http.biodiv.org>.

44 Chidi Oguamanam, "Localizing Intellectual Property in the Globalization Epoch: The Integration of Indigenous Knowledge" (2004) 11 Indiana Journal of Global Legal Studies 135 at 168.

45 Louis Henkin et al., eds., *International Law: Cases and Materials* (St. Paul, MN: West Publishing, 1987) at 439.

46 *Vienna Convention on the Law of Treaties,* opened for Signature May 23, 1969. Entered into force, 27 January 1988. 8 I.L.M. 679 (1969).

47 TRIPs agreement, *supra* note 39.

48 Graham Dutfield, "TRIPs-Related Aspects of Traditional Knowledge" (2001) 33 Case Western Reserve Journal of International Law 233.

49 United Nations Conference on Trade and Development (UNCTAD), *The TRIPs Agreement and Developing Countries* (New York and Geneva: UNCTAD, 1996) at 34.

50 But see T. Jlosvay, "Scientific Property" (1953) 2 American Journal of Comparative Law 180. For further analysis of this issue, see J. Soltysinski, "New Forms of Protection for Intellectual Property in the Soviet Union and Czechoslovakia" (1969) 32 Modern Law Review 408.

51 Harold Pott, "The Definition of Invention in Patent Law" (1944) 7 Modern Law Review 113.

52 35, U.S.C. 161-4 (1982), *supra* note 11 (emphasis added).

53 Nicholas Seay, "Protecting the Seeds of Innovation: Patenting Plants" (1988-9) 16 AIPLA Quarterly Journal 418 at 420.

54 35, U.S.C. 102 (1994), *supra* note 11 (emphasis added). Ironically, the 1790 and 1793 US Patent Acts did not have any geographic distinctions on the matter of invention. Indeed, the case of *Dawson* v. *Follen,* 7 F. Cas. 216 (C.C.D. Pa. 1808) held that the patentee must be the original inventor "in relation to every part of the world."

55 51 US (10 How.) 477 (1850). For a fuller discussion of this geographical dichotomy and its overall implications, see Donald Chisum, "Foreign Activity: Its Effect on Patentability

under United States Law" (1980) 11 International Review of Industrial Property and Copyright Law 26.
56 Emphasis added.
57 Shayana Kadidal, "Subject-Matter Imperialism? Biodiversity, Foreign Prior Art and the Neem Patent Controversy" (1996-77) 37 IDEA: The Journal of Research and Technology 371.
58 *Ibid.*
59 Michel Goulet, "Novelty under Canada's Patent Act: A European Accent" (1998) 13 Intellectual Property Journal 83. See also *Patent Act of Canada*, R.S.C. 1985, c. P-4, as amended.
60 John Sinnott, *World Patent Law and Practice* Vols. 2b, 2c, 2d, 2e, 2f, 2g (New York: Matthew Bender, 1977).
61 Gerald Rose, "Do You Have a 'Printed Publication'? If Not, Do You Have Evidence of Prior 'Knowledge or Use'?" (1979) 61 Journal of Patents and Trademark Office Society 643.
62 Kadidal, *supra* note 57 at 391.
63 For example, in *Carter Prods, Inc.* v. *Colgate-Palmolive Co.*, 130 F. Supp. 557, 104 U.S.P.Q. (BNA) 314 (D. Md. 1955), it was held that a typewritten patent document from the patent office in Argentina was not "printed matter" and therefore could not invalidate a patent application on the same subject in the United States.
64 Kadidal, *supra* note 57 at 392. See also Meetali Jain, "Global Trade and the New Millennium: Defining the Scope of Intellectual Property Protection of Plant Genetic Resources and Traditional Knowledge in India" (1999) 22 Hastings International and Comparative Law Review 777.
65 Josef Henry Vogel, "An Economic Analysis of the Convention on Biological Diversity: The Rationale for a Cartel" (on file with the author); Rosemary Coombe, "Intellectual Property, Human Rights and Sovereignty: New Dilemmas in International Law Posed by the Recognition of Indigenous Knowledge and the Conservation of Biodiversity" (1998) 6 Indiana Journal of Global Legal Studies 59.
66 For a detailed discussion of this case, see "Mexican Bean Piracy" RAFI Geno-Types, 17 January 2000.
67 An Act to Approve and Implement the Trade Agreements Concluded in the Uruguay Round of Multilateral Trade Negotiations, The Uruguay Round Agreements Act 1994, 8 December 1994, P.L. 103-465, 108 Stat. 4809.
68 *Ibid.* Emphasis added.
69 George Francis Takach, *Patents: A Canadian Compendium of Law and Practice* (Edmonton: Juriliber, 1993).
70 van Overwalle, *supra* note 9 at 155.
71 *Consolboard Inc.* v. *MacMillan Bloedel (Sask.) Ltd.*, (1981) 56 C.P.R. (2d) 145 at 154.
72 *Pioneer Hi-Bred.* v. *Commissioner of Infants* (1989) 1 S.C.R. 1623 at 1630.
73 *Ibid.* Compare with Tetraploide Kamille, (BGH GRUR 1993, 651).
74 Joseph Rossman, "Plant Patents" (1931) 13 Journal of Patent Office Society 7.
75 Robert Allyn, "Plant Patent Questions" (1933) 15 Journal of Patent Office Society 180.
76 Robert Cook, "Applying the Plant Patent Law" (1931) 13 Journal of Patent Office Society 22.
77 *Diamond* v. *Chakrabarty, supra* note 14 at 198.
78 35 U.S.C (1988), *supra* note 11 at sections 161-4. Note that the section 112 referred to deals with written specifications.
79 For example, *Plant Varieties and Seeds Act of 1964* (UK).
80 Hearings Before the Subcommittee on Departmental Operations of the Committee on Agriculture, 91st Congress, 2nd Sess. 7 (1970) (statement by Allenby White, Chairman, Breeder's Rights Study Committee, American Seed Trade Association, quoting S. Rep. No. 315, 71st Congress., 2nd Sess., (1930) (as quoted in Scalise and Nugent, *supra* note 17 at 91-92); Martin Adelman and Sonia Badia, "Prospects and Limits of the Patent Provisions in the TRIPs Agreement: The Case of India" (1996) 29 Vanderbilt Journal of Transnational Law 507.
81 Scalise and Nugent, *supra* note 17 at 92.
82 Jerome Reichman, "From Free Riders to Fair Followers: Global Competition under the TRIPs Agreement" (1996-97) International Law and Politics 11.
83 Pott, *supra* note 51 at 113.

84 Stephen Gratwick, "Having Regard to What Was Known and Used" (1972) 88 Law Quarterly Review 341; Stephen Gratwick, "Having Regard to What Was Known and Used" Revisited" (1986)102 Law Quarterly Review 403.
85 A Commission appointed by the German government in 1886 to examine the possibility of forming a definition of invention failed to come up with a definition. See Pott, *supra* note 51 at 117.
86 *Ibid.*, at 114.
87 46 R.P.C. (1929) at 248.
88 Richard Gardiner, "Language and the Law of Patents" (1994) 47 Current Legal Problems 32.
89 Pott, *supra* note 51 at 117. That is to say, the judge is called to evaluate the quantum of difference from the prior art by making a subjective estimate of its effect on the mind of the hypothetical person skilled in the art.
90 Section 161, 35 U.S.C. (1982), *supra* note 11.
91 *Yoder Bros., Inc.,* v. *California-Florida Plant Corp.,* 537 F.2d 1347, 1378, 193 U.S.P.Q. 264, 291 n.34 (5th Cir. 1976).
92 S. Rep. No. 315, 71st Cong., 2d Sess. (1930).
93 *Ibid.*
94 Kloppenburg, *supra* note 8 at 239.
95 (1994) Plant Variety Protection Office Journal 13.
96 Kloppenburg, *supra* note 8 at 144. Further, under the UPOV conventions "novelty" refers more to "commercial or market" novelty than to technological novelty, and it requires that a variety must not have been sold or marketed in the country where the application is filed earlier than one year before the filing date.
97 Jain, *supra* note 64 at 780.
98 Vandana Shiva, *The Violence of the Green Revolution* (London: Zed Books, 1991) at 259; Vandana Shiva, *Staying Alive: Women, Ecology and Development* (London: Zed Books, 1988).
99 C.L. McDougall, and R. Hall, *Intellectual Property Rights and the Biodiversity Convention: The Impact of GATT* (Bedfordshire: 1995).
100 Gurdial Singh Nijar, "Towards a Legal Framework for Protecting Biological Diversity and Community Intellectual Rights: A Third World Perspective" Third World Network Discussion Paper, Penang, Malaysia, 1997 (on file with the author).
101 C.L. McDougall, and R. Hall, *supra* note 99 at 4.
102 *Ibid.*
103 Shiva, *Staying Alive, supra* note 98 at 259.
104 The Crucible Group I, *People, Plants, and Patents: The Impact of Intellectual Property on Trade, Plant Biodiversity, and Rural Society* (Ottawa: IDRC, 1994) at xviii.
105 Van Overwalle, *supra* note 9 at 145.
106 Victor Renier, "Vegetable Novelties and Inventor's Rights" (1960) [Belgium] Annals of Law and Political Science 253.
107 *President and Fellows of Harvard College* v. *Commissioner of Patents,* [1998] 79 C.P.R. (3d) 98.
108 *Merck and Co.* v. *Apotex Inc.,* (1994) 59 C.P.R. (3d) 133 (Fed. TD) at 178.
109 As Van Overwalle explains: "The technical problems do not occur at the stage of the introduction of genetic information. The technology for making a DNA segment is manageable, and assuming that a description is available, a person skilled in the art should be able to copy the DNA segment to be inserted, on the basis of the description. The difficulties occur in the stage between the gene construct and the plant genome. The cause of these difficulties is that the DNA rapidly enters the nucleus of the plant cell and begins integrating at a random site resulting in the insertion of the DNA segment at a different site in every cell transformed. *The underlying implication of this difficulty is that a person skilled in the art who applies the method as specified, can still arrive at another transgene plant that has different DNA."* See Van Overwalle, *supra* note 9 at 157-9.
110 *Supra* note 107 at 112.
111 67 GRUR 577 (1962) [as cited in Van Overwalle, *supra* note 9 at 185]. But see Red Dove (Rote Taube, BGHZ 52, 74, 72 GRUR 772) (1969). See also Robert Allyn, "More about Plant Patents" (1933) 15 Journal of Patent Office Society 963.

112 "Australians Abandon 2 Plant 'Patent' Claims" RAFI News Release, 21 January 1998. In this case, Australian crop development agencies were forced to abandon their claims on two chickpea varieties obtained from an international public research institute (International Crops Research Institute for the Semi-Arid Tropics-ICRISAT) based in India. See also "International Research Centre (ICARDA) Breaks Trust: Allows Australians to Patent Plants Supposedly Held in Trust for Farmers" RAFI News Release, 2 February 1998. This type of questionable patent has given rise to recent calls for a restructuring of CGIAR, which was the umbrella organization responsible for the twelve international gene banks spread all over the word. See "Structure and Goals of the CGIAR Need Redefining," <http://users.ox.ac.uk/wgtrr/grairafi.htm>. However, as already noted, by an agreement made in October 1994 between the CGIAR and the FAO, the twelve gene banks have been put under the legal and intergovernmental control of the latter.

113 *Supra* note 14.

114 *Diamond v. Chakrabarty, ibid.*

115 The literature on this is legion. But see, for example, Craig Jacoby and Charles Weiss, "Recognizing Property Rights in Traditional Bio-Cultural Contribution" (1997) 16 Stanford Environmental Law Journal 74; Curtis Horton, "Protecting Biodiversity and Cultural Diversity under Intellectual Property Law: Toward a New International System" (1995) 10 Journal of Environmental Law and Litigation 1; Darrel Posey, *Traditional Resource Rights: International Instruments for Protection and Compensation for Indigenous Peoples and Local Communities* (IUCN, 1996); Elias Carreno Peralta, "A Call for Intellectual Property Rights to Recognize Indigenous Peoples' Knowledge of Genetic and Cultural Resources" in Anatole Krattiger et al., eds., *supra* note 4; Dan Perlman and Glen Adelson, *Biological Diversity: Exploring Values and Priorities in Conservation* (Malden, MA: Blackwell Publishing, 1997).

116 Michael Hochberg, ed., *Aspects of the Genesis and Maintenance of Biological Diversity* (Oxford: Oxford University Press, 1996).

117 Robert Merges, "Intellectual Property in Higher Life Forms: The Patent System and Controversial Technologies" (1988) 47 Maryland Law Review 1051.

118 Horton, *supra* note 115.

119 Daniel Jenks, "The Convention on Biological Diversity: An Efficient Framework for the Preservation of Life on Earth?" (1995) 15 Northwestern Journal of International Law and Business 636.

120 Michael Balick and Paul Alan Cox, *Plants, People and Culture: The Science of Ethnobotany* (New York: Freeman and Company, 1996).

121 Naomi Roht-Arriaza, "Of Seeds and Shamans: The Appropriation of the Scientific and Technical Knowledge of Indigenous and Local Communities" (1996) 17 Michigan Journal of International Law 940.

122 R.S. Crespi, *Patents: A Basic Guide to Patenting Biotechnology* (Cambridge, MA: Cambridge University Press, 1988).

123 Kathryn Rackleff, "Preservation of Biological Diversity: Toward a Global Convention" (1992) 3 Colorado Journal of International Environmental Law and Policy 405.

124 Saponins are various sugar compounds that form colloidal solutions when mixed and agitated with water. They are often used in detergents and foaming agents.

125 Cocaine, derived from the coca plant, has served as a model for the synthesis of many local anesthetics.

126 Related species with peculiar chemical qualities are explored to discover whether they also possess similar pharmacological qualities.

127 Mataatua Declaration on Cultural and Intellectual Property Rights of Indigenous Peoples, <http://users.ox.ac.uk/wgtrr/mataatua.htm>.

128 As the International Society of Ethnobiology recently noted in its newly formulated Code of Ethics: "It is acknowledged that much research has been undertaken in the past without the sanction or prior consent of indigenous and traditional peoples and that such research has resulted in wrongful expropriation of cultural and intellectual heritage rights of the affected peoples causing harm and violation." See International Society of Ethnobiology (ISE) Code of Ethics, <http://users.ox.ac.uk/wgtrr/isecode.htm>.

129   Declaration and Draft Model Law by the OAU/STRC Task Force on Community Rights and Access to Biological Resources, March 1998, <http://users.ox.ac/uk/wgtrr/OAU-decl.htm>.
130   Ex parte Lartimer, 1889 Dec. Comm'r Pat. 123.
131   *Kuehmsted* v. *Farbenfabriken of Eberfield Co.,* 179 F. 701, 704-05 (7th Cir.1910).
132   Harold Wegner, "Purified Protein Patents: A Legal Process Gone Berserk?" (1990) 12 European Intellectual Property Review 187.
133   Michael Sanzo, "Patenting Biotherapeutics" (1991) 20 Hofstra Law Review 387.
134   Donald Holland, "Can Product-By-Process Patents Provide the Protection Needed for Proteins Made by Recombinant DNA Technology?" (1992) 74 Journal of Patents and Trademark Office Society 902.
135   Thomas Kiley, "Patents on Random Complementary DNA Fragments?" (1992) 14 Science 915.
136   In re Bergstrom, 427 F.2d 1394, 166 U.S.P.Q. (BNA) 256 (C.C.P.A. 1970).
137   Howard Forman, *The Law of Chemical, Metallurgical and Pharmaceutical Patents* (New York: Central Book, *1967).* Compare with, K.K. Beier, R.S. Crespi, and J. Strauss, *Biotechnology and Patent Protection: An International Review* (Paris: OECD, 1985).
138   "Biopiracy Update: A Global Pandemic" RAFI Update Sep/Oct., 1995.
139   *Merck and Co.* v. *Olin Mathieson Chem. Corp.,* 253 F. 2d 156 (4th Cir. 1958).
140   *Ibid.*
141   Shiva, "Bio-Piracy: Need to Change Western IPR Systems" *supra* note 2.
142   Steve Connor, "African Root Could Be Cure for Athlete's Foot," *The Independent,* London, 21 February 2000 at 3. It is equally significant that six of the trees had to be destroyed to produce fifty grams of the anti-fungal agent.
143   See the decision of the German Federal Patent Court in "Menthonthiole" [1978] G.R.U.R. 702.
144   *General Electric Co.* v. *DeForest Radio Co.,* 28 F. 2d 641 (3d Cir. 1928) cert. denied., 278 US 656 (1929).
145   Michael Davis, "The Patenting of Products of Nature" (1995) 21 Rutgers Computer and Technology Law Journal 293.
146   *Ibid.*
147   In re Bergstrom, *supra* note 136.
148   *Parke-Davis* v. *Mulford Co.,* 189 F. 95 (C.C.S.D.N.Y. 1911), modified, 196 F.496 (2d Cir. 1912).
149   *Ibid.*
150   *Diamond* v. *Chakrabarty, supra* note 14 at 135.
151   Golden, *supra* note 13 at 127.
152   Emily Marden, "The Neem Tree Patent: International Conflict Over the Commodification of Life" (1999) 22 Boston College International and Comparative Law Review 27.
153   RAFI Communique, Nov./Dec. 1995; R.K. Gupta and Balasubramaniam, "The Turmeric Effect" (1998) 20 World Patent Information 185; R.V. Anuradha, "IPRs: Implications for Biodiversity and Local and Indigenous Communities" (2001) 10 Review of European Community and International Environmental 27.
154   The Tirpir seed is used by the Indians as a contraceptive, antibiotic, and abortifacient. The cunani is used for fishing as it kills fish without poisoning them or polluting the rivers. See "Indians Want Patent: Chief Prepare International Law Suit Against Scientist Who Registered Indigenous Knowledge" ISTOE Magazine, No. 1581, Sao Paulo, 19 January 2000. The Wapishana nation was arbitrarily split by the Brazilian and British governments into Brazil and Guyana during a border dispute in 1904.
155   *Ibid.*
156   *Ibid.*
157   "No Cure for Patents: Biotech Patents Distort and Discourage Innovation and Increase Costs for Dubious Drugs" RAFI Genotypes, July 1997 (emphasis added). But see Michael Malinowski, "Globalization of Biotechnology and the Public Health Challenges Accompanying It" (1996) 60 Albany Law Review 119. For an in-depth analysis of the methods of tinkering with and rearranging the molecular structure of TKUP for patent purposes, see Shayana Kadidal, "Plants, Poverty, and Pharmaceutical Patents" (1993) 103 Yale Law Journal 223.

158 Golden, *supra* note 13 at 131.

159 *Ibid.,* at 111.

160 Samuel Oddi, "TRIPs: Natural Rights and a 'Polite Form of Economic Imperialism'" (1996) 29 Vanderbilt Journal of Transnational Law 415. Emphasis added.

161 Vandana Shiva, *Monocultures of the Mind: Perspectives on Biodiversity and Biotechnology* (London: Zed Books, 1993). For further analysis of the sociology and cultural life of science, see the dated but brilliant work of Thomas Kuhn, *The Structure of Scientific Revolutions* (Chicago: University of Chicago Press, 1972); R. Horton, "African Traditional Thought and Western Science" (1976) 37 Africa 2.

162 For a sound analysis of this aspect of intellectual property, see Rosemary Coombe, *The Cultural Life of Intellectual Property: Authorship, Appropriation, and the Law* (Durham: Durham University Press, 1998).

163 Barry Mandelker, "Indigenous People and Cultural Appropriation: Intellectual Property Problems and Solutions" (2000) 16 Canadian Intellectual Property Review 367.

164 In the United States, for example, the Federal Circuit has consistently stated that it is the policy of the court to "adapt" and expand the patent scope in favour of the pharmaceutical and biotechnology industry. In the philosophy of the court, the more patents issued, the better. See *Pioneer Hi-Bred Int.* v. *J.E.M. AG Supply,* 200 F.3d 1374, 1376 (Fed. Cir. 2000); in re Lundak, 773 F.2d 1216, 1220 n.1 (Fed. Cir. 1985); Robert Merges and Glenn Reynolds, "The Proper Scope of the Copyright and Patent Power" (2000) 37 Harvard Journal on Legislation 45.

165 United Nations Conference on Trade and Development (UNCTAD), *The TRIPs Agreement and Developing Countries, supra* note 49 at 4.

166 Thomas Cottier, "The Protection of Genetic Resources and Traditional Knowledge in International Law: Past, Present and Future" in Susette Biber-Klemm, ed., *Legal Claims to Biogenetic Resources* (unpublished papers presented at the workshop held at the Institute for European and International Economic Law, University of Berne, June 1997; on file with the author).

167 Ewens, *supra* note 37 at 305; Keith Aoki, "Neocolonialism, Anti-Commons Property, and Biopiracy in the (Not-so-Brave) New World Order of International Intellectual Property Protection" (1998) 6 Indiana Journal of Global Legal Studies 11.

168 Donald Richards, *Intellectual Property Rights and Global Capitalism: The Political Economy of the TRIPs Agreement* (New York: M.E. Sharpe, 2004).

169 Michael Halewood, "Indigenous and Local Knowledge in International Law: A Preface to Sui Generis Intellectual Property Protection" (1999) 44 McGill Law Journal 953.

170 *Ibid.*

171 Edward McWhinney, "Towards an Empirically-Based New International Economic Order" (1989) 27 Canadian Yearbook of International Law 309 (arguing that "classical international law is too narrowly Eurocentrist or Western in its historical origins"); see also Kurt Taylor Gaubatz and Matthew MacArthur, "How International Is 'International Law'?" (2001) 22 Michigan Journal of International Law 239.

172 Thomas Franck and Steven Hawkins, "Justice in the International System" (1989) 10 Michigan Journal of International Law 127.

173 It may be naive to assume that the lax patent system merely appropriates and privatizes TKUP and plants and that, as such, it poses a problem only for the South. As a matter of fact, there are legitimate concerns that the lax patent system restricts access to information that ordinarily should be in the public domain. This concern is of particular interest to scientists and other researchers, especially with regard to dubious patents on gene fragments. Mario Biagioli, "The Instability of Authorship: Credit and Responsibility in Contemporary Biomedicine" (1998) 12 Life Science Forum 3; Declan Butler and Paul Smaglik, "Celera Genome Licensing Spark Concerns over 'Monopoly'" (2000) 403 Nature 231; Eliot Marshall, "Patent on HIV Receptor Provokes an Outcry" (2000) 287 Science 1375; Roberto Mazzoleni and Richard Nelson, "The Benefits and Costs of Strong Patent Protection: A Contribution to the Current Debate" (1998) 27 Research Policy 273.

174 The Crucible Group II, *Seeding Solutions: Policy Options for Genetic Resources – People, Plants, and Patents Revisited* (Ottawa: IDRC, 2000).

175 See generally the dated but useful work of Clyde Eagleton, *The Responsibility of States in International Law* (New York: New York University Press, 1970).

176 R.H. Reichman, "Intellectual Property in International Trade: Opportunities and Risks of A GATT Connection" (1989) 22 Vanderbilt Journal of Transnational Law 747.

177 *Ibid.*

178 John Mugabe et al., eds., *Access to Genetic Resources: Strategies for Sharing Benefits* (Bonn: IUCN-ELC, 1997); The Crucible Group, Vol. 2, *supra* note 174. *Seeding Solutions: Options for National Laws Governing Access to and Control over Genetic Resources* (pre-publication version – not for quotation) (Ottawa: IDRC, 2001) (on file with the author).

179 Republic of the Philippines Executive Order No. 247, <htpp://users.ox.ac.uk/wgtrr/rp.htm>. The complexities of the mechanisms and suggestions on access to plants and TKUP are beyond the scope of the paper. For a recent compendium of suggestions, see, Crucible Group II, *ibid.*

180 Convention on Biological Diversity, opened to signature 6th June 1992 [1993] 35 I.L.M. 813; entered into force Dec. 29, 1993.

181 CBD, *ibid.*

182 *Ibid.*

183 Coombe, *supra* note 65.

184 International Covenant on Economic, Social and Cultural Rights, 16 December 1966, 993 U.N.T.S. 3.

185 The UN Draft Declaration on the Rights of Indigenous Peoples, UN ESCOR, Commission on Human Rights, 11th Sess., Annex 1, UN Doc. E/CN.4/Sub.2 (1993); Draft United Nations Declaration on the Rights of Indigenous Peoples, Sub-Commission on Prevention of Discrimination and Protection of Minorities on its Forty-Sixth Session, Geneva, 1-26 August 1994, E/CN.4/SUB.2/1994/56, 28 October 1994; Draft of the Inter-American Declaration on the Rights of Indigenous Peoples, 19 September 1995, <http://users.ox.ac.uk/wgtrr.oas.htm>.

186 UN Report, *The Role of Patents in the Transfer of Technology to Developing Countries Report of the Secretary-General, United Nations* (New York: Martinus Nijhoff, 1964); S. Vedaram, "The New Indian Patents Law" (1972) 3 International Review of Industrial Property and Copyright Law 39; M. Bruce Harper, "TRIPs Article 27.2: An Argument for Caution" (1997) 21 William and Mary Environmental Law and Policy Review 381; National Aeronautics and Space Act of 1958, 42 U.S.C. 2457.

187 Krista Singleton-Cambage, "International Legal Sources and Global Environmental Crises: The Inadequacy of Principles, Treaties, and Custom" (1995) 2 ILSA Journal of International and Comparative Law 171; Paul Brietzke, "Insurgents in the 'New' International Law" (1994) 13 Wisconsin International Law Journal 1; Oscar Schacter, "Recent Trends in International Law Making" (1992) 12 Australian Yearbook of International Law 1.

188 Thomas Franck, *Fairness in International Law and Institutions* (Oxford: Clarendon Press, 1995).

189 Soft law is an assortment of "legal" instruments, including UN General Assembly resolutions, declarations of intergovernmental organizations, declarations of governmental meetings, guidelines, resolutions, recommendations and action programs, indeterminate provisions of treaties, unratified conventions, dissenting opinions of individual judges of the International Court of Justice or the various human rights courts, and so on.

190 C.M. Chinkin, "The Challenge of Soft Law: Development and Change in International Law" (1989) 38 International and Comparative Law Quarterly 850; R.R. Baxter, "International Law in "Her Infinite Variety"" (1980) 29 International and Comparative Law Quarterly 549.

191 Patricia Birnie, "International Environmental Law: Its Adequacy for Present and Future Needs" in A. Hurrell and B. Kingsbury, eds., *The International Politics of the Environment* (Oxford: Clarendon Press, 1992).

192 T. Gruchalla-Wisierski, "A Framework for Understanding 'Soft Law'" (1984) 30 Revue de Droit de McGill 37.

193 Pierre-Marie Dupuy, "Soft Law and the International Law of the Environment" (1991) 12 Michigan Journal of International Law 420.

194 Sumudu Atapattu, "Recent Trends in International Environmental Law" (1998) 10 Sri Lanka Journal of International Law 47.

195 Birnie, *supra* 191 at 59. Be that as it may, the legal status of each General Assembly Resolution has to be tested individually. However, any abstention or negative vote has to be examined to ascertain the underlying opinio juris. The same tests should also apply to declarations of principles by the United Nations.

196 D.N. Saraf, "Resolutions of International Organizations: Binding Norms?" (1990) 14 Cochin University Law Review 1 at 9. For an excellent analysis of the varied effects on resolutions of international organizations, see Jochen Frowein, "The Internal and External Effects of International Organizations" (1987) 49 Heidelberg Journal of International Law 778. There is not much disagreement as to the internal bindingness of international resolutions on the officials and subordinate organs of the affected international organizations.

197 *Ibid.*

198 ICJ Reports (1986) at 100.

199 Michael Bothe, "Legal and Non-Legal Norms: A Meaningful Distinction in International Relations?" (1980) 11 Netherlands Yearbook of International Law 65.

200 Jan Klabbers, "The Undesirability of Soft Law" (1998) 67 Nordic Journal of International Law 381.

201 Jan Klabbers, "The Redundancy of Soft Law " (1996) Nordic Journal of International Law 167.

202 Jan Klabbers, *The Concept of Treaty in International Law* (London: Kluwer Law International, 1996).

203 Nico Schrijver, *Sovereignty over Natural Resources: Balancing Rights and Duties* (Cambridge: Cambridge University Press, 1997).

204 The COICA Statement, <http://users.ox.ac.uk/wgtrr/coica.htm>.

205 "An Appeal from SRISTI (the Society for Research and Initiatives for Sustainable Technologies and Institutions) to Ethnobiologists," <http://users.ox.ac.uk/-wgtrr/ethnape2.htm>. See also Anthony Cunningham "Guidelines for Equitable Partnerships in New Natural Products Developments: Recommendations for A Code of Practice," <http://users.ox.ac.uk/wgtrr/cunning.htm>; Final Statement from the UNDP Consultation on Indigenous Peoples' Knowledge and Intellectual Property Rights, Suva, April 1995, <http://users.ox.ac.uk/wgtrr/suva.htm>.

206 Nihal Jayawickrama, "The Right of Self-Determination: A Time for Reinvention and Renewal" (1993) 57 Saskatchewan Law Review 1.

207 International Society of Ethnobiology (ISE) Code of Ethics, <http://users.ox.ac.uk/wgtrr/isecode.htm>.

208 Peter Drahos, "Indigenous Knowledge and the Duties of Intellectual Property Owners" (1997) 11 Intellectual Property Journal 179. Drahos calls such declarations the softest of soft law.

209 See *Knowledge, Innovations and Practices of Indigenous and Local Communities: Implementation of Article 8 (j)*, Note by the Executive Secretary of the CBD for the Third COP, 4-15 November 1996, UNEP/CBD/COP/3/19, <http//:www.biodiv.org> (last visited 17 May 2000; site now discontinued).

210 See *Recommendations of the Permanent Forum on Indigenous Issues to the Convention on Biological Diversity*, UN Doc. UNEC/CBD/WG8J/3/8, at 2 (September 2003).

211 Oguamanam, *supra* note 44.

212 World Trade Organization, Ministerial Conference, 4th Session, Doha, 9-14 November 2001, Ministerial Declaration, Adopted on 14 November 2001, WT/MIN (01)/DEC/1 at para. 19.

213 Ewens, *supra* note 37 at 307.

214 Drahos, *supra* note 208 at 182.

215 William Lesser, *Sustainable Use of Genetic Resources under the Convention on Biological Diversity: Exploring Access and Benefit Sharing Issues* (Oxford: CAB International, 1997) at 129. For example, Article 5 (1) (b) of the FAO International Treaty on Plant Genetic Resources for Food and Agriculture obliges contracting parties to "promote the collection of plant genetic resources for food and agriculture and relevant associated information on those plant genetic resources that are under threat or are of potential use."

216   R.V. Anuradha, "In Search of Knowledge and Resources: Who Sows? Who Reaps?" (1997) 6 Review of European Community and International Law 263.

217   Lyle Glowka, *A Guide to Designing Legal Frameworks to Determine Access to Genetic Resources* (Gland: IUCN, 1998).

218   *Draft Report of the World Intellectual Property Organization (WIPO) Fact-Finding Missions on Intellectual Property and Traditional Knowledge (1998-99),* Geneva, Switzerland.

219   *IUCN: Inter-Commission Task Force on Indigenous Peoples and Sustainability – Cases and Actions* (Utrecht: IUCN and International Books, 1997).

220   There have been regional legislative initiatives in this regard. For example, Article 75 of Decision 486 of the Andean Community on a Common Industrial Regime, which entered into force on 1 December 2000, nullifies any such patent. See Manuel Ruiz, "The Andean Community's New Industrial Regime: Creating Synergies between the CBD and Intellectual Property Rights" (2000) Bridges 12.

221   Crucible Group I, *supra* note 104 at 17.

222   For example, since time immemorial the Ayurvedic system has been codified in fifty-four authoritative texts, the Siddha system in twenty-nine authoritative texts, and the Unani Tibb system in thirteen. In India, the First Schedule of the *Drugs and Cosmetics Act,* No. 23 of 1940, as amended by the *Drugs and Cosmetics (Amendment) Act* no. 71 of 1986, specifies the authoritative books of the three systems. See WIPO Report, *supra* note 218 at 73.

223   For example, many traditional healing methods in Southern Nigeria have been transmitted over the millennia in secret signs and symbols known only to the initiates of the traditional healing cults.

224   Articles 8, 10, 15, and the preamble of the CBD.

225   COICA Statement, *supra* 204; Mataatua Declaration, *supra* note 127.

226   WIPO Report, *supra* note 218.

227   As earlier noted, the idea of intellectual property among non-Western societies is already well established in anthropological circles. See R.H. Lowie, *Primitive Society* (New York: Routledge, 1920) at 235-43; *Intellectual Property Needs and Expectations of Traditional Knowledge Holders: WIPO Report on Fact-Finding Missions on Intellectual Property and Traditional Knowledge (1998-1999)* (Geneva: WIPO, 2001).

228   R.H. Coase, "The Problem of Social Cost" (1960) 3 Journal of Law and Economics 1.

229   Drahos, *supra* note 208 at 183.

230   *Ibid.,* at 197.

231   Lester Yano, "Protection of the Ethnobiological Knowledge of Indigenous Peoples" (1993) 41 UCLA Law Review 443 at 486.

232   Naomi Ariazza, "Of Seeds and Shamans: The Appropriation of the Scientific and Technical Knowledge of Indigenous and Local Communities" (1996) 17 Michigan Journal of International Law 919.

233   Cottier, *supra* note 166 at 11.

234   Ikechi Mgbeoji, "Patents and Traditional Knowledge of the Uses of Plants: Is a Communal Patent Regime Part of the Solution to the Scourge of Biopiracy?" (2001) 9 Indiana Journal of Global Legal Studies 163; Ashish Kothari and R.V. Anuradha, "Biodiversity and Intellectual Property Rights: Can the Two Co-Exist" (1999) 2 Journal of International Wildlife Law and Policy 204; Gurdial Nijar Singh, *In Defence of Local Community and Biodiversity: A Conceptual Framework and the Essential Elements of A Rights Regime* (Penang, Malaysia).

235   Cottier, *supra* note 166 at 12.

236   Halewood, *supra* note 169 at 993 (emphasis added). See also Annex No. 6: Guiding Principles for the Consultative Group on International Agricultural Research Centres on Intellectual Property and Genetic Resources, in CGIAR, Centres Position Statement on Genetic Resources, Biotechnology and Intellectual Property Rights (19 May 1998).

237   Jorge Caillaux, "Biological Resources and the Convention on Biological Resources" (1994) 1 Journal of Environmental Law and Policy in Latin America and the Caribbean 9.

238   Charles McManis, "The Interface between International Intellectual Property and Environmental Protection: Biodiversity and Biotechnology" (1998) 76 Washington University Law Quarterly 255.

239 Ruth Gana, "Has Creativity Died in the Third World? Some Implications of the Internationalization of Intellectual Property" (1995) 24 Denver Journal of International Law and Policy 109.

240 Kirsten Petersen, "Recent Intellectual Property Trends in Developing Countries" (1992) 33 Harvard International Law Journal 277; Mark Hannig, "An Examination of the Possibility to Secure Intellectual Property Rights for Plant Genetic Resources Developed by Indigenous Peoples of the NAFTA States: Domestic Legislation Under the International Convention for New Plant Varieties" (1996) 13 Arizona Journal of International and Comparative Law 175.

241 Stephen Brush, "Is Common Heritage Outmoded?" in Stephen Brush and Doreen Stabinsky, eds., *Valuing Local Knowledge: Indigenous People and Intellectual Property Rights* (Washington/ Covelo: Island Press, 1996).

242 Kuhn, *supra* note 161 at 66; Fritz Machlup, *An Economic Review of the Patent System,* Study No. 15 of the Subcommittee on Patents, Trademarks, and Copyrights, of the Committee on the Judiciary, US Senate 85th Session, 2nd Session, Washington, 1958.

243 David Safran, "Protection of Inventions in the Multinational Marketplace: Problems and Pitfalls in Obtaining and Using Patents" (1983) 9 North Carolina Journal of International Law and Commercial Regulation 117.

244 Virtually all patent law jurisdictions the world over have provisions on an employer's right in an employee's invention. See David Vaver, *Intellectual Property Law: Copyright, Patents, and Trademarks* (Toronto, ON: Irwin Law, 1997) at 147-9.

245 Brush, *supra* note 241 at 145. In any event, individual rights can and do co-exist with communal rights. See, for example, Ronald Ganet, "Communality and Existence: The Rights of Groups" (1983) 56 Southern California Law Review 1001; Leighton McDonald, "Can Collective and Individual Rights Coexist?" (1998) 22 Melbourne University Law Review 310.

246 See Justice Von Dossa in *Milpurrurru* v. *Indofurn (Pty) Ltd.,* (1995) 30 I.P.R. 209 at 216.

247 Article 8 (j) of the CBD, *supra* note 43.

248 The *Kuwaiti Patent Law* provides as follows:

An invention shall not be considered new, whether wholly or partially, in the following two cases:
1. Where during the twenty years preceding the date of submission of the application for patent, the invention had been used openly in Kuwait.
2. Where, during the twenty years preceding the date of submission of the application for patent, a patent had been granted in respect of the invention or part thereof to other than the inventor or his assignee.

See Article 3, *Law No. 4 (Kuwait) of 1962 Governing Patent Designs and Industrial Models,* reproduced in John Sinnot, *Patent Laws of the World,* Vol. 2E at 3.

249 The Libyan Patent Law provides that:

A patent is not considered to be new, in whole or in part, in the following two cases:
(i) If during the 50 years preceding the date of submission of the application, the invention had already been publicly used in Libya, or its description or design had been advertised in publications in Libya, in such a manner as to render possible the exploitation thereof by experts.
(ii) If within the fifty years preceding the date of submission of the application for patent, a patent of the invention or a part thereof had already been granted to persons other than the inventor or his assigns.

See Article 1 (b) i, *Libya Law No. 8 of 1959 Relating to Patents, Designs, and Industrial Models,* reproduced in Sinnott, Vol. 2E, *ibid.*

250 WIPO Report, *supra* 218 at 125.

251 John Frow, "Public Domain and Collective Rights in Culture" (1998) 13 Intellectual Property Journal 39 at 51.

252 For further discussion of the various secrecy regimes, see WIPO Report, *supra* note 218.

253   Eugenio da Costa e Silva, *Biodiversity-Related Aspects of Intellectual Property Rights (IPRs)* (UNU/ IAS Working Paper No. 17, July 1996).

254   *Ibid.*

255   Although there are theoretical differences between utility patents on TKUP and specialized plant patent systems, the excessive dilution of the conditions of patentability has blurred this distinction. Therefore, the proposals here should be considered as being applicable to both utility patents and plant patents.

256   *Romanian Law No. 62 of the Grand National Assembly of 30 October 1974 on Inventions and Innovations,* reproduced in Sinnot, *supra* note 248, Vol. 2F.

257   Sinnott, *supra* note 248, Vols. 2F and 2G.

258   For example, section 116 of the US *Patent Act* provides that: "when two or more persons have made inventive contributions to subject matter claimed in an application, they shall apply for a patent jointly ... they shall be named as the inventors." See also Rivka Monheit, "The Importance of Correct Inventorship" (1999) 7 Journal of Intellectual Property Law 191.

259   *Community Intellectual Rights Protection Act of 1994,* S. 1841, 9th Congress of the Republic of Philippines [as cited in, Jacoby and Weiss, *supra* note 115 at 80].

260   PL is the acronym for "projeto de Lei" (legislative bill).

261   Silva, *supra* note 253 at 42.

262   Lise Osterborg, "Patent Term a la Carte?" (1986) 17 International Review of Industrial Property and Copyright Law 60.

263   *Report of the Intergovernmental Committee for a Convention on Biological Diversity,* UN Environmental Program, 7th Negotiating Sess., 5th Session of the International Negotiating Committee., UN Doc. UNEP/Bio.Div/N7-INC.5/4 (1992).

264   *Ibid.*

265   See Articles 15 and 16 of the CBD, *supra* note 43.

266   Decision 486. The Community of Andean Nations consists of Bolivia, Columbia, Ecuador, Peru, and Venezuela. It enjoys regulatory authority through decisions and resolutions. As a rule, decisions need no internal approval processes and become national law automatically upon their publication in the community's official journal. This decision came into effect on 1 December 2000 but was adopted on 14 September 2000. See Ruiz, *supra* note 220 at 12.

267   Decision 486, *ibid.*

268   *Ibid.,* Article 21 of the Brazilian proposed bill referred to above also makes similar provisions.

269   *Charan Lal Sahu v. Union of India* (2000) Vol. 118 International Law Reports 451.

270   See, for example, *Alfred Dunhill and Son, Inc., v. Puerto Rico,* (1982) 458 US 592.

271   International Workshop on Indigenous Peoples and Development. Ollantaytambo, Qosqo, Peru, 21-26 April 1997, <http://users.ox.ac.uk/wgtrr.ollan.htm>.

272   Farhana Yamin, The Biodiversity Convention and Intellectual Property Rights (a WWF International Discussion Paper), October 1995 at 4. The concept of farmers rights is entrenched in the FAO Treaty, as per Article 9.

273   Tshimanga Kongolo, "Towards a More Balanced Coexistence of Traditional Knowledge and Pharmaceuticals Protection in Africa" (2001) Vol. 35 No. 2 Journal of World Trade 349.

274   Oddi, *supra* note 160.

275   Kerry Ten Kate and Sarah Laird, *The Commercial Use of Biodiversity: Access to Genetic Resources and Benefit Sharing* (London: Earthscan Publications Ltd, 1999) at 62.

276   Agreement Concerning the Establishment of an International Patents Bureau, The Hague, 6 June 1947, reproduced in Treaty Series No. 84 (1965); Agreement Revising the Agreement Signed at the Hague on 6 June 1947, Concerning the Establishment of an International Patents Bureau, with Protocol and Resolution, The Hague, 16 February-31 December, 1961, reproduced in J.W. Baxter, ed., *World Patent Law and Practice,* Vol. 2A (London: Sweet and Maxwell, 1976) at 310.

277   Article 1, Agreement Concerning the Establishment of an International Patents Bureau, The Hague, 6 June 1947, reproduced in Treaty Series No. 84 (1965), *ibid.*

278   Peter Drahos, "Indigenous Knowledge, Intellectual Property and Biopiracy: Is a Global Bio-Collecting Society the Answer? [Opinion] (2000) 6 European Intellectual Property Review 245.

279   *Supra* note 39.
280   This legislation limited protectable plants to asexually reproduced species. Furthermore, unlike the standard utility patent, this legislation lowered the standard of patentability by excluding utility. Therefore, once the purported new plant invention was distinct and new, it qualified for patentability.
281   *Supra* note 13.
282   Kloppenburg, *supra* note 8.
283   Philippe Cullet, "Plant Variety Protection in Africa: Towards Compliance with TRIPs Agreement" (2001) 45 Journal of African Law 97.
284   M.B. Rao and M. Guru, *Understanding TRIPs: Managing Knowledge in Developing Countries* (New Delhi: Response Books, 2003).
285   Donald Richards, *Intellectual Property Rights and Global Capitalism: The Political Economy of the TRIPs Agreement* (New York: M.E. Sharpe, 2004) at 178.
286   Kloppenburg, *supra* note 8.
287   *Ibid.*, at 132.
288   *Supra* note 23.
289   Kloppenburg, *supra* note 8 at 143.
290   *Ibid.*
291   Dan Leskien and Michael Flitner, "Intellectual Property Rights and Plant Genetic Resources: Options for A Sui Generis System" [Issues in Genetic Resources No. 6] (International Plant Genetic Resources Institute, Rome, June 1997) at 62.
292   Cullet, *supra* note 283 at 100.
293   *Ibid.*

Chapter 6: Conclusion
1   Ikechi Mgbeoji, "Patent First, Litigate Later! The Scramble for Speculative and Overly Broad Genetic Patents: Implications for Access to Health Care and Biomedical Research" (2003) Vol. 2, No. 2. Canadian Journal of Law and Technology 83.
2   V.H. Heywood, ed., *Global Biodiversity Assessment* (Cambridge: Cambridge University Press, UNEP, 1996). See also Norman Myers, "The Hamburger Connection: How Central America's Forests Become North America's Hamburgers" (1981) 10 Ambio 3-8.
3   Since the industrialization of agriculture, plant genetic erosion has intensified. See Holly Saigo, "Agricultural Biotechnology and the Negotiation of the Biosafety Protocol" (2000) 12 The Georgetown International Environmental Law Review 779.
4   Mark Ritchie et al., "Intellectual Property Rights and Biodiversity: The Industrialization of Natural Resources and Traditional Knowledge" (1996) 11 St. John's Journal of Legal Comment 431.
5   Annie Patricia Kameri-Mbote and Phillipe Cullet, "Agro-Biodiversity and International Law: A Conceptual Framework" (1995) 11 Journal of Environmental Law 257.
6   As earlier noted, this statement needs to be qualified. For example, the so-called Green Revolution, which emphasized seeds that need high doses of chemical fertilizers and pesticides, may have produced high yields; however, as in the Philippines, it also led to a 358 percent increase in farm expenses due to chemical inputs. Farmers there suffered a 52 percent drop in farm income. See Rebecca Margulies, "Protecting Biodiversity: Recognizing International Intellectual Property Rights in Plant Resources" (1993) 14 Michigan Journal of International Law 322.
7   For instance, the Irish potato famine reduced the population of Ireland by one-third. In the 1970s, United States farmers suffered a 20 percent loss of their corn crop to the corn blight.
8   The CBD defines biotechnology as "any technological application that uses biological systems, living organisms or derivatives thereof, to make, modify products or processes for specific use." See, Article 1, Convention on Biological Diversity (1992) I.L.M. 813. (Entered into force 29 December 1993); Michael Doane, "TRIPs and International Intellectual Property Protection in an Age of Advancing Technology" (1994) 9 American University Journal of International Law and Policy 465.

The actual and potential commercial usefulness of this industry is quite remarkable. For instance, total sales for the biotechnology industry in 1991 in the United States alone was approximately $4billion; in 2000 it is estimated at $20 billion to $50 billion. Similarly, between 1985 and 1990 the number of biotechnology patents filed in the United States grew by 15 percent annually. M. Kenney, *Biotechnology: The University-Industrial Complex* (New Haven: Yale University Press, 1986); Rebecca Eisenberg, "Proprietary Rights and the Norms of Science in Biotechnology Research" (1989) 21 Intellectual Property Law Review 29.

Some scientists claim that the rise of the biotechnology industry and the pursuit of patent rights by academics retard communications among scientists. See *Commercialisation of Academic Biomedical Research: Hearings Before the Subcommittee on Investigations and Oversight and the Subcommittee on Science, Research and Technology of the House Committee on Science and Technology,* 97th Congress, 1st Session (1981).

9   RAFI Communiqué, February/March 2000.
10   The term "genetic modification" is used to indicate the introduction of DNA segments into an organism through recombinant DNA technology. All living organisms contain the molecule deoxyribonucleic acid (DNA), which itself looks a like a woven strand and is comprised of thousands of genes. Each of these genes controls the existence of a particular trait or characteristic. The combination of these genes distinguishes one organism from another and/or makes them similar to one another. See Robert Merges, "Intellectual Property in Higher Life Forms: The Patent System and Controversial Technologies" (1988) 47 Maryland Law Review 1051; Cheryl Bardales, "A Primer of Genetic Engineering 1: Basic Structural Components of the Cell" (1994) 4 Dickinson Journal of Environmental Law and Policy 7. The potential risks involved and the need for informed consent in dealing with modified life forms has engendered calls for the labelling of such organisms. See Karen Goldman, "Labelling of Genetically Modified Foods: Legal and Scientific Issues" (2000) 12 Georgetown International Environmental Law Review 717.

The scope and implications of genetic modification of life forms, genetic testing of human beings, and patents on the products of genetically modified organisms (including human cell lines) is complex and virtually borderless. Most of the issues fall outside the purview of this book. However, for a brief overview of the varied dimensions of genetic engineering, especially its propensity to politicization, genetic discrimination, eugenics, and racism, see Lori Andrews, "Past as Prologue: Sobering Thoughts on Genetic Enthusiasm" (1997) 27 Seton Hall Law Review 893 (arguing that the claims for a genetic cure-all for human ailments is largely oversold and exaggerated); Lisa Ikemoto, "The Racialization of Genomic Knowledge" (1997) 27 Seton Hall Law Review 937; James Bowman, "Genetics and African Americans" (1997) 27 Seton Hall Law Review 919; UNESCO Universal Declaration on the Human Genome and Human Rights 1997, 11 November 1997. UNGA Res. 53/152, reprinted in IHRR Vol. 6, No. 3 (1999).

Similar concerns have also been raised with respect to the problematic question of patent and proprietary rights to human parts and body tissue. See Jane Churchill, "Patenting Humanity: The Development of Property Rights in the Human Body and the Subsequent Evolution of Patentability of Living Things" (1994) 8 Intellectual Property Journal 249; Roger Magnusson, "The Recognition of Proprietary Rights in Human Tissue in Common Law Jurisdictions" (1992) 18 Melbourne University Law Review 601. For a judicial construction of some of these complex issues, particularly patents on body parts and the possibility of the emergence of a future human "spare-parts" industry, see Paul Matthews, "Whose Body? People as Property" (1983) 36 Current Legal Problems 193; "The Human Tissue Trade," RAFI Communiqué, January/February, 1997(noting that trade in human tissue may be worth over US$1billion by 2002).

11   Saigo, *supra* note 3 at 790; The Rural Urban Advancement Foundation, "Terminator: So Bad, Even Monsanto Can't Put a Spin on It" Geno-Types, 8 October 1998.
12   "Biopatenting and the Threat to Food Security: A Christian and Development Perspective" CIDSE Press Release, 10 February 2000, <http://www.cidse.be/pubs/tg1ppcon.htm>.
13   Saigo, *supra* note 3 at 797; *Agribusiness Consolidation: Hearings before the Subcommittee on International Trade of the House Committee on Agriculture,* 106th Congress 65-70 (1999) (state-

ment of Leland Swenson, President of National Farmers Union). On 3 April 2000 Monsanto, Pharmacia, and Upjohn merged to become the Pharmacia Corporation.

14 A few companies control 30 percent of the global seed trade. For a detailed listing of corporate controllers of the global seed industry, see "The Seed Giants: Who Owns Whom?" RAFI Update, September 1999. See also, The Crucible Group, Vol. 1, *Seeding Solutions: People, Plants, and Patents Revisited* (Ottawa: IDRC, 2000) at 16-17.

15 Neil Hamilton, "Why Own the Farm If You Can Own the Farmer (And the Crop)? Contract Production and Intellectual Property Protection of Grain Crops" (1994) 73 Nebraska Law Review 91.

16 Mark Ritchie et al., *supra* note 4 at 440.

17 "The State of Food Insecurity in the World," available at <http://www.fao.org>; Jane Burgermeister, "Number of Chronically Hungry People Is Rising by 5m a year" (2003) 327 British Medial Journal 1303; Seymour Rubin, "Economic and Social Rights and the New International Economic Order" (1986) 1 American University Journal of International Law and Policy 67.

18 Jeroen Van Wijk, "Broad Biotechnology Patents Hamper Innovation" (1995) 25 Biotechnology and Development Monitor 15-17; "Plant Variety Rights Sparks Debate in Asia," <http://www.cgiar.org/irri>.

19 Jeremy Oczek, "In the Aftermath of the 'Terminator' Technology Controversy: Intellectual Property Protections for Genetically Engineered Seeds and the Right to Save and Replant Seed" (2000) 41 Boston College Law Review 627.

20 Ikechi Mgbeoji, "The Terminator Patent and Its Discontents: Rethinking the Normative Deficit in Utility Test of Modern Patent Law" (2004) 17 St. Thomas Law Review 96; "US Patent on New Genetic Technology Will Prevent Farmers from Saving Seed," <http://www.rafi.org/genotypes/980311seed.html>.

21 In a decision handed down in Saskatoon, Saskatchewan, in March 2001 Judge Andrew Mackay of the Canadian Federal Court ordered a farmer to pay $15,000 in damages for patent infringement to Monsanto for genetically modified canola found on his farm. The allegation of gene escape was not proven. However, according to the court: "A farmer whose field contains seed or plants originating from seed spilled into them, or blown as seed, in swaths from a neighbor's land or even growing from germination by pollen carried into his field from elsewhere by insects, birds, or by the wind, may own the seeds or plants on his land even if he did not set about to plant them. He does not, however, own the right to the use of the patented gene, or of the seed or plant containing the patented gene or cell." See *Monsanto Canada Inc.* v. *Schmeiser*, [2001] F.C.J. No. 436 at para 92. On appeal, the Supreme Court of Canada (in a split decision [5-4]) upheld the judgment of the trial court.

22 For some insight into this question, see Stephen Lewis, "Attack of the Killer Tomatoes? Corporate Liability for the International Propagation of Genetically Altered Agricultural Products" (1997) 10 Transnational Law 178-88; Judy Kim, "Out of the Lab and Into the Field: Harmonization of Deliberate Release Regulations for Genetically Modified Organisms" (1993) 16 Fordham International Law Journal 1170.

23 Safe Food News (2000 edition): Scientists Explain Health and Environmental Risks, at 5, <http://www.safe-food.org>.

24 *Ibid.*

25 Judson Berkey, "The Regulation of Genetically Modified Foods" (1999) ASIL Insights (October) 1.

26 For example, within four years of placing genetically altered seeds in the market, about 45 million acres of US farmland have been planted with biotech crops. About one-half of US cotton fields, 40 percent of soybean fields, and 20 percent of corn fields had been genetically altered in the US by 1998. See Kurt Buechle, "The Great Global Promise of Genetically Modified Organisms: Overcoming Fear, Misconceptions, and the Cartagena Protocol on Biosafety" (2001) 9 Indiana Journal of Global Legal Studies 283.

27 Sean Murphy, "Biotechnology and International Law" (2001) 42 Harvard International Law Journal 47.

28 Amartya Sen, *The Political Economy of Hunger* (Oxford: Clarendon Press 1990); Amartya Sen, *The Standard of Living* (Cambridge, UK: Cambridge University Press, 1987). Amartya Sen's

path-breaking work in this field, which shattered the myth of lack of food in the world, earned him a Nobel prize.

29 Safe Food News, *supra* note 23 at 17.
30 Amartya Sen, *Poverty and Famine: An Essay on Entitlement and Deprivation* (Oxford: Clarendon, 1981).
31 Karen Graziano, "Biosafety Protocol: Recommendations to Ensure the Safety of the Environment" (1995) 7 Colorado Journal of International Environmental Law and Policy 179.
32 See, for example, Heather Scoffield, "British Scientists Slam Approvals for Genetically Modified Foods" Globe and Mail (Toronto) 7 October 1999, at A5; Kevin Bastian, "Biotechnology and the United States Department of Agriculture: Problems of Regulation in a Promotional Agency" (1990) 17 Ecology Law Quarterly 413.
33 Saigo, *supra* note 3 at 790 and the authorities cited therein.
34 Safe Food News, *supra* note 23 at 4.
35 For example, wild rice in Taiwan was destroyed when a massive gene flow from cultivated rice swamped out traits such as the perennial lifecycle from dormant seeds. See Graziano, *supra* note 31. Other dangers include groundwater and surface water pollution, health risks related to mutation of organisms, toxicity, and allergies. For example, the outbreak of Eosinophilia-myalgia syndrome (EMS) in 1992 was caused by exposure to a particular batch of a synthetic amino acid. The outbreak affected more than 1,500 people and resulted in 38 deaths in the United States. Denmark was the first European country to impose a law specifically dealing with the deliberate release of transgenic organisms. There are similar laws in Germany that impose restrictions on the deliberate release of harmful genetically modified organisms into the environment. See Claus-Joerg Ruetsch and Terry R. Broderick, "New Biotechnology Legislation in the European Community and Federal Republic of Germany" (1990) 18 International Business Law 408.

    On the other hand, countries such as Taiwan, China, India, Mexico, Brazil, the Philippines, and South Korea do not regulate biotechnology. See Peter Huber, "Biotechnology and the Regulation Hydra" (1987) 90 Technology Review 8. In March 1990 the Council of the European Communities approved two directives on genetically modified organisms: Directive 90/219, Council Directive of 23 April 1990 on the contained use of genetically modified micro-organisms, O.J. L117/1 (1990) otherwise known as Directive 90/219; Directive 90/220, Council Directive of 23 April 1990 on the deliberate release into the environment of genetically modified organisms, O.J. L117/15 (1990), otherwise known as Directive 90/220.
36 Safe Food, *supra* note 23 at 16.
37 GM Science Review, First Report, July 2003, <http://www.gmsciencedebate.org.uk>. Part of the emerging concern deals with the issue of transgenic pollution. Recent studies tend to confirm this fear as genes from transgenic plants might be up to twenty times more likely to "outcross" into relative species than the plant's natural genetic material. See J. Bergelson et al., "Promiscuity in Transgenic Plants" (1998) 25 Nature 25; H.J. Rogers and H.C. Parkes, "Transgenic Plants and the Environment" (1995) 46 Journal of Experimental Botany 467; Reid Adler, "Controlling the Applications of Biotechnology: A Critical Analysis of the Proposed Moratorium on Animal Patenting" (1988) 1 Harvard Journal of Law and Technology 1.
38 States do not grant patents on inventions deemed to be harmful to society or on products that endanger national security.
39 See Chapter 1.
40 James Buchanan, "Between Advocacy and Responsibility: The Challenge of Biotechnology for International Law" (1994) 1 Buffalo Journal of International Law 221.
41 See James Chalfant et al., "Recombinant DNA: A Case Study in Regulation of Scientific Research" (1979) 8 Ecology Law Quarterly 55.
42 Most countries in Europe have a policy of labelling genetically modified plants. The Cartagena Protocol, however, only provides for the labelling of international food ship-

ments. See *Cartagena Protocol on Biosafety, 23 February 2000, Report of Panel IV: Consideration of the Need for Modalities of Protocol Setting Out Appropriate Procedures Including, in Particular, Advance Informed Agreements in the Field of the Safe Transfer, Handling and Use of any Living Modified Organism Resulting from Biotechnology Diversity, UNEP Arguments for a Protocol Pursuant to Article 19.3 of the Convention,* UN Doc. UNEP/Bio.Div/Panels/Int.4 (1993). For a negotiating history of the protocol, see Jonathan Adler, "Cartagena Protocol: Biosafe or Bio-sorry?" (2000) 12 Georgetown International Environmental Law Review 772.

43 This procedure is the backbone of the protocol. However, it applies to only a small percentage of traded LMOs. Some of the excluded LMOs include most pharmaceuticals for humans, LMOs in transit to a third party, LMOs destined for contained use, and LMOs that have been declared safe by a meeting of the parties. See Aaron Crosbey and Stas Burgiel, "The Cartagena Protocol on Biosafety: An Analysis of Results" (an IISD briefing note), <http://iisd.ca/pdf/biosafety>.

44 For an exhaustive analysis of the status of the precautionary principle in international law, see Harald Hohmann, *Precautionary Legal Duties and Principles of Modern International Environmental Law* (Dordrecht: Martinus Nijhoff; 1994).

45 Paul Hagen and John Barlow Weiner, "The Cartagena Protocol on Biosafety: New Rules for International Trade in Living Modified Organisms" (2000) 12 Georgetown International Environmental Law Review 697; Ved Nanda, "Genetically Modified Food and International Law: The Biosafety Protocol and Regulations in Europe" (2000) 28 Denver Journal of International Law and Policy 235.

46 James Nickel, "The Human Right to a Safe Environment: Philosophical Perspectives on Its Scope and Justification" (1993) 18 Yale Journal of International Law 281. The constitutions of many countries have recognized a right to a safe environment. For further analysis of this trend, see Melissa Thorme, "Establishing Environment as a Human Right" (1991) 19 Denver Journal of International Law and Policy 301; Luis Rodriguez-Rivera, "Is the Human Right to Environment Recognized under International Law? It Depends on the Source" (2001) 12 Colorado Journal of International Law and Policy 1.

47 33 I.L.M. 1226 (1994).

48 Peter-Tobias Stoll, "Controlling the Risks of Genetically Modified Organisms: The Cartagena Protocol on Biosafety and the SPS Agreement" (1999) 10 Yearbook of International Environmental Law 82.

49 For an overview and analysis of the nature of the concept of precaution at international environmental law, see David VanderZwaag, "The Precautionary Principle in Environmental Law and Policy: Elusive Rhetoric and First Embraces" (1999) 8 Journal of Environmental Law and Practice 355.

50 See Article 10 (6) of the protocol. It seems there is a semantic debate as to whether it is a principle or an approach. See Ellen Hey, "The Precautionary Concept in Environmental Policy and Law: Institutionalising Caution" (1992) Georgetown International Environmental Law Review 303. Both terms are used interchangeably, and it seems there is no juridical difference arising therefrom. See D. Freestone and E. Hey, eds., *The Precautionary Principle and International Law: The Challenge of Implementation* (The Hague: Kluwer Law International, 1996); Ludwig Kramer, *E.C. Treaty and Environmental Law* (London, Sweet and Maxwell, 1995).

51 James Cameron and Juli Abouchar, "The Precautionary Principle: A Fundamental Principle of Law and Policy for the Protection of the Global Environment" (1991) 14 Boston College International and Comparative Law Review 1.

52 James Hickey and Vern Walker, "Refining the Precautionary Principle in International Environmental Law" (1995) 14 Virginia Environmental Law Journal 423.

53 Lothar Gundling, "The Status in International Law of the Principle of Precautionary Action" (1990) 5 Journal of Estuarine and Coastal Law 23.

54 For a brief but effective analysis of the arguments for and against legislative regulation of biotechnology, see David Rosenblatt, "The Regulation of Recombinant DNA Research: The Alternative of Local Control" (1982) 10 Boston College Environmental Affairs Law Review 37.

55   Clifford Russell, "Two Propositions about Biodiversity" (1995) 28 Vanderbilt Journal of Transnational Law 689; Chris Backes, "The Precautionary Principle in International, European, and Dutch Wildlife Law" (1998) 9 Colorado Journal of International Environmental Law and Policy 43.

56   William Andrew Shutkin, "International Human Rights Law and the Earth: The Protection of Indigenous Peoples and the Environment" (1991) 31 Virginia Journal of International Law 479.

57   Rosemary Coombe, "Intellectual Property, Human Rights and Sovereignty: New Dilemmas in International Law Posed by the Recognition of Indigenous Knowledge and the Conservation of Biodiversity" (1998) 6 Indiana Journal of Global Legal Studies 59.

58   International Covenant on Economic, Social and Cultural Rights, 16 December 1966, 993 U.N.T.S. 3.

59   *Ibid.*

60   Coombe, *supra* note 57 at 60-62; Scott Leckie, "Another Step towards Indivisibility: Identifying the Key Features of Violations of Economic, Social and Cultural Rights" (1998) 20 Human Rights Quarterly 81.

61   Articles 16 and 17 of the International Covenant on Economic, Social and Cultural Rights require states to submit reports of the legislative steps taken to implement its objectives and obligations.

62   Anja Meyer, "International Environmental Law and Human Rights: Towards the Explicit Recognition of Traditional Knowledge" (2001) 10 Review of European Community and International Environmental 37.

63   Benjamin Richardson, "Indigenous Peoples, International Law and Sustainability" (2001) 10 Review of European Community and International Environmental Law 1; Russell Barsh, "Is the Expropriation of Indigenous Peoples' Land GATT-able?" (2001) 10 Review of European Community and International Environmental 13.

64   R.V. Anuradha, "IPRs: Implications for Biodiversity and Local and Indigenous Communities" (2001) 10 Review of European Community and International Environmental 27.

65   But see Brendan Tobin, "Redefining Perspectives in the Search for Protection of Traditional Knowledge: A Case Study from Peru" (2001) 10 Review of European Community and International Environmental 27.

66   Vandana Date, "Global Development and Its Environmental Implications: The Interlinking of Ecologically Sustainable Development and Intellectual Property Rights" (1997) 27 Golden Gate University Law Review 630.

67   *Guinea/Guinea Bissau Maritime Delimitation Case* (cited in Ian Brownlie, *The Human Right to Development* [London: Commonwealth Secretariat, 1989], at 16-17).

68   K. Mbaye, "Le droit au development comme un droit de l'homme" (1975) 5 (2-3) Human Rights Journal 53; Jose Zalaquett, "An Interdisciplinary Approach to Development and Human Rights" (1983-85) 4-5 Boston College Third World Law Journal 1.

69   Ignaz Seidl-Hohenveldern, *International Economic Law* (Dordrecht: Martinus Nijhoff, 1989) at 4.

70   Zalaquett, *supra* note 68 at 6.

71   Obiora Chinedu Okafor, "The Status and Effect of the Right to Development in Contemporary International Law: Towards a South-North 'Entente'" (1995) 7 African Journal of International and Comparative Journal 865.

72   Philip Alston, "Development and the Rule of Law: Prevention versus Cure as a Human Rights Strategy" in *Development, Human Rights and the Rule of Law: Report of a Conference Held in the Hague on 27 April -1 May 1981, Convened by the International Commission of Jurists* (Oxford: Pergamon Press, 1981) at 101.

73   See Declaration on the Right to Development, UN GAOR, 41st Sess., Resolutions and Decisions, Agenda Item 101, at 3-6, 9th plenary meeting, 4 December 1986 UN Doc. A/Res/41/ 128. Resolution adopted by the General Assembly [on the report of the 3d Committee (A/ 41/925 and Corr. 1)]. For a detailed compendium and multidimensional analysis of the right to development, see Asbjorn Eide et al., eds., *Economic, Social and Cultural Rights: A Textbook* (Dordrecht: Martinus Nijhoff, 1995).

74 *Question of the Realization of the Right to Development, Global Consultation on the Right to Development as a Human Right,* Report Prepared by the Secretary-General Pursuant to Commission on Human Rights Resolution 1989/45; E/CN.4/1990/9/rev.1 (26 September 1990); Phillip Alston, "Making Space for New Human Rights: The Case of the Right to Development" (1988) 1 Harvard Human Rights Yearbook 3. The right to development was an integral part of the now defunct clamour for a New International Economic Order (NIEO). The North-South divide on the issue is by no means a function of the nationality of the various scholars working on the concept of a right to development.

75 Okafor, *supra* note 71 at 868.

76 P.J.I.M. Waart, "Implementing the Right to Development: The Perfection of Democracy" in Subtrata Chowdhury et al., eds., *The Right to Development in International Law* (Dordrecht: Martinus Nijhoff, 1992).

77 Okafor, *supra* note 71 at 73; Jacques Chonchol, "The Declaration on Human Rights and the Right to Development: The Gap between Proposals and Reality" in *Development, Human Rights and the Rule of Law: Report of a Conference Held in the Hague on 27 April-1 May 1981, Convened by the International Commission of Jurists* (Oxford: Pergamon Press, 1981) at 109.

78 Some scholars argue that there is no right to development; rather, what exists is the need for such a right. See Karel de Vey Mestdagh, "The Right to Development: From Evolving Principle to Legal Right – In Search of Its Substance" in *Development, Human Rights and the Rule of Law: Report of a Conference Held in the Hague on 27 April-1 May 1981, Convened by the International Commission of Jurists* (Oxford: Pergamon Press, 1981) at 156.

79 Oscar Schachter, "The Evolving International Law of Development" (1976) 15 Columbia Journal of Transnational Law 9.

80 Phillip Alston, "The United Nations' Specialized Agencies and Implementation of the International Covenant on Economic, Social and Cultural Rights" (1979) 18 Columbia Journal of Transnational Law 79; Phillip Alston, "Revitalizing United Nations Work on Human Rights and Development" (1991) 18 Melbourne University Law Review 216. For an excellent analysis of the nature and concept of human right to development, see James Paul, "The Human Right To Development: Its Meaning and Importance" (1992) 25 John Marshall Law Review 235; J.S. Warioba, "The Reform of the United Nations in the Context of the Law of the Sea and the United Nations Conference on the Environment and Development" (1995) 7 RADIC 426; Seymour Rubin, "Economic and Social Human Rights and the New International Economic Order" (1986) 1 American University Journal of International Law and Policy 67.

81 Tom Allen, "Commonwealth Constitutions and Implied Social and Economic Rights" (1994) 6 RADIC 555. For an excellent analysis of the interdependent nature of human rights, see Craig Scott, "Reaching Beyond (Without Abandoning) the Category of 'Economic, Social and Cultural Right'" (1999) 21 Human Rights Quarterly 633; Gregorio Peces-Barba, "Reflections on Economic, Social and Cultural Rights"(1981) 2 Human Rights Law Journal 281; Antonio Trindade, "Environment and Development: Formulation and Implementation of the Right to Development as a Human Right" (1993) 3 Asian Yearbook of International Law 15.

82 *UN Declaration on the Right to Development,* UN GA Resolution 41/128 of December 4, 1986.

83 R.N. Kiwanuka, "Developing Rights: The UN Declaration on the Right to Development" (1988) 35 Netherlands International Law Review 265.

84 *Guinea/Guinea Bissau Maritime Delimitation Case, supra* note 67.

85 Schachter, *supra* note 79.

86 Inamul Haq, "The Problem of Global Economic Inequity: Legal Structures and Some Thoughts on the Next Years" (1979) 9 Georgia Journal of International and Comparative Law 507.

87 Ivan Head, "The Contribution of International Law to Development" (1987) Canadian Yearbook of International Law 29; but see Louis Henkin, "Economic-Social Rights as 'Rights': A United States Perspective" (1981) 2 Human Rights Law Journal 223; Mitchell Ginsberg

and Leonard Lesser, "Current Developments in Economic and Social Rights: A United States Perspective" (1981) 2 Human Rights Law Journal 237.

88 "Anglo-Norwegian Fisheries Case," (1951) ICJ Reports 131, ILR 18 at 73.

89 Seidll-Hohenveldern, *supra* note 69 at 7.

90 *Ibid.*

91 Sarah Dillon, "Fuji-Kodak, the WTO, and the Death of Domestic Political Constituencies" (1999) 8 Minnesota Journal of Global Trade 197; Paul Stephan, "The New International Law: Legitimacy, Accountability, Authority and Freedom in the New Global Order" (1999) 70 University of Colorado Law Review 1555.

92 Robert Blomquist, "Protecting Nature 'Down Under': An American Law Professor's View of Australia's Implementation of the CBD: Laws, Policies, Progress, Institutions, and Plans, 1992-2000" (2000) 9 Dickinson Journal of Environmental Law and Policy 237.

93 Jim Chen, "Globalization and Its Losers" (2000) 9 Minnesota Journal of Global Trade 157.

94 For a critique of Jim Chen's treatise on the losers in the age of globalization, see Elizabeth Larsch-Quinn, "Commentary: Democracy Should Not Have Losers" (2000) 9 Minnesota Journal of Global Trade 589; John Miller, "Globalization and Its Metaphors" (2000) 9 Minnesota Journal of Global Trade 594; Paul Thompson, "Globalization, Losers, and Property Rights" (2000) 9 Minnesota Journal of Global Trade 602.

95 Alex Geisinger, "Sustainable Development and the Domination of Nature: Spreading the Seed of the Western Ideology of Nature" (1999) 27 Environmental Affairs 43.

96 Although it may be argued that memberships of globalizing institutions such as the WTO are discretionary, given the structure and process of global trade and politics, the reality is that weak states are practically coerced into membership. Whether the element of coercion would be conducive to or frustrate the compliance quotient of such members with the legal norms established by the WTO and similar organizations is a matter of interest (especially to compliance theorists). A preliminary view would support the conclusion that such legal norms may not command much compliance. Indeed, the well-known cases of public revolt against such institutions and the legal norms they establish seem to indicate that, sooner rather than later, those institutions and norms would have to wear a "human face" (or at least make some concessions towards social distributive justice). For a recent examination of these and related issues, see Peter Gerhart, "Reflections: Beyond Compliance Theory – TRIPs as a Substantive Issue" (2000) 32 Case Western Reserve Journal of International Law 357; Kal Raustiala, "Compliance and Effectiveness in International Regulatory Cooperation" (2000) 32 Case Western Reserve Journal of International Law 387; J.H. Reichman, "The TRIPs Agreement Comes of Age: Conflict or Cooperation with the Developing Countries?" (2000) 32 Case Western Reserve Journal of International Law 441.

97 *Draft Report of the World Intellectual Property Organization (WIPO) Fact-Finding Missions on Intellectual Property and Traditional Knowledge (1998-99)*, Geneva, Switzerland.

98 Dan Leskien and Michael Flitner, "Intellectual Property Rights and Plant Genetic Resources: Options for a Sui Generis System" [Issues in Genetic Resources No. 6], Jan Engels, vol. editor (International Plant Genetic Resources Institute, June 1997) (on file with author and available on request).

99 Leskien and Flitner, *supra* note 98 at 46.

100 Carlos Correa, "Sovereign and Property Rights over Plant Genetic Resources" [FAO Background Study Paper No. 2 Commission on Plant Genetic Resources-First Extraordinary Session, Rome, 7-11 November 1994].

101 Agreement Related to Trade-Related Aspects of Intellectual Property Rights, 33 I.L.M. 1197 (1994).

102 "No Cure for Patents: Biotech Patents Distort and Discourage Innovation and Increase Costs for Dubious Drugs" RAFI Genotypes, July 1997; but see Michael Malinowski, "Globalization of Biotechnology and the Public Health Challenges Accompanying It" (1996) 60 Albany Law Review 119.

103 Edgar Asebey and Jill Kempenaar, "Biodiversity Prospecting: Fulfilling the Mandate of the Biodiversity Convention" (1995) 28 Vanderbilt Journal of Transnational Law 703. India also threatened to do the same, as did Brazil, Indonesia, and Malaysia.

104 United States' Declaration Made at the United Nations Environment Programme for the Adoption of the Agreed Text of the Convention on Biological Diversity, issued 22 May 1992, 31 I.L.M. 848 (1992). Compare with the interpretative statement attached to the CBD. See Melinda Chandler, "The Biodiversity Convention: Selected Issues of Interest to the International Lawyer" (1993) 4 Colorado Journal of International Environmental Law and Policy 141.

105 International Convention for the Protection of New Varieties of Plants, 2 December 1961, as revised at Geneva on 10 November 1972, on 23 October 1978, and on 19 March 1991; 815 U.N.T.S. 89. This convention is administered by WIPO. At present forty-four countries are party to the UPOV, and the member states are predominantly from the North.

106 Andean Group: Commission Decision 313, 6 February 1992, reproduced in 32 I.L.M. 180 (1993).

107 *Ibid.* Emphasis added.

108 Martine de Koning, "Biodiversity Prospecting and the Equitable Remuneration of Ethnobiological Knowledge: Reconciling Industry and Indigenous Interests" (1998) 12 Intellectual Property Journal 261.

109 Crucible Group I, *supra* note 14 at 74.

110 For a succinct history of the legitimation process of the claims of non-Western paradigms of knowledge, see Johanna Sutherland, "Representations of Indigenous Peoples' Knowledge and Practice in Modern International Law and Politics" (1995) Vol. 2, No. 1 Australian Journal of Human Rights 4.

111 Crucible Group I, *supra* note 14 at 74.

112 *Ibid.*, at 9.

113 International Covenant on Economic, Social and Cultural Rights, 16 December 1966, 993 U.N.T.S.3; International Labour Organization Convention Concerning Indigenous and Tribal Peoples in Independent Countries; Reprinted in 28 I.L.M. 1382 (1989); Charter of Economic Rights and Duties of States of 1974 (UNGAOR 3281, XXIV); *Commission on Human Rights, Preliminary Report on the Study of the Problem of Discrimination against Indigenous Populations,* UN Doc.E/CN.4/sub.2/L.566 [1972]; The United Nations Draft Declaration on the Rights of Indigenous Peoples, UN ESCOR, Commission on Human Rights, 11th Sess., Annex 1, UN Doc. E/CN.4/Sub.2 (1993); UN Declaration on the Rights of Indigenous Peoples, UN Docs.E/CN.4/1995/2; reprinted in 34 I.L.M. 541 (1995).

114 Patricia Thompson, "Philippines Indigenous Peoples Rights Act" (1998) Human Rights and the Environment 12.

115 Statement by President Fidel Ramos of the Philippines as cited in Thompson, *ibid.*

116 *Ibid.*

117 Janet McDonnell, *The Dispossession of the American Indian, 1887-1934* (Indianapolis: Indiana University Press, 1991); Chris Cunneen and Terry Libesman, *Indigenous People and the Law in Australia* (Sydney: Butterworths, 1995) at 3.

118 Leskien and Flitner, *supra* note 98 at 70. See also Laura Campbell, "The Role of International Trade and Economics in Developing Multilateral Environmental Agreements" in *The Effectiveness of Multilateral Agreements* (Nord: Helsinki, Workshop Proceedings and Study Reports, 1996) at 10.

119 Experts Group on Environmental Law of the World. *Environmental Protection and Sustainable Development: Experts Group on Environmental Law of the World Commission on Environment and Development* (London: Graham and Trotman/Martinus Nijhoff Publishers, 1987).

120 Geir Lundestad, *East, West, North and South: Major Developments in International Politics, 1945-1986* (Oslo: Norwegian University Press, 1988) at 264.

121 Erica-Irene Daes, "Intellectual Property and Indigenous Peoples" (2001) 95 A.S.I.L. Proceedings 143.

Selected Bibliography

Books

Ackerknecht, Erwin. *Medicine and Ethnology, Selected Essays*. (Baltimore, MD: Johns Hopkins Press, 1991).

Adamantopoulos, Konstantinos. *An Anatomy of the World Trade Organization* (The Hague: Kluwer Law International, 1997).

Adams, Nassau. *Worlds Apart: The North-South Divide and the International System* (London: Zed Books, 1993).

Allen, R., and Christine Prescott-Allen. *Genes from the Wild: Using Genetic Resources for Food and Raw Materials* (London: Earthscan Publications, 1988).

Anderfelt, Ulf. *International Patent Legislation and Developing Countries* (The Hague: Martinus Nijhoff, 1971).

Ashton, T.S. *The Industrial Revolution, 1760-1830* (Oxford: Oxford University Press, 1948).

Balick, Michael, and Paul Alan Cox. *Plants, People and Culture: The Science of Ethnobotany* (New York: Freeman and Company, 1996).

Baslar, Kemal. *The Concept of the Common Heritage of Mankind in International Law* (The Hague: Martinus Nijhoff, 1998).

Baxter, J.W. *World Patent Law and Practice*, Vol. 2 (London: Sweet and Maxwell, 1976).

Beier, F.K., R.S. Crespi, and J. Strauss. *Biotechnology and Patent Protection: An International Review* (Paris: OECD, 1985).

Bentley, Lionel, and Brad Sherman. *The Making of Modern Intellectual Property Law: The British Experience, 1760-1911* (Cambridge: Cambridge University Press, 1999).

Birnie, P., and A.E. Boyle. *International Law and the Environment* (Oxford: Clarendon Press, 1992).

Bodenhausen, G.H.C. *Guide to the Application of the Paris Convention for the Protection of Industrial Property* (Geneva: BIRPI, 1968).

Bookchin, Murray. *The Ecology of Freedom: The Emergence and Dissolution of Hierarchy* (Palo Alto, CA: Cheshire Books, 1982).

Brockway, Lucile. *Science and Colonial Expansion: The Role of the British Botanic Gardens* (New York: Academic, 1979).

Brownlie, Ian. *Principles of Public International Law* (Oxford: Clarendon Press, 1982).

Burhenne, W.E., and W.A. Irwin. *The World Charter for Nature* (Berlin: Erich Schmidt Verlag, 1986).

Castaneda, Jorge. *Legal Effects of United Nations Resolutions* (New York: Columbia University Press, 1969).

Chakrabarti, Kalyan. *Conservation Versus Development* (Calcutta: Darbari Prokashan, 1994).

Claude, Inis. *Swords into Ploughshares: The Problems and Progress of International Organization* (Toronto: Random House, 1971).

Conforti, Benedetto. *International Law and the Role of Domestic Legal Systems* (Dordrecht: Martinus Nijhoff, 1993).

Coombe, Rosemary. *The Cultural Life of Intellectual Property: Authorship, Appropriation, and the Law* (Durham: Durham University Press, 1998).

Coulter, Moureen. *Property in Ideas: The Patent Question in Mid-Victorian Britain* (Missouri: Thomas Jefferson University Press, 1991).

Crawford, James. *The Creation of States at International Law* (Oxford: Clarendon Press, 1979).

Crucible Group. *People, Plants, and Patents: The Impact of Intellectual Property on Biodiversity, Conservation, Trade and Rural Society* (Ottawa: IDRC, 1994).

–. *Seeding Solutions: People, Plants, and Patents Revisited*, Vol. 1 (Ottawa: IDRC, 2000).

–. *Seeding Solutions: Options for National Laws Governing Access to and Control over Genetic Resources*, Vol. 2 (pre-publication version, not for quotation). (Ottawa: IDRC, 2001).

Cunneen, Chris, and Terry Libesman. *Indigenous People and the Law in Australia* (Sydney: Butterworths, 1995).

D'Amato, Anthony. *The Concept of Custom as a Source of International Law* (Ithaca: Cornell University Press, 1971).

Deane, Phyllis. *The First Industrial Revolution* (Cambridge: Cambridge University Press, 1965).

De Klemm, Cyrille, and Clare Shine. *Biological Diversity Conservation and the Law: Legal Mechanisms for Conserving Species and Ecosystems* (Gland: IUCN, 1993).

De Silva, Padmasiri. *Environmental Philosophy and Ethics in Buddhism* (London: Macmillan Press, 1998).

Devall, B., and G. Sessions. *Deep Ecology: Living as if Nature Mattered* (Salt Lake City: G.M. Smith, 1985).

Dorn, Harold. *The Geography of Science* (Baltimore: Johns Hopkins University Press, 1971).

Drahos, Peter. *A Philosophy of Intellectual Property* (Dartmouth: Aldershot, 1996).

Ducor, Phillipe. *Patenting the Recombinant Products of Biotechnology and Other Molecules* (London: Kluwer Law International, 1998).

Dutfield, G. *Intellectual Property Rights and the Life Science Industries: A Twentieth-Century History* (London: Ashgate, 2003).

Dutton, H.I. *The Patent System and Inventive Activity during the Industrial Revolution, 1750-1852* (Manchester: Manchester University Press, 1984).

Easlea, Brian. *Witch-Hunting, Magic, and the New Philosophy: An Introduction to Debates of the Scientific Revolution, 1450-1750* (Brighton, Sussex: Haverfield Press, 1980).

Ekins, Paul. *A New World Order: Grassroots Movements for Global Change* (London: Routledge, 1992).

Ezeabasili, N. *African Science: Myth or Reality* (New York: Vantage Press, 1977).

Firestone, O.J. *Economic Implications of Patents* (Ottawa: University of Ottawa Press, 1971).

Foster, George. *Traditional Societies and Technologies Changes* (Delhi: Allied Publishers, 1973).

Franck, Thomas. *Fairness in International Law and Institutions* (Oxford: Clarendon Press, 1995).

Gaski, Andrea, and Kurt Johnson. *Prescription for Extinction: Endangered Species and Patented Oriental Medicines in Trade* (Washington, DC: Traffic USA, 1996).

Gervais, Daniel. *The TRIPs Agreement: Drafting History and Analysis* (London: Sweet and Maxwell, 2004).

Gillespie, Alexander. *International Environmental law, Policy and Ethics* (Oxford: Clarendon Press, 1997).

Glowka, Lyle. *A Guide to Designing Legal Frameworks to Determine Access to Genetic Resources* (Gland: IUCN, 1998).

Good, Byron. *Medicine, Rationality, and Experience: An Anthropological Perspective* (Cambridge: Cambridge University Press, 1994).

Harding, Sandra. *Whose Science? Whose Knowledge? Thinking From Women's Lives* (Ithaca, NY: Cornell University Press, 1991).

Higgins, Rosalyn. *The Development of International Law through the Political Organs of the United Nations* (London: Oxford University Press, 1963).

Hohmann, Harald. *Precautionary Legal Duties and Principles of Modern International Environmental Law* (Dordrecht: Martinus Nijhoff, 1994).

International Commission of Jurists. *Development, Human Rights and the Rule of Law: Report of a Conference held in The Hague on 27 April-1 May 1982, Convened by the International Commission of Jurists* (Oxford: Pergamon Press, 1981).

International Union for Conservation of Nature and Natural Resources (IUCN). *Indigenous Peoples and Sustainability: Cases and Actions* (Utrecht: International Books, 1997).

Isaac, Erich. *Geography of Domestication* (Englewood Cliffs, NJ: Prentice-Hall, 1970).

Jewkes, John, D. Sawers, and R. Stillerman. *The Sources of Invention* (London: Macmillan, 1969).

Kate, Kerry Ten, and Sarah Laird. *The Commercial Use of Biodiversity: Access to Genetic Resources and Benefit Sharing* (London: Earthscan Publications, 1999).

Keller, Evelyn. *Reflections on Gender and Science* (New Haven: Yale University Press, 1985).

Kenney, M. *Biotechnology: The University-Industrial Complex* (New Haven: Yale University Press, 1986).

Kiss, A.C. *Survey of Current Developments in International Environmental Law* (Morges: IUCN, 1976).

Klabbers, Jan. *The Concept of Treaty in International Law* (London: Kluwer Law International, 1996).

Kloppenburg, Jack Jr. *First the Seed: The Political Economy of Plant Biotechnology, 1492-2000* (Cambridge: Cambridge University Press, 1988).

Kuhn, T. *The Structure of Scientific Revolutions* (Chicago: University of Chicago Press, 1972).

Ladas, Stephen. *The International Protection of Industrial Property* (Cambridge: Harvard University Press, 1930).

–. *Patents, Trademarks, and Related Rights: National and International Protection* (Cambridge: Harvard University Press, 1975).

Landes, David. *The Unbound Prometheus: Technological Change and Industrial Development in Western Europe From 1750 to the Present* (Cambridge: Cambridge University Press, 1969).

Lesser, William. *Institutional Mechanisms Supporting Trade in Genetic Materials: Issues under the Biodiversity Convention and GATT/TRIPS* (Geneva: UNEP, 1994).

–. *Sustainable Use of Genetic Resources under the Convention on Biological Diversity: Exploring Access and Benefit Sharing Issues* (Oxford: CAB International, 1997).

Lindley, Mark. *The Acquisition and Government of Backward Territory in International Law: Being a Treatise on the Law and Practice Relating to Colonial Expansion* (New York: Negro Universities Press, 1969).

Lowie, R.H. *Primitive Society* (New York: Routledge, 1920).

Luxemberg, Rosa. *The Accumulation of Capital* (London: Routledge, 1951).

Macer, Darryl R.J. *Shaping Genes: Ethics, Law, and Science of Using Genetic Technology in Medicine and Agriculture* (Christchurch, NZ: Eubios Ethics Institute, 1990).

Macleod, Christine. *Inventing the Industrial Revolution: The English Patent System, 1660-1800* (Cambridge: Cambridge University Press, 1988).

Mathias, Peter. *The First Industrial Nation: An Economic History of Britain* (New York: Charles Scribner's Son, 1969).

McDonnell, Janet. *The Dispossession of the American Indian, 1887-1934* (Bloomington: Indiana University Press, 1991).

McElroy, Ann, and Patricia Townsend. *Medical Anthropology in Ecological Perspective* (Boulder, CO: Westview Press, 1989).

McNeely, Jeffrey. *Conservation and the Future: Trends and Options toward the Year 2025* (Gland: IUCN, 1997).

–. *Economics and Biological Diversity: Developing and Using Economic Incentives to Conserve Biological Resources* (IUCN: Switzerland, 1988).

Merchant, Carolyn. *The Death of Nature: Women, Ecology and the Scientific Revolution* (New York: Harper and Row, 1980).

–. *Radical Ecology: The Search for a Liveable World* (London: Routledge, 1992).

Mgbeoji, Ikechi, Nasrul Islam, Isabel Martinez, and Wang Xi. *Environmental Law In Developing Countries: Selected Issues* (Cambridge: IUCN, 2001).

Mooney, Pat. *The Seeds of the Earth: A Private or Public Resource?* (Ottawa: Inter Pares, 1979).

Mooney, Pat, and Cary Fowler. *Shattering: Food, Politics and the Loss of Genetic Diversity* (Tucson: The University of Arizona Press, 1990).

Morrell, J., and A. Thackray. *Gentlemen of Science: Early Years of the British Association for the Advancement of Science* (London: Royal Historical Society, 1986).

Needham, J. *The Grand Titration: Science and Society in East and West* (London: Allen and Unwin, 1969).

–. *Science and Civilization in China*, Vols. 1-2 (Cambridge: Cambridge University Press, 1954).

Panjabi, Ranee. *The Earth Summit at Rio: Politics, Economics, and the Environment* (Boston: Northeastern University Press, 1997).

Parry, Clive. *The Sources and Evidences of International Law* (Manchester: Manchester University Press, 1965).

Patent Law of the United Kingdom by the Chartered Institute of Patent Agents (London: Sweet and Maxwell, 1975).

Penrose, Edith. *The Economics of the International Patent System* (New Haven, CN: Greenwood Press, 1974).

Perlman, Dan, and Glenn Adelson. *Biodiversity: Exploring Values and Priorities in Conservation* (Malden, MA: Blackwell Publishing, 1997).

Qureshi, Asif. *The World Trade Organization: Implementing International Trade Norms* (Manchester: Manchester University Press, 1996).

Rao, M.B., and M. Guru. *Understanding TRIPs: Managing Knowledge in Developing Countries* (New Delhi: Response Books, 2003).

Reid, Walter V., and Kenton Miller. *Keeping Options Alive: The Scientific Basis for Conserving Biodiversity* (Washington, DC: World Resources Institute, 1989).

Richards, D. *Intellectual Property Rights and Global Capitalism: The Political Economy of the TRIPs Agreement* (New York: M.E. Sharpe, 2004).

Schiff, Eric. *Industrialization without National Patents: The Netherlands, 1869-1912, Switzerland, 1850-1907* (New Jersey: Princeton University Press, 1971).

Schrijver, Nico. *Sovereignty over Natural Resources: Balancing Rights and Duties* (Cambridge: Cambridge University Press, 1997).

Seidl-Hohenveldern, Ignaz. *International Economic Law* (Dordrecht: Martinus Nijhoff, 1989).

Sen, Amartya. *The Political Economy of Hunger* (Oxford: Clarendon Press, 1990).

–. *Poverty and Famine: An Essay on Entitlement and Deprivation* (Oxford: Clarendon Press, 1981).

Sen, Amartya, and John Muellbauer. *The Standard of Living* (Cambridge, UK: Cambridge University Press, 1987).

Shah, S. *Ecology and the Crisis of Overpopulation: Future Prospects for Global Sustainability* (Cheltenham, UK: Edward Elgar, 1998).

Shaw, Malcolm. *International Law*, 2nd ed. (Cambridge: Grotius Publications, 1986).

Sherwood, Robert. *Intellectual Property and Economic Development* (Boulder, CO: Westview Press, 1990).

Shiva, Vandana. *Monocultures of the Mind: Perspectives on Biodiversity and Biotechnology* (New Jersey: Zed Books, 1993).

–. *Staying Alive: Women, Ecology and Development* (London: Zed Books, 1988).

–. *The Violence of the Green Revolution* (London: Zed Books, 1991).

Simmonds, Norman. *Principles of Crop Improvement* (New York: Longman, 1979).

Sinha, Rajiv. *Ethnobotany: The Renaissance of Traditional Herbal Medicine* (Jaipur, India: Ina Shree Publishers, 1996).

Stanley, Autumn. *Mothers and Daughters of Inventions* (New Jersey: Rutgers, 1993).

Stenson, Anthony, and Tim Gray. *The Politics of Genetic Resource Control* (London: Macmillan Press, 1999).

Stevens, Henry Bailey. *The Recovery of Culture* (New York: Harper and Brothers Publishers, 1949).

Takacs, David. *The Idea of Biodiversity: Philosophies of Paradise* (Baltimore: Johns Hopkins University Press, 1996).

Tinbergen, Jan. *Rio: Reshaping the International Order – A Report of the Club of Rome* (New York: E.P. Dutton, 1976).

United Nations. *Global Outlook 2000* (New York: United Nations, 1990).

United Nations Conference on Trade and Development (UNCTAD). *The TRIPs Agreement and Developing Countries* (New York and Geneva: UNCTAD, 1996).

Van Heijnsbergen, P. *International Legal Protection of Wild Fauna and Flora* (Amsterdam: Ohmsha Press, 1997).

Vavilov, Nikolai. *The Origin, Variation, Immunity and Breeding of Cultivated Plants*, trans. K. Chester (New York: Ronald Press, 1951).

Venne, Sharon Helen. *Our Elders Understand Our Rights: Evolving International Law Regarding Indigenous Rights* (Penticton, BC: Theytus Books, 1998).

Ward, Barbara. *The Rich Nations and the Poor Nations* (Toronto: Canadian Broadcasting Corporation, 1961).

–. *Spaceship Earth* (New York: Columbia University Press, 1966).

Wilson, Edward. *The Diversity of Life* (Cambridge, MA: Harvard University Press, 1992).

World Development Report: Development and the Environment (Washington, DC: The World Bank Group, 1992).

WTO Annual Report, 1999 (Geneva: WTO Publications, 1999).

Yamin, Farhana. *The Biodiversity Convention and Intellectual Property Rights* (WWF International Discussion Paper) October 1995.

Young, John. *Sustaining the Earth: The Story of the Environmental Movement: Its Past Efforts and Future Challenges* (Cambridge, MA: Harvard University Press, 1990).

Collections

Barnes, R.H., Andrew Gray, and Benedict Kingsbury, eds. *Indigenous Peoples of Asia* (Ann Arbour, MI: Association for Asian Studies, 1995).

Bhagwati, Jagdish, and Mathias Hirsch, eds. *The Uruguay Round and Beyond: Essays in Honor of Arthur Dunkel* (Ann Arbor, Michigan: The University of Michigan Press, 1999).

Bilderbeek, Simone, Ankie Wijgerde, and Netty van Schaik, eds. *Biodiversity and International Law: The Effectiveness of International Environmental Law* (Amsterdam: IOS Press, 1992).

Brolman, Catherine, René Lefeber, and Marjoleine Zieck, eds. *Peoples and Minorities in International Law* (Dordrecht: Martinus Nijhoff, 1993).

Brush, Stephen, and Doreen Stabinsky, eds. *Valuing Local Knowledge: Indigenous People and Intellectual Property Rights* (Washington/Covelo: Island Press, 1996).

Campbell, Dennis, ed. *International Environmental Law and Regulations*, Vol. 2 (Toronto: John Wiley and Sons, 1997).

Charney, Jonathan I., Donald K. Anton, and Mary Ellen O'Connell, eds. *Politics, Values and Functions: International Law in the 21st Century* (The Hague: Martinus Nijhoff Publishers, 1997).

Chowdhury, Subrata Roy, Erik M.G. Denters, and Paul J.I.M. de Waart, eds. *The Right to Development in International Law* (Dordrecht: Martinus Nijhoff, 1992).

Clarke, W.C., and R.E. Munn, eds. *Sustainable Development of the Biosphere* (Cambridge: Cambridge University Press, 1986).

Correa, Carlos, and Abdulqawi Yusuf, eds. *Intellectual Property and International Trade* (The Hague: Kluwer Law International, 1998).

Crawford, James, ed. *The Rights of Peoples* (Oxford: Clarendon Press, 1988).

Ellen, Roy, Peter Parkes, and Alan Bicker, eds. *Indigenous Environmental Knowledge and Its Transformations: Critical Anthropological Perspectives* (Amsterdam: Harwood Academic Publishers, 2000).

Freestone, D., and E. Hey, eds. *The Precautionary Principle and International Law: The Challenge of Implementation* (The Hague: Kluwer Law International, 1996).

Gadbaw, Michael, and Timothy Richards, eds. *Intellectual Property Rights: Global Consensus, Global Conflict?* (Boulder, CO: Westview Press, 1988).

Glowka, Lyle, Françoise Burhenne-Guilmin, and Hugh Synge, eds. *A Guide to the Convention on Biological Diversity* (Gland, Swit: IUCN, 1994).

Goldman, Michael, ed. *Privatizing Nature: Political Struggles for the Global Commons* (London: Pluto Press, 1998).

Greaves, Tom, ed. *Intellectual Property Rights for Indigenous Peoples: A Source Book* (Oklahoma: Society for Applied Anthropology, 1994).

Hoppers, Catherine A. Odora, ed. *Indigenous Knowledge and the Integration of Knowledge Systems: Towards a Philosophy of Articulation* (Cape Town, South Africa: New Africa Books, 2003).

Hurrell, A., and B. Kingsbury, eds. *The International Politics of the Environment* (Oxford: Clarendon Press, 1992).

Jarvis, Mark, ed. *The Influence of Religion on the Development of International Law* (Dordretch: Martinus Nijhoff Publishers, 1991).

Johnson, Martha, ed. *Lore: Capturing Traditional Environmental Knowledge* (Ottawa: Dene Cultural Institute and the International Development Research Centre, 1992).

Juma, Calestous, and J.B. Ojwang, eds. *In Land We Trust: Environment, Private Property and Constitutional Change* (Nairobi: Initiatives Publishers, 1996).

Kaufman, Les, and Kenneth Mallory, eds. *The Last Extinction,* 2nd ed. (Cambridge, MA: MIT Press, 1993).

Krattiger, Anatole F., Jeffrey A. McNeely, William H. Lesser, and Kenton R. Miller, eds. *Widening Perspectives on Biodiversity* (Gland, Swit.: IUCN, 1994).

Mendelsohn, Oliver, and Upendra Baxi, eds. *The Rights of Subordinated Peoples* (Delhi: Oxford University Press, 1994).

Meyer, George G., Kenneth Blum, and John G. Cull, eds. *Folk Medicine and Herbal Healing* (Springfield, IL: Thomas, 1981).

Moldan, Bedrich, and Suzanne Billharz, eds. *Sustainability Indicators: Report on Indicators of Sustainable Development* (Toronto: John Wiley and Sons, 1997).

Morley, Peter, and Roy Wallis, eds. *Culture and Curing: Anthropological Perspectives on Traditional Medical Beliefs and Practices* (Pittsburgh, PA: University of Pittsburgh Press, 1978).

Shiva, Vandana, ed. *Biodiversity Conservation: Whose Resource? Whose Knowledge?* (New Delhi: India National Trust for Art and Cultural Heritage, 1995).

Singh, Rana, ed. *Environmental Ethics: Discourses, and Cultural Traditions* (Varanasa: The National Geographical Society of India, 1993).

Swanson, Timothy. *Intellectual Property Rights and Biodiversity Conservation: An Interdisciplinary Analysis of the Values of Medicinal Plants* (Cambridge: Cambridge University Press, 1995).

VanderZwaag, David, and Bob Kapanen, eds. *Readings in International Environmental Law* (Halifax: Dalhousie Law School, 1994).

Waart, Paul de, ed. *International Law and Development* (Dordrecht: Martinus Nijhoff Publishers, 1988).

Warren, D.M., Jan Slikkerveer, David Brokensha; and technical editor, Wim H.J.C. Dechering. *The Cultural Dimension of Development: Indigenous Knowledge Systems* (London: Intermediate Technology Publications, 1995).

Weeramantary, C.G., ed. *Human Rights and Scientific and Technological Development* (New York: United Nations Press, 1990).

–. *The Impact of Technology on Human Rights* (New York: United Nations Press, 1993).

Articles from Journals

Abbott, Frederick. "Commentary: The International Intellectual Property Order Enters the 21st Century" (1996) 29 Vanderbilt Journal of Transnational Law 471.

–. "GATT and the European Community: A Formula for Peaceful Coexistence" (1990) 12 Michigan Journal of International of International Law 1.

–. "Protecting First World Assets in the Third World: Intellectual Property Negotiations in the GATT Multilateral Framework" (1989) 22 Vanderbilt Journal of Transnational Law 689.

Ackiron, Evan. "Patents for Critical Pharmaceuticals: The AZT Case" (1991) 17 American Journal of Law and Medicine 145.

Adelman, Martin, and Sonia Badia. "Prospects and Limits of the Patent Provisions in the TRIPS Agreement: The Case of India" (1996) 29 Vanderbilt Journal of Transnational Law 507.

Adler, Jonathan. "The Cartagena Protocol and Biological Diversity: Biosafe or Bio-sorry?" (2000) 12 Georgetown International Environmental Law Review 772.

–. "More Sorry than Safe: Assessing the Precautionary Principle and the Proposed International Biosafety Protocol" (2000) 35 Texas International Law Journal 173.

Afifi, Faryan. "Unifying International Patent Protection: The World Intellectual Property Organization Must Coordinate Regional Patent Systems" (1993) 15 Loyola L.A. International and Comparative Law Journal 453.

Akehurst, Michael. "Custom as a Mere Source of International Law" (1974-75) 47 British Yearbook of International Law 12.

–. "The Hierarchy of the Sources of International Law" (1974-75) 47 British Yearbook of International Law 273.

Alford, W.P. "Don't Stop Thinking About ... Yesterday: Why There Was No Indigenous Counterpart to Intellectual Property Law in Imperial China" (1993) 7 Journal of Chinese Law 3.

–. "Making the World Safe for What? Intellectual Property Rights, Human Rights and Foreign Economic Policy in the Post-European Cold War World" (1996-97) 29 International Law and Politics 135.

Alfredsson, Dudmundur. "The Rights of Indigenous Peoples with a Focus on the National Performance of the Nordic Countries" (1999) 59 Heidelberg Journal of International Law 529.

Allen, Ted. "The Philippine Children's Case: Recognizing Standing for Future Generations" (1993) 6 Georgetown Environmental Law Review 713.

Allot, Phillip. "State Responsibility and the Unmaking of International Law" (1988) 29 Harvard International Law Journal 1.

Allyn, Robert. "More about Plant Patents" (1933) 15 Journal of Patent Office Society 963.

–. "Plant Patent Questions" (1933) 15 Journal of Patent Office Society 180.

Alston, Philip. "Conjuring Up New Human Rights: A Proposal for Quality Control" (1984) 78 American Journal of International Law 607.

–. "Making Space for New Human Rights: The Case of the Right to Development" (1988) 1 Harvard Human Rights Yearbook 3.

Anaya, James. "Environmentalism, Human Rights and Indigenous Peoples: A Tale of Converging and Diverging Interests" (2000) 7 Buffalo Environmental Law Journal 1.

Anuradha, R.V. "In Search of Knowledge and Resources: Who Sows? Who Reaps?" (1997) 6 Review of European Community and International Law 263.

–. "IPRs: Implications for Biodiversity and Local and Indigenous Communities" (2001) 10 Review of European Community and International Environmental 27.

Aoki, Keith. "(Intellectual) Property and Sovereignty: Notes toward a Cultural Geography of Authorship" (1996) 48 Stanford Law Review 1293.

–. "Neocolonialism, Anticommons Property, and Biopiracy in the (Not-so-Brave) New World Order of International Intellectual Property Protection" (1998) 6 Indiana Journal of Global Legal Studies 11.

Archer, Heather. "Effect of United Nations Draft Declaration on Indigenous Rights on Current Policies of Member States" (1999) 5 Journal of International Legal Studies 205

Asebey, Edgar, and Jill Kempenaar. "Biodiversity Prospecting: Fulfilling the Mandate of the Biodiversity Convention" (1995) 28 Vanderbilt Journal of Transnational Law 703.

Ayittey, George. "How the Multilateral Institutions Compounded Africa's Economic Crisis" (1999) 30 Law and Policy in International Business 585.

Backes, Chris, and Jonathan Verschuuren. "The Precautionary Principle in International, European, and Dutch Wildlife Law" (1998) 9 Colorado Journal of Environmental Law and Policy 43.

Baker, Katharine. "Consorting with Forests: Rethinking Our Relationship to Natural Resources and How We Should Value Their Loss" (1995) 22 Ecology Law Quarterly 677.

Bakhashab, Omar. "Islamic Law and the Environment: Some Basic Principles" (1988) 3 Arab Law Quarterly 287.

Bale, Harvey. "Patent Protection and Pharmaceutical Innovation" (1996-7) 29 International Law and Politics 95.

Barron, Brian. "Chinese Patent Legislation in Cultural and Historical Perspective" (1991) 6 Intellectual Property Journal 313.

Bassiouni, Cherif. "A Functional Approach to General Principles of International Law" (1990) 11 Michigan Journal of International Law 768.

Baxter, R.R. "International Law in 'Her Infinite Variety'" (1980) 29 International and Comparative Law Quarterly 549.

–. "Multilateral Treaties as Evidence of Customary International Law" (1965-66) 41 British Yearbook of International Law 278.

Beier, Friedrich-Karl. "The European Patent System"(1981) 14 Vanderbilt Journal of Transnational Law 1.

–. "One Hundred Years of International Cooperation: The Role of the Paris Convention in the Past, Present and Future" (1984) 15 International Review of Industrial Property and Copyright Law 1.

Bently, L., and B. Sherman. "The Ethics of Patenting: Towards a Transgenic Patent System" (1995) 3 Medical Law Review 275.

Bergelson, Joy, Colin B. Purrington, and Gale Wichmann. "Promiscuity in Transgenic Plants" (1998) 395 Nature 25.

Berkey, Judson. "The Regulation of Genetically Modified Foods" (October 1999) ASIL Insights 1.

Boozang, Kathleen. "Western Medicine Opens the Door to Alternative Medicine" (1998) 24 American Journal of Law and Medicine 12.

Bosselman, Klaus. "Plants and Politics: The International Legal Regime Concerning Biotechnology and Biodiversity" (1995) 7 Colorado Journal of International Environmental Law and Policy 111.

–. "The Right to Self-Determination and International Environmental Law: An Integrative Approach" (1997) 1 New Zealand Journal of Environmental Law 1.

Bothe, Michael. "Legal and Non-Legal Norms: A Meaningful Distinction in International Relations?" (1980) 11 Netherlands Yearbook of International Law 65.

Boyle, Alan. "State Responsibility and International Liability for Injurious Consequences of Acts Not Prohibited by International Law: A Necessary Distinction?" (1990) 39 International and Comparative Law Quarterly 1.

Bradley, Curtis. "Biodiversity and Biotechnology" (1995) 7 Colorado Journal of International Environmental Law and Policy 107.

Brenner-Beck, Dru. "Do As I Say, Not As I Did" (1992) 11 Pacific Basin Law Journal 84.

Brownlie, Ian. "Legal Status of Natural Resources in International Law" (1979) 162 Recueil de Cours 245.

Brunee, Jutta. "'Common Interest': Echoes from an Empty Shell? Some Thoughts on Common Interest and International Environmental Law" (1987) 49 Heidelberg Journal of International Law 791.

Burhenne-Guilmin, Francoise, and Susan Casey-Lefkowitz. "The Convention on Biological Diversity: A Hard Won Global Achievement" (1992) 3 Yearbook of International Environmental Law 44.

Butterton, Glenn. "Pirates, Dragons and US Intellectual Property Rights in China: Problems and Prospects of Chinese Enforcement" (1996) 38 Arizona Law Review 1081.

Caillaux, Jorge. "Biological Resources and the Convention on Biological Resources" (1994) 1 Journal of Environmental Law and Policy in Latin America and the Caribbean 9.

Cameron, James, and Juli Abouchar. "The Precautionary Principle: A Fundamental Principle of Law and Policy for the Protection of the Global Environment" (1991) 14 Boston College International and Comparative Law Review 1.

Carroll, Amy. "Not Always the Best Medicine: Biotechnology and the Global Impact of US Patent Law" (1994-95) 44 American University Law Review 2433.

Catanese, Adrienne. "Paris Convention, Patent Protection, and Technology Transfer" (1985) 3 Boston University International Law Journal 209.

Cate, Fred. "Sovereignty and the Globalization of Intellectual Property" (1998) 6 Indiana Journal of Global Legal Studies 1.

Chandler, Melinda. "The Biodiversity Convention: Selected Issues of Interest to the International Lawyer" (1993) 4 Colorado Journal of International Environmental Law and Policy 141.

Chapman, Bruce. "Rational Environmental Choice: Lessons for Economics from Law and Ethics" (1993) 6 Canadian Journal of Law and Jurisprudence 63.

Charney, Jonathan. "Biodiversity: Opportunities and Obligations" (1995) 28 Vanderbilt Journal of Transnational Law 613.

Chen, Jim. "Globalization and Its Losers" (2000) 9 Minnesota Journal of Global Trade 157.

Chibundu, Maxwell. "Law in Development: On Tapping, Gourding, and Serving Palm-Wine" (1997) 29 Case Western Reserve Journal of International Law 167.

Chinkin, Christine. "The Challenge of Soft Law: Development and Change in International Law" (1989) 38 International and Comparative Law Quarterly 580.

Chisum, Donald. "Foreign Activity: Its Effect on Patentability under United States Law" (1980) 11 International Review of Industrial Property and Copyright Law 26.

Christie, Edward. "The Eternal Triangle: The Biodiversity Convention, Endangered Species Legislation and the Precautionary Principle" (1993) 10 Environmental and Planning Law Journal 470.

Christol, Carl. "The Common Heritage of Mankind Provision in the 1979 Agreement Governing the Activities of States on the Moon and Other Celestial Bodies" (1980) 14 International Lawyer 429.

Churchill, Jane. "Patenting Humanity: The Development of Property Rights in the Human Body and the Subsequent Evolution of Patentability of Living Things" (1994) 8 Intellectual Property Journal 250.

Cocca, A. A. "Mankind as the New Legal Subject: A New Juridical Dimension Recognized by the United Nations" (1971) Proceedings of the 13th Colloquium on the Law of Outer Space 211.

Cohen, Morris. "Property and Sovereignty" (1927-28) 13 Cornell Law Quarterly 8.

Coombe, Rosemary. "Challenging Paternity: Histories of Copyright" (1994) 6 Yale Journal of Law and the Humanities 407.

–. "Critical Cultural Legal Studies" (1998) 10 Yale Journal of Law and the Humanities 463.

–. "The Cultural Life of Things: Anthropological Approaches to Law and Society in Conditions of Globalization" (1995) 10 American University Journal of International Law and Policy 791.

–. "Intellectual Property, Human Rights and Sovereignty: New Dilemmas in International Law Posed by the Recognition of Indigenous Knowledge and the Conservation of Biodiversity" (1998) 6 Indiana Journal of Global Legal Studies 59.

–. "The Properties of Culture and the Politics of Possessing Identity: Native Claims in the Cultural Appropriation Controversy" (1993) 6 Canadian Journal of Law and Jurisprudence 249.

Crawford, James. "Revising the Draft Articles on State Responsibility" (1999) 10 European Journal of International Law 435.

Cullet, Philippe. "Plant Variety Protection in Africa: Towards Compliance with TRIPS Agreement" (2001) 45 Journal of African Law 97.

Currier, Andrew. "To Publish or to Patent, That Is the Question" (2000) 16 Canadian Intellectual Property Review 337.

Date, Vandana. "Global 'Development' and Its Environmental Ramifications: The Interlinking of Ecologically Sustainable Development and Intellectual Property Rights" (1997) 27 Golden Gate University Law Review 631.

Davies, Seaborne. "The Early History of the Patent Specification" (Part 1) (1934) 50 Law Quarterly Review 86.

–. "The Early History of the Patent Specification" (Part 2) (1934) 50 Law Quarterly Review 92.

Davis, Michael. "The Patenting of Products of Nature" (1995) 21 Rutgers Comparative Technology Law Journal 293.

De Klemm, Cyril. "The Convention on Biological Diversity: State Obligations and Citizens' Duties" (1989) 19 Environmental Policy and Law 50.

Dominice, Christian. "The International Responsibility of States for Breach of Multilateral Obligations" (1999) 10 European Journal of International Law 353.

Doorman, Gerald. "Patent Law in the Netherlands: Suspended in 1869 and Re-established in 1910" (1948) 30 Journal of the Patent Office Society 225.

Doremus, Holly. "Patching the Ark: Improving Legal Protection of Biological Diversity" (1991) 18 Ecology Law Quarterly 265.

Drahos, Peter. "Global Property Rights in Information: The Story of TRIPs at the GATT" (1995) 13 Prometheus 12.

–. "Indigenous Knowledge and the Duties of Intellectual Property Owners" (1997) 11 Intellectual Property Journal 179.

–. "Indigenous Knowledge, Intellectual Property and Biopiracy: Is a Global Bio-Collecting Society the Answer?" [Opinion] (2000) 6 European Intellectual Property Review 245.

Dupuy, Pierre-Marie. "The International Law of State Responsibility: Revolution or Evolution?" (1989) 11 Michigan Journal of International Law 105.

–. "Reviewing the Difficulties of Codification: On Ago's Classification of Obligations of Means and Obligations of Result in Relation to State Responsibility" (1999) 10 European Journal of International Law 371.

–. "Soft Law and the International Law of the Environment" (1991) 12 Michigan Journal of International Law 420.

Dutfield, G. "TRIPs-Related Aspects of Traditional Knowledge" (2001) 33 Case Western Reserve Journal of International Law 233.

Dyzenhaus, David. "The Legitimacy of Legality" (1996) 46 University of Toronto Law Journal 129.

Edgar, Craig. "Patenting Nature: GATT on a Hot Tin Roof" (1994) 34 Washburn Law Journal 76.

Eisenberg, R. "Patenting the Human Genome" (1990) 39 Emory Law Journal 721.

–. "Proprietary Rights and the Norms of Science in Biotechnology Research" (1989) 21 Intellectual Property Law Review 29.

Erstling, Jay. "The Protection of Intellectual Property: Of Metaphysics, Motivation, and Monopoly" (1991) 3 Sri Lanka Journal of International Law 51.

–. "The Role of Licensing and Technology Transfer in the Protection and Management of Intellectual Property: A Developing Country Perspective" (1993) 5 Sri Lanka Journal of International Law 21.

Ezejiofor, Gaius. "The Law of Patents: A Review" (1973-4) 9 African Law Studies 39.

Falk, Richard. "Toward a World Order Respectful of the Global Ecosystem" (1991-92) 19 Boston College Environmental Affairs Law Review 711.

Franck, Thomas, and Steven Hawkins. "Justice in the International System" (1989) 10 Michigan Journal of International Law 127.

Frowein, Jochen. "The Internal and External Effects of Resolutions by International Organizations" (1989) 49 Heidelberg Journal of International Law 778.

Fuller, Lon. "Human Interaction and the Law" (1969) 14 American Journal of Jurisprudence 1.

–. "Law as an Instrument of Social Control and Law as a Facilitation of Human Interaction" (1975) 1 Brigham Young University Law Review 89.

Gadbaw, Michael. "Intellectual Property and International Trade: Merger or Marriage of Convenience?" (1989) 22 Vanderbilt Journal of Transnational Law 223.

Gana, Ruth. "Has Creativity Died in the Third World? Some Implications of the Internationalization of Intellectual Property" (1995) 24 Denver Journal of International Law and Policy 109.

–. "Prospects for Developing Countries under the TRIPs Agreement" (1996) 29 Vanderbilt Journal of Transnational Law 735.

Gardner, R.C. "Diverse Opinions on Biodiversity" (1999) 6 Tulsa Journal of Comparative and International Law 303.

Gathii, James Thuo. "The Legal Status of the Doha Declaration on TRIPs and Public Health under the Vienna Convention of the Law of Treaties" (2002) 15 Harvard Journal of Law and Technology 291.

–. "A Market Based Approach to Construing Articles 7 and 8 of TRIPs in the Context of AIDS Drugs" (2001) 53 Florida Law Review 727.

–. "Rights, Patents, Markets and the Global AIDS Pandemic" (2002) 14 Florida Journal of International Law 261.

Gaubatz, Kurt Taylor, and Matthew MacArthur. "How International Is 'International Law?'" (2001) 22 Michigan Journal of International Law 239.

Geisinger, Alex. "Sustainable Development and the Domination of Nature: Spreading the Seed of the Western Ideology of Nature" (1999) 27 Environmental Affairs 43.

Ghosh, Subha. "Globalization, Patents and Traditional Knowledge" (2003) 17 Columbia Journal of Asian Law 73.

Golden, John. "Biotechnology, Technology Policy, and Patentability: Natural Products and Invention in the American System" (2001) 50 Emory Law Journal 101.

Gollin, Michael. "Using Intellectual Property to Improve Environmental Protection" (1991) 4 Harvard Journal of Law and Technology 193.

Goodhart, A.L. "The Migration of the Common Law" (1960) 76 Law Quarterly Review 39.

Gorove, Stephen. "The Concept of 'Common Heritage of Mankind': A Political, Moral or Legal Innovation?" (1972) 9 San Diego Law Review 390.

Gratwick, Stephen. "Having Regard to What Was Known and Used" (1972) 88 Law Quarterly Review 341.

–. "'Having Regard to What Was Known and Used': Revisited" (1986) 102 Law Quarterly Review 403.

Graziano, Karen. "Biosafety Protocol: Recommendations to Ensure the Safety of the Environment"(1995) 7 Colorado Journal of International Environmental Law and Policy 179.

Greenfield, Michael. "Recombinant DNA Technology: A Science Struggling with the Patent Law" (1993) 25 Intellectual Property Law Review 135.

Greif, Siegfried. "Patents and Economic Growth" (1987) 18 International Review of Industrial Property and Copyright Law 191.

Grote, Rainer. "The Status and Rights of Indigenous Peoples in Latin America" (1999) 59 Heidelberg Journal of International Law 497.

Gruchalla-Wisierski, T. "A Framework for Understanding 'Soft Law'" (1984) 30 Revue de Droit de McGill 37.

Guest, Richard. "Intellectual Property Rights and Native American Tribes" (1995-96) 20 American Indian Law Review 111.

Gundling, Lothar. "The Status in International Law of the Principle of Precautionary Action" (1990) 5 Journal of Estuarine and Coastal Law 23.

Gutowski, Robert. "The Marriage of Intellectual Property and International Trade in the TRIPs Agreement: Strange Bedfellows or a Match Made in Heaven?" (1999) 47 Buffalo Law Review 713.

Gutterman, Alan. "The North-South Debate Regarding the Protection of Intellectual Property Rights" (1993) 28 Wake Forest Law Review 89.

Halewood, Michael. "Indigenous and Local Knowledge in International Law: A Preface to Sui Generis Intellectual Property Protection" (1999) 44 McGill Law Journal 953.

Hamilton, Marci. "The TRIPs Agreement: Imperialistic, Outdated, and Overprotective" (1996) 29 Vanderbilt Journal of Transnational Law 613.

Hamilton, Neil. "Why Own the Farm If You Can Own the Farmer (and the Crop)? Contract Production and Intellectual Property Protection of Grain Crops" (1994) 73 Nebraska Law Review 91.

Hey, Ellen. "The Precautionary Concept in Environmental Policy and Law: Institutionalizing Caution" (1992) IV Georgetown International Environmental Law Review 303.

Hickey, James Jr., and Vern Walker. "Refining the Precautionary Principle in International Environmental Law" (1995) 14 Virginia Environmental Law Journal 423.

Mgbeoji, Ikechi. "Beyond Rhetoric: State Sovereignty, Common Concern, and the Inapplicability of the Common Heritage Concept to Plant Genetic Resources" (2003) 16 Leiden Journal of International Law 821.

–. "The Juridical Origins of the International Patent System: Towards a Historiography of the Role of Patents in Industrialization" (2003) Vol. 5, No. 2 Journal of the History of International Law 403.

–. "Patents and Traditional Knowledge of The Uses of Plants: Is a Communal Patent Regime Part of the Solution to the Scourge of Biopiracy?" (2001) 9 Indiana Journal Of Global Legal Studies 163.

–. "Patent First, Litigate Later! The Scramble for Speculative and Overly Broad Genetic Patents: Implications for Access to Health Care and Bio-medical Research" (2003) Vol. 2, No. 2 Canadian Journal of Law and Technology 83.

–. "The 'Terminator' Patent and its Discontents: Rethinking the Normative Deficit in Utility Test of Modern Patent Law" (2004) 17 St. Thomas Law Review 96.

–. "(Under)Mining the Seabed? Between the International Seabed Authority Mining Code and Sustainable Bioprospecting of Hydrothermal Vent Ecosystems in the Seabed Area: Taking Precaution Seriously" (2004) 18 Ocean Yearbook 413.

Mutua, Makau wa. "Savages, Victims and Saviours: The Metaphor of Human Rights" (2001) 42 Harvard International Law Journal 201.

Myers, Norman. "The Hamburger Connection: How Central America's Forests become North America's Hamburgers" (1981) 10 Ambio 3.

Naganathan, Nirmala. "What Constitutes Custom in International Law?" (1971) 1 Colombo Law Review 68.

Oddi, Samuel. "Beyond Obviousness: Invention Protection in the Twenty-First Century" (1989) 38 American University Law Review 1097.

–. "The International Patent System and Third World Development: Reality or Myth?" (1987) 63 Duke Law Journal 831.

–. "TRIPs: Natural Rights and a 'Polite' Form of Economic Imperialis" (1996) 29 Vanderbilt Journal of Transnational Law 415.

–. "Un-Unified Theories of Patents: The Not-Quite-Holy Grail" (1996) 71 Notre Dame Law Review 267.

Oguamanam, Chidi. "Between Reality and Rhetoric: The Epistemic Schism in the Recognition of Traditional Medicine in International Law" (2003) 16 St. Thomas Law Review 59.

–. "Localizing Intellectual Property in the Globalization Epoch: The Integration of Indigenous Knowledge" (2004) 11 Indiana Journal of Global Legal Studies 135.

–. "The Protection of Traditional Knowledge: Towards a Cross-Cultural Dialogue on Intellectual Property Rights" (2004) 15 Australian Intellectual Property Journal 34.

Okafor, Obiora Chinedu. "The Status and Effect of the Right to Development in Contemporary International Law: Towards a South-North 'Entente'" (1995) 7 RADIC 865.

Onwuekwe, Chika. "The Commons Concept and Intellectual Property Rights Regime: Whither Plant Genetic Resources and Knowledge?" (2004) 2 Pierce Law Review 65.

Onyejekwe, Kele. "GATT, Agriculture, and Developing Countries" (1993) 17 Hamline Law Review 77.

Plant, Arnold. "The Economic Theory Concerning Patents for Inventions" (1934) 1 Economic 67.

Reichman, J.H. "From Free Riders to Fair Followers: Global Competition under the TRIPS Agreement" (1996-97) 29 International Law and Politics 11.

–. "Intellectual Property in International Trade: Opportunities and Risks of a GATT Connection" (1989) 22 Vanderbilt Journal of Transnational Law 747.

–. "The TRIPs Agreement Comes of Age: Conflict or Cooperation With the Developing Countries" (2000) 32 Case Western Reserve Journal of International Law 441.

Richardson, Benjamin. "Environmental Law in Post-Colonial Societies: Straddling the Local-Global Institutional Spectrum" (2000) 11 Colorado Journal of International Environmental Law and Policy 1.

–. "Indigenous Peoples, International Law and Sustainability" (2001) 10 Review of European Community and International Environmental Law 1.

Roht-Arriaza, Naomi. "Of Seeds and Shamans: The Appropriateness of the Scientific and Technical Knowledge of Indigenous and Local Communities" (1996) 17 Michigan Journal of International Law 940.

Sagoe, T. Ekua. "Industrial Property Law in Nigeria" (1992) 14 Comparative Law Yearbook of International Business 312.

Saigo, Holly. "Agricultural Biotechnology and the Negotiation of the Biosafety Protocol" (2000) 12 Georgetown International Environmental Law Review 779.

Sands, Philippe. "International Law in the Field of Sustainable Development" (1994) 55 British Yearbook of International Law 303.

Sarma, Lakshmi. "Biopiracy: Twentieth Century Imperialism in the Form of International Agreements" (1999) 13 Temple International and Comparative Law Journal 107.

Schachter, Oscar. "The Evolving International Law of Development" (1976) 15 Columbia Journal of Transnational 1.

–. "Recent Trends in International Law Making" (1992) 12 Australian Yearbook of International Law 1.

Schermers, Henry. "Some Recent Cases Delaying the Direct Effect of International Treaties in Dutch Law" (1989) 10 Michigan Journal of International Law 266.

Scott, Craig. "Reaching Beyond (without Abandoning) the Category of 'Economic, Social and Cultural Rights'" (1999) 21 Human Rights Quarterly 633.

Seidl-Hohenveldern, Ignatius. "Transformation or Adoption of International Law into Municipal Law" (1963) 12 International and Comparative Law Quarterly 88.

Seita, Alex. "Globalization and the Convergence of Values" (1997) 30 Cornell International Law Journal 429.

Sherwood, Robert. "Human Creativity for Economic Development: Patents Propel Technology" (2000) 33 Akron Law Review 1.

–. "The TRIPs Agreement: Implications for Developing Countries" (1996-7) 37 Journal of Law and Technology 491.

Sherwood, Robert M., Vanda Scartezini, and Peter Dirk Siemsen. "Promotion of Inventiveness in Developing Countries through a More Advanced Patent Administration" (1999) 39 IDEA: The Journal of Research and Technology 473.

Slater, David. "Contesting Occidental Visions of the Global: The Geopolitics of Theory and North-South Relations" (1994) 4 Beyond Law (Bogota) 97.

Spaulding, Norman III. "Commodification and Its Discontents: Environmentalism and the Promise of Market Incentives" (1997) 16 Stanford Environmental Law Journal 293.

Stedman, John. "The Employed Inventor, the Public Interest, and Horse and Buggy Law in the Space Age" (1970) 45 New York University Law Review 1.

Tejera, Valentina. "Tripping over Property Rights: Is it Possible to Reconcile the Convention on Biological Diversity with Article 27 of the TRIPS Agreement?" (1999) 33 New England Law Review 976.

VanderZwaag, David. "The Precautionary Principle in Environmental Law and Policy: Elusive Rhetoric and First Embraces" (1999) 8 Journal of Environmental Law and Policy 355.

Van Heijnsbergen, P. "Biodiversity and International Law" (1991) 45 International Spectator 681.

Vaver, David. "Intellectual Property Today: Of Myths and Paradoxes" (1990) 69 Canadian Bar Review 98.

Vavilov, Nikolai. "Studies on the Origin of Cultivated Plants" (1925) Vol. 16, No. 2 Bulletin of Applied and Plant Breeding 1.

Vazquez, Carlos Manuel. "The Four Doctrines of Self-Executing Treaties" (1995) 89 AJIL 695.

Wolfrun, Rudiger. "The Principle of the Common Heritage of Mankind" (1983) 43 ZaorRv-Heidelberg Journal of International Law 312.

–. "The Protection of Indigenous Peoples in International Law" (1999) 59 ZaorRv-Heidelberg Journal of International Law 369.

Yusuf, Abdulqawi. "International Law and Sustainable Development: The Convention on Biological Diversity" (1995) 1 African Yearbook of International Law 109.

Zimmerman, K. "The Deforestation of the Brazilian Amazon: Law, Politics, and International Cooperation" (1990) 21 Inter-American Law Review 513.

Decided Cases and Arbitral Rulings

Advisory Opinion No. 17, Interpretation of the Convention between Greece and Bulgaria Respecting Reciprocal Emigration (1930) P.C.I.J (ser. B) No. 17, at 32 (July 31).

Aegean Sea Continental Shelf Case, (1978) ICJ Reports 3.

Alfred Snapp and Son, Inc. v. *Puerto Rico* (1982) 458 U.S. 592.

Anglo-Norwegian Fisheries Case, ICJ Reports (1951) 131.

Application for Review of Judgment No. 273 of the United Nations Administrative Tribunal (Advisory Opinion of 20 July 1982) ICJ Reports (1982) 386.

Arab Monetary Fund v. *Hashim* [1991] 1 All E.R. 316.

Asylum Case, ICJ Reports (1950) 266.

Atlantic Works v. *Brady* (1882) 107 U.S. 192, 200.

Attorney-General for Canada v. *Attorney-General for Ontario* [1937] A.C. 326.

Attorney-General v. *Adelaide Steamship Co.,* [1913] A.C. 781.

Bainbridge v. *Wigley* (1810) 171 ER 636.

Barcelona Traction (Belgium v. *Spain),* (1970) I.C.J. 273-74.

Belilos v. *Switzerland,* 132 Eur. Ct. H.R. (Ser. A) (1988).

Bonito Boats Inc. v. *Thunder Craft Boats, Inc.,* 489 U.S. 141.

Boulton and Watt v. *Bull* (1795) 126 ER 651.

Brenner v. *Manson,* 383, U.S. 519, 148 USPQ 689 (1966).

Carter Prods, Inc., v. *Colgate-Palmolive Co.,* 130 F. Supp557, 104 U.S.P.Q. (BNA) 314 (D. Md. 1955).

Case Concerning the Gabcikovo-Nagymaros Project (Hungary v. *Slovakia)* (1996) ICJ Reports 241 at para 29.

The Case of Monopolies, 77 E.R. 1263.

Case of Monopolies, Darcy v. *Allin* (1602) 77 ER 1260 at 1266 (KB).

The Charming Betsy [1804] 2 Cranch 64.

Certain Expenses of the United Nations (Advisory Opinion), ICJ Reports (1962) 151.

Charan Lal Sahu v. *Union of India* (2000) Vol. 118 International Law Reports 451.

Chorzow Factory Case, 1928 P.C.I.J. (Ser. A) No. 17, at 27-8 (13 September).

Ciba-Geigy, Technical Board of Appeal 3.3.1., 26 July 1983 (T 49/83).

Clothesworkers of Ipswich Case (1615) Godbolt 252.

Competence of the General Assembly for Admission of a State to the United Nations (1950) ICJ Reports 4.

Consolboard Inc. v. *MacMillan Bloedel (Sask.) Ltd.,* (1981) 56 C.P.R. (2d) 145 at 154.

Council Decision Concerning the Conclusion on Behalf of the European Community, as Regards Matters within Its Competence, of the Agreements Reached in the Uruguay Round of Multilateral Negotiations (1986-94) (94/800/EC), OJ No L 336/ 23 December 1994 at 2.

Dawson v. *Follen,* 7 F. Cas. 216 (C.C.D. Pa. 1808).

Delimitation of the Continental Shelf between the United Kingdom and France, 18 United Nations Report of International Arbitral Awards 3 (1979).

Diamond v. *Chakrabarty,* 447 U.S. 303 (1980).

Diversion of Water from the Meuse Case (Netherlands v. *Belgium),* 1937 P.C.I.J. (Ser. A/B) No. 70, at 4. (Reproduced in Lakshman Guruswamy et al., eds., *Supplement of Basic Documents to International Environmental Law and World Order* (Saint Paul, MN: West Publishing Co., 1994) at 1232.

Donaldson v. *Beckett* (1774) 17 *Parliamentary History* col. 999.

Ex parte Hibbard, 227 U.S.P.Q. 443 (1985).

Ex parte Lartimer, 1889 December Comm'r Pat. 123.

Fender v. *St. John-Mildmay* [1938] A.C. 1.

Foster v. *Neilson,* 27 US 253 (1829).

GA Mayras Case, [1972] ECR 1219.

Gayler v. *Wilder* 51 U.S. (10 How.) 477 (1850).

General Electric Co. v. *DeForest* Radio Co. 28 F. 2d 641 (3d Cir. 1928) cert. denied, 278 U.S. 656 (1929).

General Tire and Rubber Co. v. *Phillips Petroleum Co.,* [1967] SCR 664 at 671.

Guinea/Guinea Bissau Maritime Delimitation Case. (Cited in Ian Brownlie, *The Human Right to Development,* London, Commonwealth Secretariat (Occasional Paper Series) 1989, at 16-17.

Icelandic Fisheries Case, ICJ Reports (1974) at 3.

In re Bergstrom, 427 F.2d 1394, 166 USPQ 256, 262 (C.C.P.A. 1970).

In re Lundak, 773 F.2d 1216, 1220 n.1 (Fed. Cir. 1985).

In re The Uruguay Round Treaties Opinion [1995] 1 C.M.L.R 205 of 15 November 1994.

International Fruit Company v. *Produktschap voor Groenten en Fruit* (Case 21-24/72); Preliminary Ruling of 12 December 1972; [1972] ECR 1219.

Interpretation of the Agreement of 25 March 1951 Between the WHO and Egypt (1980) ICJ Reports 67.

Jean-Baptiste Caire v. *United Mexican States,* (1929) 5 R.I.A.A. 516.

Jefferys v. *Boosey* (1854) 10 ER 702.

Juan Antonio Oposa, et al. v. *The Honourable Fulgencio Factoran Jr., Secretary of the Department of the Environment and Natural Resources et. al., Supreme Court of the Philippines,* 33 I.L.M 174 (1994).

Jurisdiction of the Courts of Dantzig, PCIJ (1928), Ser. B., No.5.

Kim Bros. v. *Haglen* (1958) 120 U.S.P.Q. 210.

Kuehmsted v. *Farbenfabriken of Eberfield Co.,* 179 F. 701, 704-05 (7th Cir. 1910).

Lane Fox v. *Kensington and Knightsbridge Electric Lighting Co.,* [1892] 3 Ch. 424 at 428.

Legality of the Threat of Use of Nuclear Weapons (1996) ICJ Reports 241.

Lotus Case (1927) PCIJ Series A, No. 10.

Lubrizol [hybrid plants] Technical Board of Appeal 3.3.2., 10 November 1988 (T 320/87).

Merck and Co. v. *Apotex Inc.,* (1994), 59 CPR (3d) 133 (Fed. TD) at 178.

Merck and Co. v. *Olin Mathieson Chem. Corp.,* 253 F. 2d 156 (4th Cir. 1958).

Menthonthiole [1978] G.R.U.R. 702.

Merrell Dow Pharmaceuticals Inc. v. *HN Norton* [1996] Report of Patent Cases 76.

Milpurrurru v. *Indofurn (Pty) Ltd.,* (1995) 30 I.P.R. 209 at 216.

National Research Development Corp. v. *Commissioner of Patents* (1959) 102 C.L.R.252.

Nicaragua v. *United States,* Merits, ICJ Reports (1986) 138.

North Sea Continental Shelf, Judgment, ICJ Reports (1969) 43.

Nuclear Tests Case (1974) ICJ Reports 253.

Paquette Habana 175 U.S. 677 (1900).

Parke-Davis v. *Mulford Co.,* 189 F. 95 (C.C.S.D.N.Y. 1911), *modified,* 196 F.496 (2d Cir. 1912).

The Parlement Belge, 4 P.D. 129 (1879).

Pioneer Hi-Bred International Inc. v. *J.E.M. Supply Inc.,* 200 F.3d 1374, 1376 (Fed. Cir. 2000).

Pioneer Hi-Bred Ltd. v. *Commissioner of Patents,* (1989), 25 CPR (3d) 257.

Pitts v. *Bull* (1851) 2 Black W 237.

President and Fellows of Harvard College v. *Commissioner of Patents,* 79 CPR (3d) (FCTD) 1998.

Rainbow Warrior Incident (1985) *International Law Reports* 74.

Rayner v. *Department of Trade,* [1990] 2A.C. 419.

Red Dove (Rote Taubel), BGHZ 52, 74, 72 GRUR 772 (1969).

Reparation for Injuries Suffered in the Service of the United Nations (1949) *ICJ Reports* 174.

Reservations on the Convention on Genocide (1951) ICJ Reports 15.

Richardson v. *Mellish* 130 English Reports 294 at 303.

Rights of Passage Case (1960) ICJ Reports 6.

Rose Breeding, 67 GRUR 577 (1962).

Samuel Parkes and Co. Ltd. v. *Cocker Bros., Ltd.,* (1929) 46 R.P.C. 248.

Scripps Clinic and Research Foundation v. *Genentech Inc.,* 666 F. Supp. 1379, 3 U.S.PQ 2d 1481 (N.D. Calif. 1987).

Society for the Propagation of the Gospel in Foreign Parts v. *New Haven*, 8 Wheat. 464 (1823) at 490.

South West Africa Cases (1960) ICJ Reports 6.

South West Africa Cases, 1CJ Reports (1962) 310.

South West Africa/Namibia (Advisory Opinion) ICJ Reports (1971) 16.

Suramerica de Aleaciones Laminadas, C.A. v. *United States*, 966 F 2d. 660 (Federal Circuit, 1992).

Tetraploide Kamille, (BGH GRUR 1993, 651).

Trail Smelter Arbitration (*US* v. *Canada*) 3 UNRIAA 1905. Reprinted in Sands et al., at 87.

United States v. *Percham*, 32 US (7 Pet.) 51 (1833).

Volkwagen A.G. v. *Schlunk*, 486 U.S. 694, 700 (1988).

Voting Procedure Case, (1955) I.C.J. Reports 120.

Werner and Leifer Case, C-70/94 [1995] ECR 1-3189.

Yoder Bros. Inc. v. *California-Florida Plant Corp.*, 537 F.2d 1347, 1378, 193 U.S.P.Q. 264, 291 n.34 (5th Cir. 1976).

International Treaties, Conventions, Agreements, and Declarations

African Convention on the Conservation of Nature and Natural Resources 1968, 1001 U.N.T.S. 4.

Agreement Concerning the Activities of States on the Moon and other Celestial Bodies, December 5, 1979, (1979) 18 I.L.M. 1434.

Agreement Concerning the Establishment of an International Patents Bureau, The Hague, 6 June 1947, reproduced in Treaty Series No. 84 (1965).

Agreement Establishing the World Trade Organization, (1994) 33 ILM 1.

Agreement on Andean Subregional Integration, 26 May 1969, reprinted in 8 I.L.M. 910 (1969).

Agreement Related to Trade-Related Aspects of Intellectual Property Rights, 33 I.L.M. 1197 (1994).

Agreement Revising the Agreement Signed at the Hague on 6 June 1947, Concerning the Establishment of an International Patents Bureau, with Protocol and Resolution, The Hague, 16 February-31 December 1961, reproduced in J.W. Baxter, ed., *World Patent Law and Practice*, Vol. 2A (London: Sweet and Maxwell, 1976) at 310.

Antarctic Treaty, concluded at Washinton, 1 December 1959. Entered into force, 23 June 1961. 40 U.N.T.S. 71.

The ASEAN Agreement on the Conservation of Nature and Natural Resources 1985, 15 E.P.L. 2.

The Association of South-East Asia Nations Agreement on the Conservation of Nature and Natural Resources, 9 July 1955. Reprinted in Rummel-Bulska, I., and S. Osafo, eds, *Selected Multilateral Treaties in the Field of Environment*, Vol. 2 (Nairobi: United Nations Environment Programme, 1991).

Cartagena Protocol on Biosafety, 23 February 2000, Report of Panel IV: Consideration of the Need for Modalities of Protocol Setting Out Appropriate Procedures Including, in Particular, Advance Informed Agreements in the Field of the Safe Transfer, Handling and Use of Any Living Modified Organism Resulting from Biotechnology Diversity, UNEP Arguments for a Protocol Pursuant to Article 19.3 of the Convention, UN Doc. UNEP/Bio.Div/Panels/Int.4 (1993).

Charter of the United Nations (As Amended) Concluded at San Francisco, 26 June 1945. Entered into force, 24 October 1945. 1 U.N.T.S. 16 or (1976) Y.B.U.N.1043.

China-US: Bilateral Agreements Regarding Intellectual Property Rights. Reprinted in 34 I.L.M 881 (1995).

Constitution of the International Labour Organization (Without Annex but as Revised through 9 October 1946) Concluded at Versailles, 28 June 1919. Entered into force, 10 January 1920. 15 U.N.T.S. 35.

Convention Establishing the World Intellectual Property Organization, opened for Signature 14 July 1967. Reprinted in 828 U.N.T.S. 3.

Convention on Biological Diversity, 29 December 1993, reprinted in 31 I.L.M. 813. (1992) Entered into force 29 December 1993.

Convention on International Trade in Endangered Species of Wild Fauna and Flora, 3 March 1973, 993 U.N.T.S. 243.

Convention on Nature and Wildlife Preservation in the Western Hemisphere, (1940) 161 U.N.T.S 193.

Convention on the Conservation of European Wildlife and Natural Habitats, 19 September 1979, Berne, European Treaty Series No. 104; (1982) U.K.T.S. 56.

Convention on the Grant of the European Patents, 5 October 1973, 1160 U.N.T.S. 231.

Convention on the Protection of the Ozone Layer, 22 March 1985, Vienna. 26 I.L.M. 1529 (1987).

Convention on the Unification of Certain Points of Substantive Law on Patents for Invention, 27 November 1963, European Treaty Series No. 47.

Convention on Wetlands of International Importance, especially as Waterfowl Habitat, Ramsar, 996 U.N.T.S. 245.

Convention Relative to the Preservation of Fauna and Flora in their Natural State, 172 L.N.T.S 241.

Declaration of Principles Governing the Sea-Bed and the Ocean Floor, and the Subsoil Thereof, Beyond the Limits of National Jurisdiction. Adopted by the United Nations General Assembly, 17 December 1970. UN Doc. A/RES/2749 (XXV), 10 I.L.M. 220 (1971).

Declaration of Principles of International Law and Concerning Friendly Relations and Co-operation among States in Accordance with the Charter of the United Nations. Adopted by the UN General Assembly, 24 October 1970. G.A. Res. 2625, UN GAOR, 25th Sess., Supp. No. 28. at 121, UN Doc.A/8028 (1971) 9 I.L.M 1292 (1970).

European Convention on the International Classification of Patents for Invention, Paris, 19 December 1954, reproduced in Treaty Series No. 12 (1963).

European Convention Relating to the Formalities Required for Patent Applications, Treaty Series No. 43 (1955).

Final Act Embodying the Results of the Uruguay Round of Multilateral Trade Negotiations, Agreement Establishing the World Trade Organization, 15 April 1994, 33 I.L.M. 1125 (1994).

General Agreement on Tariffs and Trade, opened for signature 30 October 1947. 55 U.N.T.S. 187.

International Convention for the Protection of Plants, 16 April 1929, reprinted in 126 L.N.T.S. 305 (1931-32).

International Covenant on Economic, Social and Cultural Rights, 16 December 1966, 993 U.N.T.S. 3.

International Labour Organization Convention Concerning Indigenous and Tribal Peoples in Independent Countries, reprinted in 28 I.L.M. 1382 (1989).

International Plant Protection Convention, 5 December 1951, reprinted in 150 U.N.T.S. 67.

London Convention of 1933, 172 L.N.T.S. 241.

Montevideo Convention on Rights and Duties of States, 26 December 1933, 165 L.N.T.S 19.

Montreal Protocol on Substances that Deplete the Ozone Layer, 26 I.L.M. 1550 (1987).

Non-Legally Binding Authoritative Statement of Principles for a Global Consensus on the Management, Conservation, and Sustainable Development of all Types of Forests. Adopted by the UN Conference on Environment and Development at Rio de Janeiro, 13 June 1992. UN Doc. A/CONF. 151/26 (Vol. III) (1992), 31 I.L.M. 881 (1992).

North American Agreement on Environmental Cooperation, reprinted in (1993) 32 I.L.M. 1480.

North America Free Trade Agreement, 32 I.L.M. 289 (1993).

Paris Convention for the Protection of Industrial Property of March 20, 1883, as revised at Stockholm on July 14, 1967, 828 U.N.T.S.305.

Patent Cooperation Treaty, 19 June 1970, 1160 U.N.T.S. 231.

Rio Declaration on Environment and Development, 13 June 1992, 31 I.L.M. 874 (1992).
Single European Act, reproduced in 25 I.L.M. 503.
Statute of the International Court of Justice. Concluded at San Francisco, 26 June 1945. Entered into force 24 October 1945. 1976 Y.B.U.N. 1052.
Stockholm Declaration of the United Nations Conference on the Human Environment, Report of the UN Conference on the Human Environment, 5-16 July 1972, 11 I.L.M. 1416 (1972).
Treaty Establishing the European Economic Community (Without Annexes and Protocols). Concluded at Rome 25 March 1957. Entered into force 1 January 1958. 298 U.N.T.S. 11.
Treaty for Amazonian Cooperation, 3 July 1978; reprinted in 17 I.L.M. 1045 (1978).
Treaty on European Union, Maastricht, 7 February 1992, Official Journal, 1992, C 191 (29 July 1992).
Treaty on Principles Governing the Activities of States in the Exploration and Use of Outer Space, Including the Moon and Other Celestial Bodies, 610 U.N.T.S. 205.
United Nations Convention on the Law of the Sea (With Annex V). Concluded at Montego Bay 10 December 1982. Entered into force 16 November 1994. U.N. Doc. A/CONF.62/122, reprinted in 21 I.L.M 1261 (1982).
United Nations Framework Convention on Climate Change, concluded at Rio de Janeiro, 29 May 1992. Entered into force 21 March 1994. 31 I.L.M. 849 (1992).
Vienna Convention on the Law of Treaties, opened for Signature May 23, 1969. Entered into force 27 January 1988. 8 I.L.M. 679 (1969).
World Charter for Nature, 21 I.L.M. 455 (1983)/

Statutes and Legislation

An Act to Approve and Implement the Trade Agreements Concluded in the Uruguay Round of Multilateral Trade Negotiations, The Uruguay Round Agreements Act 1994, 19 USC 3512.
Atomic Energy Act of 1954, 42 U.S.C. 2011.
National Aeronautics and Space Act of 1958, 42 U.S.C. 2457.
Patent Act, R.S.C. 1985, c. P-4.
Plant Patent Act, 35 U.S.C. (1982).
Plant Varieties and Seeds Act of 1964 (U.K.).
Plant Variety Protection Act, 7 U.S.C (1982).
Statute of Monopolies (21 Jac. 1, cap 3.).

United Nations/UN Agencies' Documents

Agreed Interpretation of the International Undertaking Resolution 4/89 and Farmers Right, Resolution 5/89, of Twenty-fifth session, 1989.
Charter of Economic Rights and Duties of States of 1974 (UNGAOR 3281, XXIV)
Cobo, Jose Martinez. *Study of the Problem of Discrimination Against Indigenous Peoples* UN Doc. E/CN.4/Sub.2/1986/7 Add. 4, UN Sales No. E.86.XIV.3.3.
Commission on Human Rights, *Preliminary Report on the Study of the Problem of Discrimination against Indigenous Populations,* UN Doc.E/CN.4/sub.2/L.566[1972].
Declaration and Treaty Concerning the Reservation Exclusively for Peaceful Purposes of the Seabed and the Ocean Floor, Underlying the Seas beyond the Limits of Present National Jurisdiction, and the Use of Their Resources in the Interests of Mankind, UN Doc. A/AC.105 /C.2/SR (17 August 1967).
Declaration of Principles Governing the Sea-Bed and the Ocean Floor, and the Subsoil Thereof, beyond the Limits of National Jurisdiction. Adopted by the United Nations General Assembly. 17 December 1970. U.N. Doc. A/RES/2749 (XXV), 10 I.L.M. 220 (1971).
Declaration of The Hague. Concluded at The Hague, 11 March 1989. UN Doc. A/44/340-E/1989/120 (Annex) (1989), 28 I.L.M 1308 (1989).
Declaration on the Right to Development, UN GAOR, 41st Sess., Resolutions and Decisions, Agenda Item 101, at 3-6, 9th plenary meeting, 4 December 1986. UN Doc. A/Res/

41/128. Resolution adopted by the General Assembly [on the report of the 3d Committee (A/41/925 and Corr. 1)].

Draft Principles of Conduct in the Field of the Environment for Guidance of States in the Conservation and Harmonious Utilization of Natural Resources Shared by Two or More or States. Adopted by the UN Environment Programme Governing Council, 19 May 1978. UN Doc. UNEP/IG12/2 (1978). Reprinted in 17 I.L.M 1097 (1978).

Draft Report of the World Intellectual Property Organization (WIPO) Fact-Finding Missions on Intellectual Property and Traditional Knowledge (1998-1999), Geneva, Switzerland.

Draft Treaty on the Settlement of Disputes in the Field of Intellectual Property, International Bureau of WIPO, WIPO Doc. SD/CE/V/2 (8 April 1993).

Experts Group on Environmental Law of the World Commission on Environment and Development, Legal Principles for Environmental Protection and Sustainable Development. Adopted by the WCED Experts Group on Environmental Law, 4 August 1987. UN. Doc. WCED/86/23/Add. 1 (1986).

FAO Resolution Concerning Plant and Genetic Resources, Annex to Res. 8/83, 22nd Session of the FAO Conference, 23 November 1983.

The Impact of Intellectual Property Rights Systems on the Conservation and Sustainable Use of Biological Diversity and on the Equitable Sharing of Benefits From its Use. UNEP/CBD/COP/3/23, 1996.

Institutes Conserving Crop Germplasm: The IBPGR Global Network of Genebanks, International Board for Plant Genetic Resources (Rome: 1984).

Intellectual Property Needs and Expectations of Traditional Knowledge Holders: WIPO Report on Fact-Finding Missions on Intellectual Property and Traditional Knowledge (1998-1999) (Geneva: WIPO, 2001).

International Board for Plant Genetic Resources Annual Report 1983 (Rome: 1983).

International Law Commission Draft Articles on International Liability for Injurious Consequences Arising from Acts not Prohibited by International Law. Report of the International Law Commission on the Work of its Forty-First Session. UN GAOR, 44th Sess., Supp. No., at 222, UN Doc. A/44/10 (1989).

International Undertaking on Plant Genetic Resources, Resolution 8/83 of the Twenty-second Session of the FAO Conference, 23 November 1983.

Non-Legally Binding Authoritative Statement of Principles for a Global Consensus on the Management, Conservation and Sustainable Development of all Types of Forest, United Nations Document, A/CONF.151/6/Rev. 1, 15 June 1992. Reprinted in 31 I.L.M. 881 (1992).

Our Common Future: The World Commission on Environment and Development, U.N.G.A.O.RA/42/427, Chap. 2, (1987).

Permanent Sovereignty over Natural Resources, U.N. General Assembly Resolution 1803 (XVII).

"Plant Genetic Resources: Report of the Director General." Document C 83/25, August, Rome: FAO.

Protocol on Biosafety, Cartagena, 14-19 February 1999. UNEP/CBD/BSWG/6/L.2. Reproduced in (1999) 29 Environmental Law and Policy 138.

Question of the Realization of the Right to Development, Global Consultation on the Right to Development as a Human Right, Report Prepared by the Secretary-General Pursuant to Commission on Human Rights Resolution, 1989/45; E/CN.4/1990/9/rev.1 (26 September 1990).

The Relationship between Intellectual Property Rights and the Relevant Provisions of the Agreement on Trade-Related Aspects of Intellectual Property Rights (TRIPs Agreement) and the Convention on Biological Diversity. UNEP/CBD/ISOC/5, 11 May 1999.

Report of the Intergovernmental Committee for a Convention on Biological Diversity, UN Environmental Programme, 7th Negotiating Sess., 5th Session of the International Negotiating Committee, UN Doc. UNEP/Bio.Div/N7-INC.5/4 (1992).

Report of the International Conference on Population and Development, UN Doc. A/CONF.171/13, preamble, principle 3. Adopted at Cairo, 5-13 September 1994.

Report of the International Law Commission on the Work of its Forty-fifth Session, UN GAOR, 48th Sess., Supp. No. 10, at 79, UN Doc. A/48/10 (1993).

Report of the International Law Commission on the Work of its Forty-Second Session, UN GAOR, 42nd Sess., Supp. No. 10. UN Doc. A/42/10 (1990).

Report of the International Law Commission to the General Assembly, UN GAOR, 35th Sess., Supp. No, at 59, UN Doc. A/35/10 (1980).

Report of the Quinquennial Review of the International Board for Plant Genetic Resources, Consultative Group on International Agricultural Research (Rome, 1980).

Report of the Second Meeting of the Parties to the Conference to the CBD, Annex ii, Decision 11/10, UN. Doc. UNEP/CBD/COP/2/19 (30 November 1995).

Report of the UN. Secretary-General, Development and International and International Economic Cooperation: An Agenda for Development, 3, U.N. Doc.A/48/935 (48th Sess., agenda item 91, 6 May 1994).

Resolution 3281 (XXIX) UNGA, Charter of the Economic Rights and Duties of States, reprinted in (1975) 75 American Journal of International Law 484.

Resolution on Historical Responsibility of States for the Preservation of Nature for Present and Future Generations. Adopted by the UN General Assembly, 30 October 1980. G.A. Res. 35/48, UN GAOR, 35th Sess, Supp. No. 48, at 15, UN Doc A/35/48 (1981).

Resolution on Institutional Arrangement to Follow up the United Nations Conference on Environment and Development. Adopted by the U.N. General Assembly, 22 December 1992. UN Doc. A/47/191. Reprinted in 32 I.L.M. 238 (1993).

Resolution on Permanent Sovereignty over Natural Resources. Adopted by the UN General Assembly, 14 December 1962. G.A.Res. 1803, UNGAOR, 17th Sess., Supp. No. 17, at 15, UN Doc. A/5217 (1963). Reprinted in 2 I.L.M. 223 (1963).

Resolution on Permanent Sovereignty over Natural Resources. Adopted by the UN General Assembly, 17 December 1973. G.A. Res. 3171, UN GAOR, 28th Sess., Supp., No. 30, at 52, UN Doc. A/9030 (1974), reprinted in 13 I.L.M. 238 (1974).

Resolution on the Institutional and Financial Arrangements for International Environment Cooperation (Establishing the United Nations Environment Program, UNEP). Adopted by the U.N General Assembly, 15 December 1972. G.A. Res. 2997, UN GAOR, 27th Sess., Supp. 30, at 42, UN Doc. A/8370 (1973). Reprinted at 13 I.L.M. 234 (1974).

Revision of the International Undertaking on Plant Genetic Resources, Analysis of Some Technical, Economic and Legal Aspects for Consideration in Stage II: Access to Plant Genetic Resources and Farmer's Rights, Commission on Plant Genetic Resources, 6th Sess., FAO Doc. CPGR-6/95/8 Supp. (19-30 June 1995).

Rio Declaration on Environment and Development. Adopted by the UN Conference on Environment and Development (UNCED) at Rio de Janeiro, 13 June 1992. UN Doc. A/CONF.151/26 (vol. 1) (1992) Reprinted in 31 I.L.M 874 (1992).

The Role of Patents in the Transfer of Technology to Developing Countries, Doc. E/3861/Rev. 1, 1 March 1964. Report of the Secretary-General, United Nations (New York: Martinus Nijhoff, 1964). Report of the First Meeting of the Parties to the Conference to the Conference to the CBD, Annex ii, Decision 1/9, U.N. Doc. UNEP/CBD/COP/1/17 (January 1995).

Stockholm Declaration of the United Nations Conference on the Human Environment. Adopted by the UN Conference on the Human Environment at Stockholm, 16 June 1972. U.N. Doc. A/CONF.48/14/. Reprinted at 11 I.L.M. 1416 (1972).

UN Declaration on the Rights of Indigenous Peoples. UN Doc. E/CN.4/1995/2, reprinted in 34 I.L.M. 541 (1995).

UN Declaration on the Right to Development, UN GA Resolution 41/128 of 4 December 1986.

UNESCO Universal Declaration on the Human Genome and Human Rights 1997, 11 November 1997. UNGA Res. 53/152, reprinted in IHRR Vol. 6 No. 3 (1999).

United Nations Declaration on Social Progress and Development, UNGAR, 2542 (XXIV) of 11 December 1969.

United Nations Draft Declaration on the Rights of Indigenous Peoples, UN ESCOR, Commission on Human Rights, 11th Sess., Annex 1, UN Doc. E/CN.4/Sub.2 (1993).

United Nations Transnational Corporation and Management Division., United Nations Department of Economics and Social Development, Intellectual Property Rights and Foreign Direct Investment, UN Doc.ST/CTC/SER.A/24, UN. Sales No. E93, II.A.10 (1993).

United States' Declaration Made at the United Nations Environment Programme for the Adoption of the Agreed Text of the Convention on Biological Diversity, issued 22 May 1992, 31 I.L.M. 848.

UN Study on the International Dimensions of the Right to Development. UN Doc. E/CN.4/ 1334, para. 27.

World Charter for Nature. Adopted by the UN General Assembly, 28 October 1982. G.A. Res. 37/7 (Annex), U.N. GAOR, 37th Sess., Supp. No. 51, at 17, UN Doc. A/37/51, reprinted in 22 I.L.M 455 (1983).

World Economic Survey: Current Trends and Policies in the World Economy (United Nations: New York, 1991).

Miscellaneous National and International Documents

1077 Official Gazette Patent Office 24 (1987), announcement dated 7 April 1987.

Agribusiness Consolidation: Hearings before the Subcomm. on International Trade of the House Committee on Agriculture, 106th Congress 65-70 (1999).

Agricultural Patent Term Restoration, Hearing before the Sub-Committee on Courts, Civil Liberties, and the Administration of Justice of the Committee on the Judiciary, House of Representatives, 99th Congress, 2nd Sess., on H.R. 3897, Agricultural Patent Term Restoration, July 17, 1986 (Washington, DC: US Government Printing Office, 1986).

Commercialization of Academic Biomedical Research: Hearings before the Subcommittee on Investigations and Oversight and the Subcommittee on Science, Research and Technology of the House Committee on Science and Technology, 97th Congress, 1st Session (1981).

Council Decision Concerning the Conclusion on Behalf of the European Community, as Regards Matters within Its Competence, of the Agreements Reached in the Uruguay Round of Multilateral Negotiations (1986-94) (94/800/EC), OJ No L 336/.23.12.94.

Council Recommendation on the Implementation of the Polluter-Pays Principles, 14 November 1974, C(74) 223, reprinted in Philippe Sands, Richard Tarasofsky, and Mary Weiss, eds., *Documents in International Environmental Law* (Manchester: Manchester University Press, 1994).

Council Recommendation on the Uses of Economic Instruments in Environmental Policy, 31 January 1991, C(90) 177; reproduced in Philippe Sands, Richard Tarasofsky, and Mary Weiss, eds. *Documents in International Environmental Law* (Manchester: Manchester University Press, 1994) at 1185.

FAO Draft International Code of Conduct for the Collection and Transfer of Plant Germplasm, Doc. CPGR/91/10 of March 1991.

FAO, *Global Plan of Action for the Conservation and Sustainable Utilization of Plant Genetic Resources for Food and Agriculture and the Lepzig Declaration, Adopted by the International Technical Conference on Plant Genetic Resources* (Rome: FAO, 1996).

Hamilton, Walter. Senate Temporary National Economic Committee, 76th Sess., Investigation of Concentration of Economic Power: Patents and Free Enterprise 164 (Comm. Print 1941).

Indigenous Peoples' Statement on the Trade-Related Aspects of Intellectual Property Rights (TRIPs) of the WTO Agreement. Adopted in Geneva on 25 July 1999.

The Langkawi Commonwealth Heads of Government Declaration on Environment. Adopted at Langkawi, Malaysia, 21 October 1989, reprinted in (1990) American University Journal of International Law and Policy 589.

Machlup, Fritz. *An Economic Review of the Patent System* (Study of the Subcommittee on Patents, Trademarks, and Copyrights of the Committee on the Judiciary, United States Senate, 85th Congress, Second Session. Study No. 15).

Melman, S. *The Impact of the Patent System on Research*, Study No. 11, United States Senate, Committee on the Judiciary, Studies of the Sub-Committee on Patents, Trademarks and Copyrights (Washington, DC: Government Printing Office, 1958) at 18.

Neumeyer, Fredrik. *The Law of Employed Inventors in Europe*, Study No. 30, Senate Sub-Committee on Patents, Trademarks and Copyrights on the Committee of the Judiciary, 87th Congress, 2nd Session, Washington, 1963.

Partnership Agreement between the African, Caribbean and Pacific States and the European Community and Its Member States, as Concluded in Brussels on 3 February 2000. EU Document CE/TFN/GEN/23-OR ACP/00/0371/00.

Reid, Walter. Vice President, World Resources Institute, Testimony before Canadian House of Commons Standing Committee on the Environment, 23 November 1992, House of Commons, Issue No. 47, Third Session, 34th Parliament, 1991-92.

Report of the Conference of FAO, Genetic Resources and Biological Diversity, Annex 3, at 1, U.N. Doc. C/91/ REP (1991).

Report of the President's Commission on the Patent System, reproduced in Hearings Before Subcommittee No. 3 of the Committee on the Judiciary House of Representatives, 90th Congress on H.R. 5924, H.R. 13951, and related Bills for the General Revision of the Patent Laws, Title 35 of the United States Code, and For Other Purposes, Serial No. 11, Part 1 (Washington, DC: US Government Printing Office, 1968).

Report of the Quinquennial Review of the International Board for Plant Genetic Resources, Consultative Group on International Agricultural Research (Rome, 1980).

US Congress, Office of Technology Assessment, OTA-CIT-302 (Washington, DC: US Government Printing Office, April 1986).

United States' Declaration Made at the United Nations Environment Programme for the Adoption of the Agreed Text of the Convention on Biological Diversity, issued 22 May 1992, 31 I.L.M. 848.

World Bank, *World Development Report: Development and the Environment* (New York, NY: Oxford University Press, 1992).

Electronic Sources

Andean Pact: Common System on Access to Genetic Resources. <http://www.users.ox.ac.uk/~wgtrr/andpact.htm>.

The Bellagio Declaration. <http://users.ox.ac.uk/~wgtrr/bellagio.htm>.

Berkey, Judson. "The Regulation of Genetically Modified Foods." <http://www.asil.org.insights.htm>.

Bravo, Elisabeth. "Law for the Protection of Biodiversity in Ecuador." <http://www.users.ox.ac.uk/~wgtrr/ecuador4.htm>.

Burgiel, Stas. "Briefing Note on the Informal Consultations Regarding the Resumed Session of the Extraordinary Meeting of the Conference of the Parties for the Adoption of the Protocol on Biosafety to the Convention on Biological Diversity, 15-19 September 1999." <http://www.iisd.ca/linkages/biodiv/abs>.

Center for International Environmental Law (CIEL). "US Patent Office Admits Error, Cancels Patent on Sacred 'Ayahuasca' Plant." <http://www.grain.org.html>.

Community Intellectual Rights Act, Third World Network 1994. <http://www.users.ox.ac.uk/~wgtrr/cira.htm>.

Community Register for Documenting Local Community Uses of Biological Diversity. <http://www.users.ox.ac.uk/~wgtrr/bhatkot2.htm>.

Conclusions of the Workshop on "Drug Development, Biological Diversity and Economic Growth" Held by the NIH/National Cancer Institute (NCI) Bethesda, Maryland, March 1991. <http://www.users.ox.ac.uk/~wgtrr/nci.htm>.

Cosbey, Aaron, and Stas Burgiel. "The Cartagena Protocol on Biosafety: An Analysis of Results," IISD Briefing Note. <http://www.iisd.ca/pdf/biosafety.pdf>.

Declaration and Draft Model Law by the OAU/STRC Task Force on Community Rights and Access to Biological Resources, March 1998. <http://www.twnside.org.sg/title/oau1-cn.htm>.

Draft of the Inter-American Declaration on the Rights of Indigenous Peoples, 19 September 1995. <http://www.users.ox.ac.uk/~wgtrr/oas.htm>.

Dutfield, Graham. "The Costa Rica Biodiversity Law: A Brief Summary." <http://users.ox.ac.uk/~wgtrr/crley.htm>.

–. "The Public and Private Domains: Intellectual Property Rights in Traditional Ecological Knowledge" WP 03/99, *Oxford Electronic Journal of Intellectual Property Rights*. <http://users.ox.ac.uk/~mast0140/EJWP03/99.html>.

–. "Report and Commentary on the International Workshop on Benefit Sharing with Indigenous People: Organized by Centre for Science and Environment, New Delhi, India, 28-30 August 1996." <http://users.ox.ac.uk/~wgtrr/csework.htm>.

–. "Report on the Fourth International Congress of Ethnobiology, Lucknow, India, November 1994." <http://users.ox.ac.uk/~wgtrr/congeth.htm>.

–. "Report on the Second Conference on Cooperation of European Support Groups in the UN Decade of Indigenous Peoples, Almen, Netherlands, 3-5 May 1996. <http://users.ox.ac.uk/~wgtrr/almen.htm>.

FAO. "FAO Draft International Code of Conduct for Plant Germplasm Collecting and Transfer." <http://users.ox.ac.uk/~wgtrr/fao.htm>.

The Global Coalition for Biocultural Diversity Covenant on Intellectual, Cultural and Scientific Resources: A Basic Code of Ethics and Conduct for Equitable Partnerships between Responsible Corporations, Scientists or Institutions, and Indigenous Groups. <http://www.users.ox.ac.uk/~wgtrr/gcbcd.htm>.

Gupta, Anil. "Suggested Ethical Guidelines for Accessing and Exploring Biodiversity." <http://users.ox.ac.uk/~wgtrr/gupta.htm>.

Indigenous People's Seattle Declaration on the Occasion of the Third Ministerial Meeting of the World Trade Organization, 30 November-3 December 1999. <http://bio-ipr@cuenet.com>.

In Safe Hands: Communities Safeguard Biodiversity for Food Security, Leipzig, Germany, 14-16 June 1996. <http://users.ox.ac.uk/~wgtrr/safehands.htm>.

International Society of Ethnobiology (ISE) *Code of Ethics.* <http://users.ox.ac.uk/~wgtrr/isecode.htm>.

International Workshop on Indigenous Peoples and Development, Ollantaytambo, Qosqo, Peru, 21-26 April 1997. <http://users.ox.ac.uk/~wgtrr/ollan.htm>.

The Inuit Tapirisat of Canada Principles for Negotiating Research Relationships with the North. <http://www.users.ox.ac.uk/~wgtrr/inui.htm>.

The Manila Declaration Concerning the Ethical Utilisation of Asian Biological Resources. <http://www.users.ox.ac.uk/~wgtrr/asomps.htm>.

Plenderleith, Kristina. "FAO Plant Genetic Resources Conference: NGO Strategy Meeting in London." <http://users.ox.ac.uk/~wgtrr/gaiafeb2.htm>.

Posey, Darrell. Workshop on Indigenous Peoples and Traditional Resource Rights. <http://www.users.ox.ac.uk/~wgtrr/greencol.htm>.

Professional Ethics in Economic Botany: A Preliminary Draft of Guidelines. <http://users.ox.ac.uk/~wgtrr/seb.htm>.

Rangnekar, Dwijen. "A Comment on the Proposed Protection of Plant Varieties and Farmers' Rights Bill, 1999." <http://bio-ipr@cuenet.com>.

Rural Advancement Foundation International. "Australians Abandon Two Plant 'Patent' Claims: But Many Questions Remain." <http://www.rafi.org/genotypes/980122apbr.html>.

–. "Biopiracy Project in Chiapas, Mexico Denounced by Mayan Indigenous Groups." <http://www.rafi.org/news>.

–. "Bioprospecting/Biopiracy and Indigenous Peoples," RAFI Communiqué, Nov/Dec. 1994. <http://www.rafi.org/communique/fltxt/19944.html>.

–. "Bolivian Farmers Demand Researchers Drop Patent on Andean Food Crop." <http://www.rafi.org/genotypes/970618quin.html>.

–. "The Impact of Intellectual Property Rights on Sustainable Food Security and Farm Families Remains to Be Felt." <http://www.rafi.org/genotypes/990519afr.html>.

–. "Inter-American Foundation Strays into an Intellectual Property Minefield." <http://www.rafi.org/genotypes/980311ayah.html>.

–. "Quinoa Patent Dropped: Andean Farmers Defeat US University." <http://www.rafi.org/genotypes/980522quin.html>.

–. "Recent Australian Claims to Indian and Iranian Chickpeas Countered by NGOs and ICRISAT." <http://www.rafi.org/pr/release09/.html>.

–. "US Patent on New Genetic Technology Will Prevent Farmers from Saving Seed." <http://www.rafi.org/genotypes/980311seed.html>.

Shiva, Vandana. "Monocultures, Monopolies, Myths and the Masculinisation of Agriculture." <http://www.indiaserver.com/betas/vshiva>.

Sutherland, Joanna. "UNESCO Subregional Meeting on Access to Biological Resources." <http://www.users.ox.ac.uk/~wgtrr/unesco3.htm>.

Tobin, Brendan. "Know-How Licenses: Recognizing Indigenous Rights over Collective Knowledge." <http://users.ox.ac.uk/~wgtrr/tobin4.htm>.

–. "Protecting Traditional Knowledge: The Challenge of Respecting Rights and the Danger of Vested Interests. Consideration of the Peruvian Draft Law on Protecting Indigenous People's Collective Knowledge and a Latin American Communication to the Third Ministerial Conference of WTO." <http://www.wto.org/online.ddf.htm >.

Index

Abbott, Frederick, 25, 36, 41, 52, 62
Access and benefit sharing, 1, 83-85, 115
Africa, 74, 85-86, 108, 112-13, 120, 160,
    163, 207, 212
    intellectual property rights in, 142, 161
    and patents, 41, 123
    regional intellectual property regimes,
    13, 41, 57, 163, 224, 240
Agracetus, 198
Agribusiness, 20, 70-72, 81, 86-87, 125,
    191, 197, 202
Agricultural and Processed Food Products
    Export Development Authority
    (APEDA), 15
AIDS, 246, 306
Alford, William, 34
Amazon (South America), 51, 64-65, 71-72,
    80-81, 105-6, 121, 173
American Pharmaceutical Association, 30,
    138
American Seed Trade Association, 111-12,
    123-24, 127, 139
Andaman Islanders, 18
Andean farmers, 15, 41, 139, 172, 196
Andean Pact, 41, 196
Anderfelt, Ulf, 22-25, 34-35
Antibiotics, 51
APEC, 41
APEDA, 15
Appropriation of indigenous knowledge,
    16, 29-32, 38-46, 86-90, 95-118, 120-50
Arbitration of patent disputes, 46, 170
Argentina, 25, 41, 176
Article 8 (j) of the Convention on Biologi-
    cal Diversity (CBD), 75, 84, 116-17,
    128-29, 152, 156
Asebey, Edgar, 72
Asia, 10, 52-55

Asia-Pacific Economic Cooperation
    (APEC), 41
Aspirin, 143, 145
Assaying, 71-72, 142
AstraZeneca, 182
Australia, 10, 14, 28, 54, 61, 176
Authorial conceit, 149
Aventis, 182
AZT case, 217n184

Badalone, 16
Balick, Michael, 93-95, 109, 142
Barley, 51, 108, 123, 134
Basmati rice, 15,
Bellagio Declaration, 88
Bentham, Jeremy, 88
Biodiversity Issues, 45, 50-51, 54-62, 75-80
Biopiracy, 94-100, 118, 129-30
    concept and definition of, 9-15
    economic implications of, 12, 15, 36-37
    human rights aspects of, 42, 45, 92-96,
    154-55, 156-89
    North-South debates on, 90-92
    race theory of, 90-96
Bioprospecting, 65, 70-71, 142, 146, 159,
    162
Biosafety, 5, 180-90
Biotechnology, 12, 72
    controversies on, 35, 70-71
    definition of, 70-71, 184-86
    industry, 13, 65, 123-25
    patents and patentability issues, 87,
    123-24, 130-31, 160-81
    TRIPS provisions, 184-90
    social impact of, 70-72, 142-50
Blue-water doctrine, 10, 204n14.
Bolivia, 5, 104, 200
Botanic gardens, 95, 102-3, 112

Brazil, 25, 41, 61, 71, 73, 80, 104-6,
169-70, 190
Bretton Woods institutions, 236
Brownlie, Ian, 75
Brundtland Commission, 67, 80
Brunelleschi, Filipo, 16, 166
Budapest Treaty, 42, 225n267
Buddhism, 53-54

Calvinism, 18
Cameroon, 51
Canada, 10, 33, 61, 70, 109, 176
Capitalism, 17-18, 81
Carroll, Amy, 36
Cartagena Protocol, 186-88
CGIAR. *See* Consultative Group on
International Agricultural Research
Chakma, Kabita, 66
Charter for Nature. *See* World Charter for
Nature
Chile, 15, 101, 176, 197
Chimni, Bhupinder, 24
China, 23, 25, 28, 34, 61, 71-73
Ciba-Geigy, 258n26
Civilization, 1, 65-70, 80-82
as Westernization, 57-60, 63-66, 73, 162
Claims, 15, 124
Climate Change, 72-73
Clothesworkers of Ipswich case, 30
Coca-Cola, 92
Cocaine, 91
Cocoa, 68, 95, 104
Coffee, 95
COICA statement, 157, 163
Cohen, Joel, 70
Cohen, Morris, 67
Colgate-Palmolive, 216
Collective inventions, 24, 167
Colonialism, 82, 90-95, 104, 106, 150, 192
Colonization of indigenous peoples, 9, 28,
62-65, 82-90, 104, 106, 150, 192
Columbian Exchange, 89-90, 94-96
Commodification, 49, 52, 71, 121-22, 164,
170
Common concern of mankind (CCM),
77-78, 88
concept of, 78-79, 98-105
Common heritage of mankind (CHM), 4,
8, 77-78, 88, 179
definition of, 97-105
inapplicability to plant genetic resources
in the territory of states, 77, 97-105
origin of, 97-98
Communal patent system, 165, 169-75
Community Patent Convention (CPC),
40, 224n253

Compulsory licensing, 19, 38
and anti-patent movement, 31
Conforti, Benedetti, 45-46
Conservation International, 71
Consultative Group on International
Agricultural Research (CGIAR), 108-10,
113-15, 118, 141, 165
Consumerism, 4, 65-68, 74
Convention on Biological Diversity
(CBD), 10-13, 57, 71, 57, 71, 75-86,
102-5, 114-17, 128-30
relationship with TRIPs, 85, 129
Coombe, Rosemary, 13
Coordinating Body for the Indigenous
Organizations of the Amazon Basin
(COICA) statement, 157, 163
Copyright, 12, 36, 154
Corn, 61, 65, 84, 92, 96, 106-7, 112-13,
139
Cottier, Thomas, 164-65, 170
Cotton, 95-96, 182
Crop yields, 183
Crucible Group, 140, 161, 198
Cultural assimilation, 57-59, 73-74
relationship with biodiversity, 43-44,
62-65

Debt burden, 65-69
Debt-for-nature swaps, 241n173
Declaration of Belem, 157
Declaration of Malta, 86
Deep ecology, 60
Deforestation, 65-74, 80-81
Denevan, William, 64
Diamond v. *Chakrabarty*, 124-25, 141, 146,
195
Dillon, Sara, 66, 80-81
DNA, 118, 122, 144
Doha Declaration, 159
Drahos, Peter 18, 23, 160, 164, 170
Dualist doctrine, 47
Dutfield, Graham 10-12, 122-23

Ecofeminism, 55-57
Ecology, 63-65
Ecuador, 15, 73, 104, 133, 200
ECOSOC. *See* UN Economic and Social
Council
El Niño, 73
Enclosure, 96. *See also* Public domain
Environmental protection, 44-47, 52-70,
80-86
Epistemology, 88
eurocentricity of, 14, 55-57, 84-86, 90-94
and Indigenous peoples, 60, 88-96, 121-
22, 179

Erlich, Paul, 69-70
Ethiopia, 65, 85, 92, 104, 171-72
Ethnobiology, 9, 88, 91, 157
Eurocentrism, 37
  of international law, 57-60
  of patent laws 1-2, 38-49, 169-70
  of science, 57-62
Europe, 17, 27
European Patent Convention, 40, 125-27
European Patent Office, 125, 182
European Union, 40, 125, 195
Ex Parte Hibbard, 124
Exploitation of hired inventors, 22-25
Extinction of species, 65-74

FAO, 107-8, 111-18
  Treaty on Plant Genetic Resources, 4, 75,
    89, 111, 101-4, 111, 115-17, 174
  undertakings on plant genetic resources,
    116-17
Farmers' rights, 89, 113-16, 174-76
First Nations, 10
Flitner, Michael, 194
Folk knowledge, 9, 55-60, 88-90, 142
Food and Agricultural Organization. *See*
  FAO
Food security, 5, 45, 51-55, 117, 180-82
Foreign direct investment, 25. *See also*
  Patents
Forests. *See* Deforestation
Four Directions Council, 10
Frader-Frechette, Christine, 7
France, 30, 104, 171
French Revolution, 19
Friends of the Earth, 60
Frow, John, 6, 168
Frumkin, Maximillian 27

Gardiner, Richard, 137
GDP, 52, 66-67
Gene banks, 85, 89-90, 107-17
"Gene rush," 72
General Agreement on Tariffs and Trade
  (GATT), 44, 48, 126
Genetic engineering, 183-84
Genetic erosion, 196
Genetic materials, 14, 107, 112, 182-83,
  197
Genetic uniformity, 70-72, 109
Geographic indicators, 62
German Patent Office, 123
Germany, 31, 123, 135
Germplasm. *See* Plant genetic resources
Gervais, Daniel, 10
Ghana, 41
Global Biodiversity Forum (GBF), 76

Globalization, 1-7, 6, 26, 32-39, 42-62,
  73-74, 126, 150-53, 172-92
GNP, 66-67
Golden, John, 123, 146, 148
Greenhouse effect, 73
Greenpeace, 60
Green Revolution, 108-10
Group of 77. *See* North-South Relations
Guatemala, 107
Gutowski, Alan 33

Halewood, Michael, 150, 165
Harmonization of Patent Laws, 32, 37-41
Health care, 51, 92, 180
High yield varieties (HYVs) 98, 108-9,
  111-12
HIV, 72, 142
Human rights. *See* Biopiracy, human
  rights aspects of
Hybridization, 188

IARCs. *See* International Agricultural
  Research Centers
IBPGR. *See* International Board for Plant
  Genetic Resources
ICDPs. *See* Integrated Conservation and
  Development Projects
ICJ. *See* International Court of Justice
ILO. *See* International Labour
  Organization
India, 51, 53, 95
  and biopiracy, 14-15, 65, 91, 104, 133,
    138-39, 144-47
Indigenous knowledge, 10, 29, 90-95, 119,
  128-29, 141, 149, 151-60
  definition of, 10-11
Indigenous peoples, 9-10
  contributions to plant breeding, 11-12,
    90-92
  definitions of 9, 9-11
  protocols for protection of intellectual
    property, 129, 155-73
Individualism, 17
Indonesia, 51, 65, 73
Industrial designs, 154
Industrial Revolution, 16, 20-22, 28-29,
  34, 59
Industrialization 1, 20-23, 28-29, 63-65,
  126, 140, 180
Innovation, 18, 22-23, 164
  in traditional societies, 9-11, 18, 84-85,
    87, 136, 158, 166
  relationship with intellectual property
    regimes, 23-27, 36-38, 58, 150-51
Instrumentalism, 29, 103, 165
Integrated circuits, 160

Integrated Conservation and Development Projects (ICDPs), 64-65
Intellectual property rights (IPRs), 17, 28-29
  abuse of, 34-36
  enforcement of, 44
  expansion of, 44, 111
  free riders, 39
  and gene banks, 115, 118
  human rights issues, 188-89
  impact on biodiversity, 71-72, 86-88
  impact on indigenous peoples, 89, 99, 105-7, 120-21, 158-59
  increasing economic value of, 23, 38-39, 128
  market characteristics of, 38
  as perverse incentives, 86
  as private property, 18
  race theory of, 140, 157
  role in international economy, 11-14, 25, 34-36, 124, 150
International Agricultural Research Centers (IARCs), 89-90, 106-18
International Board for Plant Genetic Resources (IBPGR), 109-10, 114
International Chamber of Commerce, 12
International Court of Justice (ICJ), 45, 156
International Labour Organization (ILO) Convention 169, 9-10, 58-59, 64
International law, 52-65, 197
  indigenous peoples in, 9-10
  patent regimes in, 8, 33, 39-49
  responses to biopiracy, 5, 42
  and traditional knowledge, 7, 11
International Law Commission (ILC), 79
International Rice Research Institute (IRRI), 107, 112, 141
International Treaty on Plant Genetic Resources for Food and Agriculture. *See* FAO
International Tropical Timber Agreement (ITTP), 69
International Union for the Conservation of Nature (IUCN), 11, 64, 76
International Union for the Protection of New Varieties of Plants (UPOV), 114, 125-28, 176-78,196
Innovation, 18
  by indigenous peoples, 9-10, 58, 38, 85-87
  relationship with patents, 17, 22-25, 27, 36
Inventions, 32
  corporate ownership of, 23-25

  incentive theory of, 20-24
  by individuals, 18
  as national economic assets, 19, 29-30, 33-35
  national security aspects of, 19-20
  relationship with patent systems, 16, 20-24, 26-26
  social necessity of, 18, 29
Inventors, individual, 22-25, 208n65
Issue linkage, 128, 140
  intellectual property and trade, 39-40, 128, 140
  patent protection and foreign direct investment (*see* Foreign Direct Investment)
ITTA. *See* International Tropical Timber Agreement

Japan 31, 37-38, 70, 108, 118, 186
  patent system, 28, 37-38, 132, 161
Judeo-Christian philosophy, 3, 6-7, 52, 55-60

Kalyan, Chakrabarti, 50
Kempenaar, Jill, 72
Kenya, 61, 73, 92
Kitch, Edmund, 26
Kloppenburg Jack, 61-63, 90-91, 95-96, 98, 108-10, 123, 176-77
Koryak Indians, 18
Kuhn, Thomas, 167
Kurian, Priya, 80

Labour, 21, 35, 96
  employed in inventiveness, 17, 24, 84
  employed in infringing activities, 86
Land races, 63, 108, 111
Latin America, 41, 61, 68, 96, 106, 117
Law of the Sea, 101
Leskien, Dan, 194
Lesser, William, 14, 61, 94, 106
Liardet v. *Johnson*, 29
Lippert, Owen, 16
Locke, John, 17

Machlup, Fritz, 16, 26-27
Madagascar, 61, 72
Maize. *See* corn
Malaria, 91, 96
Malaysia, 61, 72
Martinez Cobo Report, 9
Masculihization of Science 18, 55-57
Mataatua Declaration, 172-73
McDonalds, 71
McNeely, Jeffrey, 64

Mendelian principles, 125
Medical research, 51, 84, 92-93, 147, 166, 198
Mercado Comun del Sur (MERCUSOR), 41
Merck-Inbio agreement, 172
Merges, Robert, 26
Mexico, 106, 116, 133
Microorganisms, 135
Middle Ages, 14-16, 27-30, 56, 134
Monist doctrine, 47-48
Monopolies, 35, 104, 182
 relationship with intellectual property rights, 15, 20, 33-35
Monsanto, 182
Montevideo Convention, 102
Mooney, Pat 12, 129, 111, 140
Morphine, 91
Multilateralism, 39-48, 115-18, 172, 196
Multinational enterprises, 68, 81, 179
 influence on patent laws, 2, 14
 involvement in biopiracy, 7, 116-17, 162, 180-81
 pressure for stronger intellectual property laws, 45, 173
Myers, Norman, 67

Naess, Arne, 60
National Cancer Institute, 142
Nature parks controversy, 59-60, 63
Needham, John 23
Neem tree, 147-48
Netherlands, 101, 104, 135, 109
 anti-patent movement in, 31
New economy, 35
New International Economic Order (NIEO), 98
Newton, Isaac, 18
NGOs, 15, 42, 120, 150, 157-60, 120, 182
Nigeria, 14, 41, 54
North American Free Trade Agreement (NAFTA), 41, 133
North-South relations, 1, 32, 36-40, 43-44, 52-53, 60-65, 68-69, 80-82, 94-96, 98-99, 126-28, 179, 199
Nwabueze, Remigius 12

OAPI, 41-42
Oats, 61, 123
Oddi, Samuel, 19, 26, 148, 174
Office of Technology Assessment (OTA), 148
Oguamanam, Chidi, 128
Okafor Obiora, 68-69
Oncomouse, 140-41
Opinio Juris, 45, 101-4

Organization of African Unity, 126
Organization on Economic Cooperation and Development (OECD), 217
Orientalism, 52-55
Overpopulation, 69-70

Panjabi, Ranee, 78
Paris Convention on the Protection of Industrial Property, 39-40
Patent Cooperation Treaty (PCT), 40-41, 168, 193
Patents
 anti-patent movement of the nineteenth century, 30-31
 benefits of, 23, 25-26, 36-7, 195
 colonial migration of, 27-28
 contract theory of, 20
 criteria for patentability, 14, 29, 44, 88, 122-61, 167-78
 criticism of US Patent System, 14-16, 19, 32-34, 36-38, 67, 130-35, 147
 definition of, 16-17
 disclosure requirement of, 29, 124-25, 134
 distinction from copyrights, 136
 domestic and foreign applicants for, 56-57, 135, 140, 158, 174
 on DNA, 117-18, 122, 144
 on DNA fragments, 117
 economic justification of, 23-25
 effects of strengthened protection, 25-31, 36-39
 enforcement of, 16, 32-33, 44-48, 50-60, 125, 148
 Eurocentrism of, 38, 17-20
 factors in levels of protection among states, 30-33, 35-36
 foreign prior art, 146
 historical development of, 28-31
 historical necessity of, 22-23
 incentive theory of, 19-24, 63, 180-82
 industrial utility, 71, 120, 136-40, 148, 195
 and industrialization, 20, 22-23, 28-29, 34
 as instruments of national policy, 29-31, 33-39, 126, 186
 natural rights theory of, 19-20
 North-South implications of, 11-12, 32-36, 127-30
 novelty requirement, 30-32, 119-20, 127, 130-35, 138, 147-48, 161, 194
 obviousness, 44, 120-27, 136-48, 194
 ordre public, 33, 44, 127
 origin of, 16-17

pharmaceuticals, 87, 127-31
piracy, 35-36
on plants, 118-20, 125-30
on plant varieties, 58-60, 107, 122-24, 135, 135, 141, 175
printed publication, 131-32
prior art (*see* Patents, novelty requirement)
on products of nature, 32, 63, 120, 130, 140-50
protectionism and, 29-30
relationship with inventiveness 21-24, 36
as rent transfers, 33-35, 36-37
reward theory of, 20-21
social costs of, 18-19
specification, 29, 125-26
theories in justification of, 17-23
United States, 19, 177, 194-95
Pharmaceutical industries, 120-28
Penrose, Edith, 19, 24, 36
Permanent Sovereignty over Natural Resources (PSNR), 105-7
Peru, 15
Philippines, 81-82, 107
Pioneer Hi-Bred v. *Commissioner of Patents*, 124, 134
Plant, Sir Arnold, 25
Plant breeding, 115, 119
  by indigenous peoples, 3, 6, 181
  by industrial concerns, 63, 70-72, 176, 200
  by women farmers, 58, 83-86, 136
Plant breeders rights (PBRs), 5, 114-16, 123
  farmers' privilege, 89, 118, 114-16, 174
  and indigenous peoples, 71, 116, 175-97
  and patents compared, 125-27, 195-95
Plant exploration, and germplasm collection, 99-101, 108-13
Plant genetic resources (PGRs), 85, 92-95
  conservation of, 110-12
  extinction of, 65-72
  global transfer of, 88-90, 95-115
  homogenization of, 16, 70-72
  improvement by indigenous peoples, 6, 62-65, 87, 121-22
  North-South relations, 9, 61-62, 75-105
  sovereignty issues, 12, 52, 84
Potatoes, 61, 71-72, 96, 107
Prance, Sir Ghilean, 64-74
Precautionary principle, 82, 185-88
Price, W.H., 22
Public domain, 6-7, 134, 162, 167-68

Quinine, 91-92, 96
Quinoa, 15

Reichman, Jerome 19, 35, 37, 136, 151
Reid, Walter, 78
Rice, 15, 51, 61, 65, 74, 107, 117
RiceTec, 15
Rifkind, Simon, 23
Rio Declaration, 56, 75
Ritchie, Mark, 126
Rockefeller Foundation, 106-8
Roht-Arriaza, Naomi, 11, 164
Rosen, Dan, 38
Rubber, 71, 95, 104-5
Rural Advancement Foundation International (RAFI), 129

Safran, David, 167
Salt-water doctrine. *See* Blue-water doctrine
Sandoz, 181
Schiff, Eric, 22
Science, 56-58, 90-92, 198
Self-Determination, 8-10, 158, 199
Serendipity, 21-22
Shamans, 93
Shiva, Vandana, 66, 129
Simmonds, Norman, 63
Soft law, 75-80, 120, 155-60
Soybeans, 61-63, 117, 124-25, 135-36
State Responsibility, 5-6, 79-80, 120, 151-52
Statute of Monopolies, 31
Stockholm Declaration, 56-57, 75-77
Sui Generis, 44, 113, 128-30, 175-76
Sustainable Development, 66-70, 80-82
Swanson, Timothy, 73-74
Switzerland, 22-28, 31, 165, 171
Syngenta, 182

Taboos, 54
Taxol, 72
Technology, 27, 30
  acquisition of, 21-25, 33-34
  ownership of, 43
  transfer of, 43, 153
Tejera, Valentina, 126
Terminator technology, 183
Thailand, 200
Third World. *See* North-South Relations
TKUP. *See* Traditional knowledge of the uses of plants
Tobacco, 95
Tomato, 51, 61, 84, 117
Trade, 39-41
Trade-Related Aspects of Intellectual Property Rights Agreement (TRIPS), 14, 48, 125-20
  administration, 44
  background, 14, 44-45

criticism of, 119-22, 125-27
economic impact of, 40
effectiveness of, 46
enforcement of, 47, 125-27
future issues, 129-30
provisions, 44
Trade secrets, 20
Traditional knowledge, 11-12
definition of, 9-10
distinction from indigenous knowledge,
10-11
Traditional knowledge of the uses of
plants (TKUP), 10-12, 87-90, 92-93,
120
Transfer of technology, 43, 153
TRIPS. *See* Trade-Related Aspects of
Intellectual Property Rights Agreement
Turmeric, 147

UN Conference on Trade and Develop-
ment (UNCTAD), 42-43, 149, 158
UN Convention on the Law of the Sea
(UNCLOS), 101, 248n287
UN Economic and Social Council
(ECOSOC), 43, 69
UN Environment Programme (UNEP), 42
UN Forum on Forests (UNFF), 69
United Kingdom, 22-30, 34, 97, 104
United States, 25, 30
and allegations of biopiracy, 13, 15, 107,
123-24
patents in, 19, 23, 31-34
Patent Act, 123
Plant Patent Act, 123
Plant Variety Protection Act, 123-24
trade policy, 5, 14, 36-39, 126-28
university-industry relations

US Congress, 34, 135
UPOV. *See* International Union for the
Protection of New Varieties of Plants
Uruguay Round, 44, 48, 126-28, 133
Usui, Chikako, 38

Vavilov Centres, 61-63, 94-95, 106-15,
123, 135
Venetian Patent Law of 1474, 16, 22-23,
123-24
Venice, Italy, 16, 22-27, 29
Vienna Convention on the Law of
Treaties, 46, 75, 128-29
Vitamin B12, 144-46
Vogel, Henry, 98

Watt, James, 20, 166
Western science, 1-3, 17, 59, 63, 86-90
Wheat, 51, 61, 65, 106, 117
Wolfrun, Rudiger, 100
Wood, David, 63
Women farmers, 3, 6, 58, 62-65
World Bank, 42, 52, 67-68, 110
World Charter for Nature, 56, 75
World Health Organization, 92, 158
World Intellectual Property Organization
(WIPO) 42-43, 120, 129
on intellectual property and genetic
resources and folklore, 159, 162, 168
traditional knowledge, 10-11, 120, 129,
150, 158-59
World Resources Institute, 57
World Trade Organization, 43-48, 126-30,
133-35, 153, 175-76, 197

Zeneca, 182
Zimbabwe, 41, 45

James B. Kelly, *Governing with the Charter: Legislative and Judicial Activism and Framers' Intent* (2006)

Dianne Pothier and Richard Devlin (eds.), *Critical Disability Theory: Essays in Philosophy, Politics, Policy, and Law* (2006)

Susan G. Drummond, *Mapping Marriage Law in Spanish Gitano Communities* (2006)

Louis A. Knafla and Jonathan Swainger (eds.), *Laws and Societies in the Canadian Prairie West, 1670-1940* (2006)

Ikechi Mgbeoji, *Global Biopiracy: Patents, Plants, and Indigenous Knowledge* (2006)

Randy K. Lippert, *Sanctuary, Sovereignty, Sacrifice: Canadian Sanctuary Incidents, Power, and Law* (2005)

Florian Sauvageau, David Schneiderman, and David Taras, with Ruth Klinkhammer and Pierre Trudel, *The Last Word: Media Coverage of the Supreme Court of Canada* (2005)

Gerald Kernerman, *Multicultural Nationalism: Civilizing Difference, Constituting Community* (2005)

Pamela A. Jordan, *Defending Rights in Russia: Lawyers, the State, and Legal Reform in the Post-Soviet Era* (2005)

Anna Pratt, *Securing Borders: Detention and Deportation in Canada* (2005)

Kirsten Johnson Kramar, *Unwilling Mothers, Unwanted Babies: Infanticide in Canada* (2005)

W.A. Bogart, *Good Government? Good Citizens? Courts, Politics, and Markets in a Changing Canada* (2005)

Catherine Dauvergne, *Humanitarianism, Identity, and Nation: Migration Laws of Australia and Canada* (2005)

Michael Lee Ross, *First Nations Sacred Sites in Canada's Courts* (2005)

Andrew Woolford, *Between Justice and Certainty: Treaty Making in British Columbia* (2005)

John McLaren, Andrew Buck, and Nancy Wright (eds.), *Despotic Dominion: Property Rights in British Settler Societies* (2004)

Georges Campeau, *From UI to EI: Waging War on the Welfare State* (2004)

Alvin J. Esau, *The Courts and the Colonies: The Litigation of Hutterite Church Disputes* (2004)

Christopher N. Kendall, *Gay Male Pornography: An Issue of Sex Discrimination* (2004)

Roy B. Flemming, *Tournament of Appeals: Granting Judicial Review in Canada* (2004)

Constance Backhouse and Nancy L. Backhouse, *The Heiress vs the Establishment: Mrs. Campbell's Campaign for Legal Justice* (2004)

Christopher P. Manfredi, *Feminist Activism in the Supreme Court: Legal Mobilization and the Women's Legal Education and Action Fund* (2004)

Annalise Acorn, *Compulsory Compassion: A Critique of Restorative Justice* (2004)

Jonathan Swainger and Constance Backhouse (eds.), *People and Place: Historical Influences on Legal Culture* (2003)

Jim Phillips and Rosemary Gartner, *Murdering Holiness: The Trials of Franz Creffield and George Mitchell* (2003)

David R. Boyd, *Unnatural Law: Rethinking Canadian Environmental Law and Policy* (2003)

Ikechi Mgbeoji, *Collective Insecurity: The Liberian Crisis, Unilateralism, and Global Order* (2003)

Rebecca Johnson, *Taxing Choices: The Intersection of Class, Gender, Parenthood, and the Law* (2002)

John McLaren, Robert Menzies, and Dorothy E. Chunn (eds.), *Regulating Lives: Historical Essays on the State, Society, the Individual, and the Law* (2002)

Joan Brockman, *Gender in the Legal Profession: Fitting or Breaking the Mould* (2001)

Printed and bound in Canada by Friesens

Set in Stone by Artegraphica Design

Copy editor: Joanne Richardson

Proofreader: Brenda Belokrinicev